THE REGULATION OF CELLULAR SYSTEMS

The Regulation of Cellular Systems

Reinhart Heinrich and Stefan Schuster

CHAPMAN & HALL

ITP® International Thomson Publishing

New York • Albany • Bonn • Boston • Cincinnati • Detroit • London • Madrid • Melbourne
Mexico City • Pacific Grove • Paris • San Francisco • Singapore • Tokyo • Toronto • Washington

Cover Design: Saïd Sayrafiezadeh, emDASH inc.
Art Direction: Andrea Meyer

Printed in the United States of America

For more information, contact:

Chapman & Hall
115 Fifth Avenue
New York, NY 10003

Chapman & Hall
2-6 Boundary Row
London SE1 8HN
England

Thomas Nelson Australia
102 Dodds Street
South Melbourne, 3205
Victoria, Australia

Chapman & Hall GmbH
Postfach 100 263
D-69442 Weinheim
Germany

International Thomson Editores
Campos Eliseos 385, Piso 7
Col. Polanco
11560 Mexico D.F.
Mexico

International Thomson Publishing-Japan
Hirakawacho-cho Kyowa Building, 3F
1-2-1 Hirakawacho-cho
Chiyoda-ku, 102 Tokyo
Japan

International Thomson Publishing Asia
221 Henderson Road #05-10
Henderson Building
Singapore 0315

1 2 3 4 5 6 7 8 9 10 XXX 01 00 99 97 96 95

Library of Congress Cataloging-in-Publication Data

Heinrich, Reinhart, 1946–
The regulation of cellular systems / by Reinhart Heinrich and Stefan Schuster.
p. cm.
Includes bibliographical references and index.
ISBN 0-412-03261-9 (alk. paper)
1. Cell metabolism—Regulation—Mathematical models. 2. Control theory.
I. Schuster, Stefan. II. Title.
QH634.5.H45 1996
574.87'6—dc20 96–1594
CIP

British Library Cataloguing in Publication Data available

To order this or any other Chapman & Hall book, please contact **International Thomson Publishing, 7625 Empire Drive, Florence, KY 41042.** Phone: (606) 525-6600 or 1-800-842-3636.
Fax: (606) 525-7778, e-mail: order@chaphall.com.

For a complete listing of Chapman & Hall's titles, send your requests to
Chapman & Hall, Dept. BC, 115 Fifth Avenue, New York, NY 10003.

Die angewandte Mathematik hat im Verlaufe der letzten Jahrhunderte eine so hohe Stufe der Ausbildung erreicht, ihre Schlüsse haben einen solchen Grad von Sicherheit erlangt, daß sie unter den Wissenschaften den ersten Rang einzunehmen berechtigt ist. Sie ist der Anfang und das Ende für den Sternkundigen, den Techniker, den Seemann, sie ist die feste Achse aller Naturforschung jetziger Zeit. Nur der Biologie haben die Entdeckungen Galileis, Newtons und Mariottes verhältnismäßig geringe Früchte getragen; für die Lebenserscheinungen wurden keine Formeln aufgefunden . . .

Throughout the last centuries, applied mathematics has attained such a high level of perfection and such a degree of certainty in its conclusions that it is entitled to take the first place among the Sciences. Mathematics is the be-all and end-all for the astronomer, the engineer, and the seaman; it is the solid basis of all natural sciences today. Only for biology, the yield of the discoveries of Galilei, Newton and Mariotte has been comparatively small; no formulae have been found for the phenomena of life . . .

ROBERT MAYER, *Die organische Bewegung in ihrem Zusammenhange mit dem Stoffwechsel* (Organic motion in its relation to metabolism), Heilbronn, 1845

Contents

Preface

There is no doubt that nowadays, biology benefits greatly from mathematics. In particular, cellular biology is, besides population dynamics, a field where techniques of mathematical modeling are widely used. This is reflected by the large number of journal articles and congress proceedings published every year on the dynamics of complex cellular processes. This applies, among others, to metabolic control analysis, where the number of articles on theoretical fundamentals and experimental applications has increased for about 15 years. Surprisingly, monographs and textbooks dealing with the modeling of metabolic systems are still exceptionally rare. We think that now time is ripe to fill this gap.

This monograph covers various aspects of the mathematical description of enzymatic systems, such as stoichiometric analysis, enzyme kinetics, dynamical simulation, metabolic control analysis, and evolutionary optimization. We believe that, at present, these are the main approaches by which metabolic systems can be analyzed in mathematical terms. Although stoichiometric analysis and enzyme kinetics are classical fields tracing back to the beginning of our century, there are intriguing recent developments such as detection of elementary biochemical synthesis routes and rate laws for the situation of metabolic channeling, which we have considered worth being included. Evolutionary optimization of metabolic systems is a rather new field with promising prospects. Its goal is to elucidate the structure and functions of these systems from an evolutionary viewpoint. This may entail important applications in bioengineering, where optimization obviously plays a fundamental role. One of our major goals is to present the state of the art in metabolic control analysis, focusing on its mathematical aspects. We would be glad if we could contribute to unifying the nomenclature in this field. Besides its theoretical implications, metabolic control analysis, like the other approaches reviewed in this book, provides a framework for the planning and conduction of experiments. In that sense, the book is also addressed to experimentalists. However, reviewing the multitudinous experimental applications of the theoretical tools presented would be beyond the scope of our monograph.

The present book is, to some extent, an outcome of our teaching mathematical biology for undergraduate and graduate students in biophysics at Humboldt University, Berlin. This biophysics program is based on comprehensive studies not only in biological disciplines but also in mathematics and physics and includes a specialized training in thermodynamics, systems theory, and computer modeling, among others. This interdisciplinary approach is reflected in this book. Nevertheless, most of the text will be instructive to all biologists and chemists having the usual mathematical training in these disciplines. As far as the biological background is concerned, it is supposed that the reader is familiar with basic features of enzyme catalysis, the main pathways and regulatory mechanisms in intermediary metabolism, and principles of membrane transport.

The theoretical presentation is illustrated by many examples. For pedagogical purpose, we made them as simple as possible. Often, they are reduced versions of more elaborate models, for example, of calcium oscillations, oxidative phosphorylation, and glycolysis, taken from the literature.

To many biochemists, the present text may appear a rather specialized and somehow sophisticated view on metabolic systems. On the other hand, in light of the recent developments in the mathematical analysis of these systems, the book must be considered as introductory. Nevertheless, we have tried to take into account a representative selection of the recent literature.

The reader will become aware of many open questions. This concerns, for example, the mathematical description of the interaction of metabolism and gene expression, the simulation of cellular metabolism on a large scale, including many interacting pathways and membrane transport, and appropriate ways of modeling the various types of metabolic channeling. One of the pending problems in metabolic control analysis is a comprehensive extension to oscillations in living cells. Although cellular metabolism is one of the best studied objects in biology, we are far from satisfactorily understanding the emergence and evolution of such a complex machinery in terms of basic theories of self-organization.

While writing this book, we benefited greatly from discussions with Dr. Milan Brumen (Maribor), Dr. David Fell (Oxford), Dr. Jannie Hofmeyr (Stellenbosch), Dr. Hermann-Georg Holzhütter (Berlin), Dr. Daniel Kahn (Toulouse), Dr. Boris N. Kholodenko (Moscow), Dr. Jean-Pierre Mazat (Bordeaux), Dr. Tom A. Rapoport (Boston), Dr. Christine Reder (Bordeaux), Dr. Enrique Meléndez-Hevia (La Laguna), Dr. Francisco Montero (Madrid), Dr. Gösta Pettersson (Lund), Johann Rohwer (Amsterdam), Dr. Thomas G. Waddell (Chattanooga), and Dr. Hans Westerhoff (Amsterdam). We discussed with these colleagus intensively a wide variety of topics relevant to cellular regulation, ranging from nonlinear dynamics to organic chemistry and from rapid biochemical equilibria to molecular evolution. The venture of writing this monograph would probably have been impossible without the stimulating and cooperative atmosphere within the scientific community of metabolic modeling, which is reflected, for example, in the large number of scientific congresses in recent years.

In our institute, many colleagues have contributed to the completion of the manuscript in different ways. Margrit Sternberg took care of the bibliography with patience and painstaking. Petra Schubert expertly typed the manuscript, drew many reaction schemes and rescued what we lost in the many different computer files of the text. Dr. Edda Klipp did several numerical simulations and produced many of the nice figures. Several colleagues and students have cross-read drafts of the manuscript and helped us eliminate some inconsistencies. In particular, we mention Stephan Frickenhaus, Ines Jentzsch, Edda Klipp, Ulrike Laitko, Amadeus Stephani, Thomas Wilhelm, and Jana Wolf. We are glad to thank all of them. We are gratefully indebted to Dr. Clas Blomberg (Stockholm) for reviewing a draft of the manuscript very carefully and giving many helpful suggestions.

We would like to express our warmest thanks to Chapman & Hall for friendly and efficient cooperation. Of particular help has been the fruitful work of the publishers Dr. Eleanor S. Riemer and Gregory Payne, who have never been out of patience when we were not able to meet the deadlines. We also remember with pleasure our discussion with Gregory Payne on our book project on a restaurant terrace above the roofs of Washington. We also thank Mary Ann Cottone and Jennifer G. Lane for expertly managing the production of the book. We are gratefully indebted to Dr. Michael Conrad (Detroit) for establishing our contact with this publishing house.

Symbols

A_j	affinity of reaction j
$\mathbf{A}$	atomic matrix
$\boldsymbol{b}_i$	eigenvector of the Jacobian
C	flux cone
C^J_{ij}	control coefficient for flux J_i and reaction j (normalized or non-normalized as indicated in the particular context)
C^S_{ij}	control coefficient for concentration S_i and reaction j (normalized or non-normalized as indicated in the particular context)
$\mathbf{C}^J$	matrix of flux control coefficients
$\mathbf{C}^S$	matrix of concentration control coefficients
D_{ijk}	second-order control coefficient
D	deviation index
$\boldsymbol{e}_k$	generating vector of a polyhedral cone
E_j	total concentration of enzyme j, when rate laws of overall enzymic reactions are considered; free concentration of enzyme j, when enzyme mechanisms are considered
$E_{T,j}$	total concentration of enzyme j, when enzyme mechanisms are considered
EP	concentration of the enzyme-product complex
ES	concentration of the enzyme-substrate complex
E_{tot}	sum of enzyme concentrations in a metabolic pathway
f	frequency of oscillations
F	Faraday constant

$\mathcal{F}$	flux cone for systems with irreversible reactions only
g_{ij}, h_{ij}	kinetic orders in the power-law approximation
ΔG	change in Gibbs free energy
G	growth rate
$\mathbf{G}$	conservation matrix
H_τ	measure of time hierarchy
$\mathbf{I}$	identity matrix
J_j	steady-state flux in kinetic modeling, or thermodynamic flow
k_{+j}, k_{-j}	rate constants of reaction j for the forward and backward directions, respectively
$\mathbf{K}$	null-space matrix
K_I	inhibition constant
K_mS	Michaelis constant of the reaction substrate
K_mP	Michaelis constant of the reaction product
K_j^+, K_j^-	Michaelis constants of the substrate and product, respectively, of an enzyme j with uni-uni mechanism
$\mathcal{K}$	cone representing all non-negative conservation relations
L	allosteric constant for cooperative enzymes
$\mathbf{L}$	link matrix in metabolic control analysis
$\mathbf{L}$	matrix of Onsager coefficients, L_{ij}, in irreversible thermodynamics
$\mathbf{M}$	Jacobian matrix
$\boldsymbol{M}$	flux mode
n	number of internal metabolites
n_H	Hill coefficient
n_{ij}	stoichiometric coefficient of substance S_i in reaction j
$\mathbf{N}$	stoichiometry matrix
$\mathbf{N}^0$	reduced stoichiometry matrix
p	parameter
P_i	concentration of ith external metabolite
q_j	equilibrium constant
${}^jO_{ik}^{S,S}$	coresponse coefficient for two concentrations S_i, S_k resulting from a perturbation of a parameter p_j; coresponse coefficients for other steady-state variables are denoted similarly
r	number of reactions
R	universal gas constant
R_{ij}^Y	response coefficient for variable Y_i and parameter p_j (normalized or non-normalized as indicated in the particular context)
$\mathbf{R}^Y$	matrix of response coefficients
S, P	substrate and product concentrations, respectively
S_i	concentration of the ith internal metabolite
S_i^0	concentration of the ith internal metabolite at a reference state
tr	trace of the Jacobian matrix

T	temperature
T_k	conservation quantities (numbered by index k)
v_j	reaction rate (as a function of concentrations and parameters)
V	cellular volume
V_L	Lyapunov function
V_m	maximal activity of an enzyme catalyzing an irreversible reaction
V_j^+, V_j^-	maximal activity in the forward and reverse directions, respectively, of an enzymatic reaction j
w_j	reaction rate of elementary steps in enzyme catalysis
$\mathbf{W}$	modal matrix
X_j	thermodynamic forces
Y_i	steady-state variable
z	elementary conserved-moiety vector
$\mathbf{Z}$	conserved-moiety matrix
α_i, β_i	rate constants of aggregate reactions in the power-law approximation
δ_{ij}	Kronecker symbol
Δ	determinant of the Jacobian matrix
ε_{ij}	elasticity (normalized or non-normalized as indicated in the particular context)
$\boldsymbol{\varepsilon}$	matrix of elasticities
ε_{ijk}	second-order elasticity coefficient
λ_i	eigenvalue of the Jacobian matrix
λ	Lagrange multiplier
$\mathbf{\Lambda}$	diagonalized Jacobian matrix
μ	small parameter
$\Delta\tilde{\mu}_{H^+}$	proton-motive force
π_{jk}	parameter elasticitiy (normalized or non-normalized as indicated in the particular context)
$\boldsymbol{\pi}$	matrix of parameter elasticities
σ	entropy production rate
τ	time constants of reactions, or transient times of enzymic systems
$\Delta\Psi$	electric transmembrane potential
Ω	sum of intermediate concentrations

NOTE: Italicized symbols of substances stand for their concentrations, e.g., *ATP* or Ca^{2+}. Boldface Roman symbols denote matrices; boldface italic symbols denote vectors. The partial derivative of a vector with respect to another vector, $\partial \boldsymbol{X}/\partial \boldsymbol{Y}$, is meant to denote the matrix $(\partial X_i/\partial Y_j)$.

1

Introduction

The increasing role of mathematics in cell biology is witnessed by the ever-increasing number of mathematical *models* representing particular processes and subsystems of living cells. This development was made possible by the exploration of multitudinous elementary processes underlying the phenomena of life at the molecular level. Nevertheless, there appears to be some lack of general formalized *theory* in biology. Physics comprises very elaborate buildings of theory for several centuries. Attempts to develop general theoretical bases for mathematical description of living organisms have been made only in the last decades, partly with the aid of the laws of physics. The fact that formalized theories in biology are still rare is not only due to difficulties arising from the enormous complexity of living matter but also to the fact that experimental quantitation in biology had begun relatively late. In biochemistry, in particular, the quantitative approach has been considerably stimulated by the identification of the main metabolic pathways and the isolation of the enzymes involved, that is, since the middle of our century. We would like to stress, however, that also rigorous formalization of classical mechanics from the beginnings with Galilei in 1590 up to Lagrange's formalism in 1788 took almost 200 years.

The present book is devoted to the theoretical description of metabolic systems, that is, networks of enzyme-catalyzed reactions proceeding in living cells. Chapter 2 outlines some *fundamentals of biochemical modeling* and is meant to be introductory to the subsequent chapters. As we do not extend the width of the treatment too much, this chapter may be skipped by advanced readers. It includes the mathematical description of single enzymes in terms of rate laws. Starting from a generalized mass-action kinetics, we give several specific rate equations which we will use in subsequent chapters. Furthermore, thermodynamic flow-force relationships and the power-law formalism are compared with classical enzyme kinetics. A major part of Chapter 2 is devoted to the systemic level. Steady states are treated, including stability analysis and multistationarity, and conditions for the occurrence of metabolic oscillations are given. Basic models of bistable behavior and of glycolytic and calcium oscillations are considered.

Chapter 3 deals with structural analysis of metabolic networks. This approach is aimed at elucidating relevant relationships among system variables (e.g., concentrations or fluxes) on the basis of the network *stoichiometry* without reference to kinetic properties. Such analysis is motivated by the extreme complexity of cell physiology. Topological properties are often difficult to recognize by mere inspection and require formalized methods.

Chapter 4 deals with the implications of *time hierarchy* (i.e., the wide separation of time constants) for model construction in the field of enzyme systems. In particular, we address the quasi-steady-state and rapid-equilibrium approximation methods, which can be applied when separate time scales are relevant. Moreover, we give an overview of modal analysis, which serves to decompose the system dynamics into motions on different time scales.

Chapter 5 is meant to present the state of the art in the mathematical analysis of metabolic control. This is a theoretical framework that has been developed for about 20 years, originally based on the problem of how to define rate limitation in metabolic pathways. Metabolic control analysis serves to quantify, in terms of control coefficients, the extent to which different enzymes limit the flux under particular conditions. This analysis has become increasingly relevant for experimental investigation of metabolism; in the present book, however, we focus on theoretical aspects with some applications to concrete pathways.

Chapter 6 deals with the mathematical analysis of *optimality properties* of metabolism and *evolutionary aspects*. As is suggested by Darwin's concept of the "survival of the fittest," optimization plays an important role in evolution. This aspect opens a further access to mathematical treatment of metabolic systems. Our presentation is far from giving a comprehensive overview of the biological aspects of evolution of metabolic pathways. First steps are made toward an analysis concerning the problem of whether the contemporary state of enzyme systems is optimal compared to other conceivable states.

As mentioned above, the main efforts in the field here considered are directed toward development of models. In science, both physical objects, such as space-filling or wire models of DNA double helices or proteins, and nonmaterial, in particular, mathematical, representations are used. The usefulness of a model is determined by the compromise between adequacy (i.e., the correctness of representation) and simplicity (tractability). Every mathematical model is based on simplifying assumptions to render possible or facilitate the analytical or computational treatment and the interpretation of results. As for models of metabolic systems, such an assumption concerns, among others, the distinction between internal and external metabolites. The latter substances are assumed to have fixed (buffered) concentrations, which can sometimes, in fact, be achieved by an appropriate experimental setup. Paradoxically, even model assumptions contradicting each other may be useful when they are favorable for tractability (e.g., the quasi-electroneutrality assumption and the existence of an electric field in models of ion distributions across and near biological membranes). A model is known from the very beginning of its development not to be correct to a certain extent. The following aphorism may fit in this context: "If we do not develop models, we do not learn why they are false." Of course, the iterative process of model building tends to gradually eliminate errors and unjustified assumptions, but a certain remainder of incorrectness is deliberately accepted for the sake of simplicity. No theory can be completely correct either; any scientific representation is a simplification and a more or less distorted picture of the object it is to reflect. So the delimitation between the terms theory and model is not sharp. A model is usually not as correct and general as a theory, and its logical basis is less rigorous.

Models of metabolic processes, as any other model, are usually developed for a certain pragmatic purpose. One may intend to give a detailed mathematical representation of all the underlying enzymic reactions, which is very important for fitting experimental data in the best way possible. This type of modeling in biochemistry was stimulated to a considerable extent by the availability of powerful computers. Therefore, rather large kinetic models were developed and solved numerically. The resulting curves are often very impressive but bear the risk of pseudo-exactness because it is often unclear how reliable the theoretical background and the parameters used in the model are. The results of detailed, very complex models are difficult to interpret owing to the high number of variables involved. Alternatively, one may be interested in explaining *specific phenomena*, such as calcium oscillations or the dependence of ATP concentration on energetic load in cellular energy metabolism, or in finding the conditions for the emergence of chaos or multistationarity. It is then suitable to develop minimal models by restricting oneself to essential features. This can be done in two ways. One can start from a real pathway and try to describe it by a model simplified as far as possible so that the phenomenon of interest is retained. This generally leads to skeleton models of metabolic pathways, in which groups of reactions are lumped into overall reactions, and simple kinetic rate laws are used (e.g., linear kinetics or power laws). The lumping of reactions may be done in an *ad hoc* way or by more sound methods based on, for example, temporal and spatial hierarchies in the system. Another approach is by focusing on a specific phenomenon and trying to find the simplest model to produce this. In the present book, we study, as an example, a minimal model of a chemical reaction scheme with mass-action kinetics showing limit cycle behavior.

Efficient dynamical simulation in biochemistry requires one to analyze the underlying structure of the system. The kinetic parameters of enzymic reaction systems are often unknown and are subject to frequent changes, even in short time periods. In contrast, the structure of these systems (i.e., the topology of connection of substances by reactions) remains virtually constant, unless evolutionary time scales are studied. Therefore, the modeling of any biochemical system should include the analysis of its structural invariants, such as conservation relations among concentrations and restrictions to fluxes imposed by balance equations. Moreover, thermodynamic aspects may be included in this analysis. For example, when some reactions are irreversible, additional sign conditions for the flux values arise. In the context of structural analysis, the repeatedly posed question of how to delimit metabolic pathways is worth being tackled. The difficulty of this question results from the fact that all reactions in the living cell are virtually interdependent. One possible way of approaching this problem is by looking for the simplest routes leading from certain substrates to some product. Under the additional condition that the pathway operates at steady state, these routes may be represented by specific vectors in the so-called null-space of the stoichiometry matrix, that is, the space of all conceivable steady-state fluxes. Furthermore, practical independence of reactions often results from special thermodynamic and kinetic properties, such as irreversibility of reactions, saturation of enzymes, and separation of time constants.

Structural (topological) analysis in many fields is often done by using graph theory. As far as biochemical networks are concerned, problems arise when reactions other than monomolecular are studied, because they cannot simply be represented by arcs. Indeed, several attempts have been made to adapt graph theory to biochemical networks by introducing auxiliary vertices. In our eyes, structural analysis of metabolic systems can be tackled more elegantly by using a matrix formalism than by graph theory.

Time hierarchy is a ubiquitous phenomenon in biology. Biological evolution, ontogenetic development, transfer of genetic information, metabolic interconversions, and elementary processes of enzyme catalysis proceed on very distinct time scales, ranging approximately from 10^{17} to 10^{-12} s. Importantly, even at a given level of biological organization, for example in one and the same metabolic pathway, processes with very different time constants are involved. Relevant changes in metabolism mostly occur in a time range from seconds to hours. Temporal hierarchies have important implications for the methodology of modeling, because it allows one to detect the changes relevant in the velocity "window" of interest. Simplifications may result by neglecting very slow processes, which cannot be observed experimentally. A third class is made up by the reactions which are so fast that they can be considered to have terminated in the time scale of interest. This has the consequence that although a metabolic system generally operates far from equilibrium, subsystems may attain quasi-equilibria.

In contrast to the thermodynamic properties of reactions, such as the standard free-energy differences, the velocities of biochemical processes are determined by the properties of enzymes catalyzing them. Therefore the question of *why* evolution has brought about large differences in time constants in one and the same metabolic pathway is intriguing. It may be supposed that quasi-equilibration of subsystems by time hierarchy serves to preclude complex behavior such as chaotic dynamics in situations where such behavior is of no functional use.

Over a long time, common belief in biochemistry had been that only one, namely the slowest enzyme in a pathway, would control the flux (in some recent textbooks, this view is still maintained). This enzyme would then be the rate-limiting step, also called a pace-maker enzyme. When *metabolic control analysis* was introduced in the early seventies, it turned out that occasionally a particular enzyme may be rate limiting, but generally there is a distribution of control among many enzymes that varies with circumstances.

A general point in the construction of models of complex systems is to describe the system behavior in terms of the properties of their constituents. In metabolic control analysis, this is achieved by equations linking the systemic properties expressed by "control coefficients" to the component properties of the enzymes expressed by "elasticity coefficients." Both types of coefficients are defined so as to refer to the response to very small perturbations of reaction rates or concentrations of reactants and effectors. The concept of control coefficients was also extended to quantify the response of other steady-state variables, such as concentrations of pathway intermediates. Restricting the mathematical analysis to infinitesimal changes, one arrives at a linear theory. This simplifies the mathematical treatment and makes possible comprehensive and general elaboration, to a large extent by the use of matrix formalism.

Metabolic control analysis provides a framework for experimental investigation in that it clearly shows that understanding of the functioning of enzyme networks is mainly achieved by measuring changes around the *in vivo* state after perturbations, rather than by only determining this state itself. The analysis indicates which quantities have to be measured to determine the response behavior of metabolic systems. For a large number of metabolic pathways, such as glycolysis, the pentose phosphate pathway, oxidative phosphorylation, and tryptophan biosynthesis, the distribution of control among the enzymes involved have been determined experimentally or theoretically for various physiological states.

It is often useful to analyze metabolic systems at a higher level by grouping enzymes into "modules." This can be done according to the existence of functional units, which are not only the particular pathways but also organelles, such as mitochondria. This leads to a modular approach to metabolic control analysis. A further generalization of the original concept is to analyze time-dependent responses, in particular the control of relaxation processes and oscillations.

As mentioned above, control analysis only provides reliable predictions when small changes are considered. However, this may not be sufficient for many applications (e.g., in biotechnology and medicine). Furthermore, it may be difficult to produce sufficiently small perturbations in experiment. First attempts have been made to cope with the nonlinear effects of larger changes. One method is based on the concept of "deviation index" and the other is a second-order approach resulting from a Taylor expansion of the system equations. In any case, control analysis should preferably be combined with construction of a simulation model, in order to obtain an integrated picture of the system behavior for small and large changes in environmental and internal conditions.

Originally, one of the goals in the development of metabolic control analysis was to provide a tool for elucidation of the principles governing regulation of intracellular processes. There have been manifold speculations about the differences between "control" and "regulation." Clearly, control coefficients describe nothing but the potential response of metabolite concentrations or fluxes to changes in a reaction rate. Whether or not such changes actually occur under physiological conditions (e.g., by action of an effector) is at present beyond the realm of control analysis. Regulation is somehow linked with the functions of metabolic systems, with the difficulty that there is no clear-cut definition of the term "function." Obviously, specific functions (in the intuitive sense of the term) can be distinguished for different pathways and different cells. Examples are the fairly constant supply of a metabolic product, the homeostasis of certain substances involved in many different pathways (e.g., ATP), transmission and amplification of intercellular signals, and maintenance of biorhythms by metabolic oscillations. Many theoretical approaches to regulation have concentrated on homeostasis in systems with feedback loops. Based on concepts of metabolic control analysis, quantities that may be useful to characterize regulation in the sense of homeostasis have recently been introduced, such as internal response coefficients and coresponse coefficients.

Practical applications of metabolic control analysis are manifold. It can be used to study diseases caused by enzyme deficiencies, thus enabling us to understand why a pathway does not function properly. Conversely, one is often interested in suppressing metabolic activity in pathogenic microorganisms. To this end, it is important to detect the enzymes with the highest flux control coefficients. It may be supposed that inhibition of these by some drugs reduces pathway flux most. Similarly, one may derive from the distribution of flux control which enzymes should be amplified by genetic manipulation to give the highest effect in increasing the synthesis rate of a target biosynthetic product. One can even derive estimates for the gain in production rate when enzymes are altered in concentration or kinetic parameters. In this way, metabolic control analysis may provide tools for optimization in biotechnology.

The number of journal papers on metabolic control analysis has increased rather rapidly. The growing interest in this field is also documented by recent congress proceedings [e.g., *Control of Metabolic Processes*, Cornish-Bowden and Cárdenas (eds.), 1990; *Modern Trends in Biothermokinetics*, Schuster *et al.* (eds.),

1993b; *What is Controlling Life?*, Gnaiger *et al.* (eds.), 1994]. However, control analysis has hitherto been dealt with in very few monographs. The present book is planned to fill the gap. We tried to cover most recent developments, inclusive control in single enzymes and control in metabolic channeling.

Metabolic systems are characterized by two distinct groups of data. One set is composed of the variables (essentially concentrations and fluxes); the other set comprises the system parameters (stoichiometric coefficients, kinetic constants, etc.). Simulation models serve to compute the system variables on the basis of given values for parameters. The question arises whether the latter quantities are also amenable to theoretical explanation. To answer this question, one should consider time scales on which the kinetic properties and stoichiometry of enzymatic properties have changed, that is, the dimension of biological evolution. In contrast to chemical reactions of inanimate nature, all the enzyme-catalyzed processes in the living cell are the outcome of natural selection which have acted over billions of years.

It may well be that we will never be able to follow the details of the emergence of the contemporary enzymes and metabolic pathways. However, a certain degree of understanding may be gained by considering evolution as an optimization process. This view implies that metabolic systems found in living cells show some *fitness properties*, which may be described by *extremum principles*. One should bear in mind, however, that biological evolution has not reached a final stage. Investigation of extremum principles in biology is not, therefore, necessarily based on the hypothesis that living organisms have attained states referring to certain global optima. However, it can be assumed that subsystems of living organisms, such as metabolic pathways, cannot be further optimized under given external conditions. Moreover, usage of extremum principles is a methodology of research, which allows one to filter out important limit situations, such as special constellations in the high-dimensional parameter space, between which, or in the vicinity of which, the real systems are situated.

Obviously, investigation of optimization principles is even meaningful at the level of individual reaction steps. Here, it is an intriguing task to understand the extremely high catalytic efficiency of enzymes. One may ask, for example, whether the parameters of enzyme kinetic mechanisms (i.e., the values of elementary rate constants or the Michaelis constants) may be explained on the assumption of maximal catalytic power. On the level of multienzyme systems it is interesting to study how far the topology of enzymatic networks, represented by the stoichiometries of the pathways or special enzyme-modifier relationships, reflect optimum properties. Only recently, theoreticians became aware of the importance of the "historical dimension" for the mathematical modeling of metabolic systems, in order to gain deeper insight into structure-function relationships.

Extremum principles have a long tradition in physics; Hamilton's Principle of Least Action and the Second Law of Thermodynamics are important examples.

However, these principles are not considered as optimization principles. Biology and physics have in common to deal with processes on very distinct time scales. Whereas classical celestial mechanics deals, for example, with the motion of planets with given values for the gravitational constant and the masses of planets and the sun, cosmology has the goal, among others, to explain such parameters. This is a situation analogous to the explanation of those biological parameters that may change on long time scales.

From the methodological point of view, evolutionary optimization is related to optimization in biotechnology. Also here, relevant objectives concern the maximization of metabolic yield, the optimization of stability, and so on. On the other hand, there are some differences in that optimization in biotechnology is aimed at the improvement of one or few specialized functions, whereas biological evolution has mainly acted to achieve a well-tuned balance between several functions. This is a reason for the relevance of multicriteria optimization in the understanding of evolution.

The reader of this book is supposed to be acquainted with basic concepts of elementary algebra, standard differential calculus, as well as operations with vectors and matrices. Moreover, for the particular chapters, additional mathematical knowledge is helpful. This mainly concerns linear algebra and nonlinear algebraic and ordinary differential equation systems. Some basic knowledge of nonlinear optimization is needed for the understanding of the treatment of significant evolutionary extremum principles. In the chapters on structural analysis and control analysis, ample use is made of matrix notation. Many relations can be formulated in this way very concisely, because the mathematical treatment of metabolic systems requires a number of variables of the same type for its constituents (e.g., concentrations of many substances, or fluxes of reactions). Linear algebra had been used in the stoichiometric analysis of chemical systems as early as at the beginning of our century. Over the last two decades, it has been realized that standard linear algebra alone is insufficient for this analysis because it does not cope with non-negativity conditions. As many relevant quantities such as concentrations of reacting species, numbers of atom groups constituting these species, and velocities of irreversible reactions are always non-negative, such constraints must be taken into account. Accordingly, mathematical tools from convex algebra have turned out to be helpful.

The topics dealt with in the present book are multidisciplinary and may be treated from different viewpoints. We chose a mathematical approach corresponding to our own research. This is meant quite in the sense of what Robert Mayer, one of the discoverers of the First Law of Thermodynamics, regretted, in the above epigraph, to be missing in biology.

2

Fundamentals of Biochemical Modeling

In this book, we deal with deterministic kinetic modeling of biochemical reaction systems. The principal notions are the concentration (i.e., the number of moles of a given substance per unit volume) and the reaction rate (expressed as concentration change per unit time). This type of modeling is sometimes referred to as macroscopic or phenomenological approach, at variance with microscopic approaches, where molecules and their interactions are considered as fundamental concepts. In the latter approaches, rate constants are calculated in terms of molecular quantities, for example, in the Transition State and Kramers Rate Theories (cf. Hänggi *et al.*, 1990).

Starting from general balance equations, we outline, in this chapter, important fundamentals of biochemical modeling concerning rate laws, steady states, and time-dependent phenomena of nonlinear enzymic systems. The section dealing with enzyme kinetics is meant to give an overview of basic concepts of a wide field which we do not wish to cover comprehensively. For systematic treatises of enzyme kinetics, the reader is referred to the books by Cornish-Bowden and Wharton (1988) and Kuby (1991). We also give several specific rate laws which we will use in the chapter devoted to metabolic control analysis (Chapter 5) and a newly derived rate law for a channeled pathway. Furthermore, thermodynamic flow-force relationships and the power-law formalism are compared with classical enzyme kinetics.

As metabolic control analysis and optimization studies on biochemical systems are usually confined to steady states, we outline, in Section 2.3, fundamental concepts of analyzing stability of such states and treating multistationarity. Out of the wide domain of oscillatory behavior of biological systems, in particular, biochemical networks, we will sketch, in Section 2.4, the basic conditions for such behavior and some exemplifying models of glycolytic and calcium oscillations. Moreover, the frequently posed problem of finding minimal models showing oscillations is addressed. A three-component model of a chemical reaction system with very simple mass-action kinetics showing Hopf bifurcation is presented. Furthermore, the possible physiological significance of oscillations is discussed.

2.1. BALANCE EQUATIONS

Chemical and biochemical kinetics are based on the postulate that the reaction rate, v, at a point $\boldsymbol{r} = (x,y,z)$ in space at a time t can be expressed as a unique (usually nonlinear) function of the concentrations, S_i, of all participating chemical species at the point $\boldsymbol{r}$ and at the time t, and possibly of time,

$$v(\boldsymbol{r},t) = v[\boldsymbol{S}(\boldsymbol{r},t),t], \tag{2.1}$$

where $\boldsymbol{S}$ denotes the vector of concentrations. This equation allows for the possibility that the rate v is explicitly dependent on time t. Furthermore, Eq. (2.1) implies that (bio)chemical reactions are not subject to memory effects nor to long-range interactions; that is, interactions over distances longer than the diameter of the volume element taken for defining concentration as average number of moles per volume. It is worth noting, however, that in other fields of biological modeling [e.g., in population dynamics (cf. Gopalsamy, 1992) and molecular biology (cf. Heinrich and Rapoport, 1980)], memory effects play a major role and are then described by delay differential equations.

Direct dependence of reaction rates on time occurs, for example, in systems with oscillating inputs (Markus and Hess, 1990). In most cases, however, autonomous systems are considered; that is, systems that do not depend on time directly. For such systems, Eq. (2.1) implies that the state of a biochemical system at some point in space is uniquely given by all the concentration variables (i.e., by a finite-dimensional vector). The state is also characterized by parameters (e.g., rate constants), which are (in contrast to variables) constant in the time span of interest. As we will only deal with isothermic and isobaric systems, we also consider temperature and pressure as parameters. Furthermore, some concentrations can be treated as parameters if they are virtually constant (external metabolites, see below).

A further simplification used in many kinetic biochemical models and also throughout this book concerns spatial homogeneity; that is, all concentrations are considered uniform in the volume under study. This assumption is substantiated by the smallness of the volume of living cells and organelles, so that usual intermediates distribute uniformly by diffusion in a very short time. As for experimental setups *in vitro*, the test volumes have to be well stirred for the assumption of spatial homogeneity to be justified.

An essential characteristics of metabolic reaction networks is their stoichiometry. It indicates the molecularity (more exactly, the proportions of molecularities) with which the reactants and products enter the reactions. For example, in the reaction

$$2H_2O_2 \rightarrow 2H_2O + O_2 \tag{2.2}$$

catalyzed by the enzyme catalase (EC 1.11.1.6), hydrogen peroxide, water, and oxygen have the stoichiometric coefficients -2, 2, and 1, respectively. The signs of stoichiometric coefficients depend on the chosen orientation of the reaction. Usually, one considers the chemicals on the left-hand side of a reaction equation as reactants and those on the right-hand side as products, with the forward reaction going from "left to right" and the reverse reaction going from "right to left." This convention is not essential. Formally, the forward and backward reactions can be interchanged by inverting the signs of stoichiometric coefficients.

The set of stoichiometric coefficients of a reaction can be considered as a vector. When analyzing systems of several reactions, it is useful to arrange the set of these vectors in a matrix. Usually, the rows of this stoichiometry matrix refer to substances, whereas the columns refer to reactions. For instance, to the system of reactions

$$\text{glucose} + \text{ATP} \rightarrow \text{glucose-6-phosphate} + \text{ADP}, \tag{2.3a}$$

$$\text{glucose-6-phosphate} \rightarrow \text{glucose-1-phosphate}, \tag{2.3b}$$

catalyzed by the enzymes hexokinase (HK, EC 2.7.1.1) and phosphoglucomutase (PGM, EC 5.4.2.2), respectively, and proceeding, for example, in liver cells, one may attach the stoichiometry matrix

$$\mathbf{N} = \begin{array}{c} \begin{array}{cc} \text{HK} & \text{PGM} \end{array} \\ \begin{pmatrix} -1 & 0 \\ 1 & -1 \\ 0 & 1 \\ -1 & 0 \\ 1 & 0 \end{pmatrix} \end{array} \begin{array}{l} \\ \text{gluc} \\ \text{G6P} \\ \text{G1P} \\ \text{ATP} \\ \text{ADP} \end{array} \tag{2.4}$$

Here, all reacting species involved have been included into **N**. For larger biochemical systems, this is neither necessary nor useful. One often excludes those substances from the analysis, the concentrations of which are constant and virtually independent of the system parameters due to presence in large excess (e.g., water) or to homeostasis of the substance as maintained by the biological organism (e.g., glucose in the blood). They are usually referred to as external substances. Another example is the influx of a species at constant rate. This species can then be considered as the product of the first-order degradation of a substance present in time-invariant concentration (see Horn and Jackson, 1972). If in the example given in Eq. (2.3), glucose and glucose-1-phosphate are considered as external, the first and third rows in **N** can be canceled.

Kinetic modeling in biochemistry has been made possible by experimental identification of the structure of metabolic pathways, resulting in highly detailed charts and metabolic maps. Therefore, the situation is somewhat different from chemical kinetics dealing with an inanimate nature. The latter is often concerned with detecting the reaction mechanism; that is, identifying the compilation of elementary steps for a multistep process (cf. Bauer, 1990; Corio and Johnson, 1991). This search may not lead to a unique adequate mechanism, as several mechanisms consistent with some overall process may not be distinguishable (cf. Vajda and Rabitz, 1994).

Things are different in biochemistry, where most reactions are catalyzed by enzymes. [Examples of nonenzymic reactions in living cells are processes involving free radicals and several glycation reactions of proteins (cf. Giardino *et al.*, 1994)]. Thus, the mechanism is normally uniquely determined by the presence of enzymes, provided they are highly specific. Therefore, the stoichiometry of the systems can be taken as a prerequisite of the analysis. In contrast, the atomic composition of the substances (e.g., proteins) is often incompletely known, unlike in "nonbiological" chemical kinetics.

Reaction rates, v, are usually given as the rate of change in the extent of reaction divided by the volume, V,

$$v(t) = \frac{1}{V}\frac{d\xi}{dt}. \tag{2.5}$$

The extent of reaction, ξ, is defined as

$$\xi(t) = \frac{1}{n_i}\,\Delta N_i(t) \tag{2.6}$$

(cf. Prigogine and Defay, 1954; Smith and Missen, 1992), where ΔN_i is the difference $N_i(t) - N_i(t_0)$ of mole numbers of substance S_i, with t_0 being some reference point in time, and n_i denotes the stoichiometric coefficient of substance S_i for the reaction under consideration.

If the stoichiometric coefficients coincide with the molecularities in the reaction, the reaction rate is uniquely defined in terms of concentration changes by

Eqs. (2.5) and (2.6). When detailed knowledge about the molecular mechanism is not available for a given overall reaction equation, stoichiometric coefficients are indeterminate up to rescaling by a common factor. Accordingly, the reaction rate can be arbitrarily scaled as well.

In the usual situation that the biochemical system encompasses (much) more than one reaction, we denote reaction rates by v_j ($j = 1, \ldots, r$) and the stoichiometric coefficients by n_{ij}, where i and j refer to the subscripts of the substance and the reaction, respectively.

When (bio)chemical reactions are the only cause of concentration changes (i.e., when there is no mass flow due to convection, diffusion, etc.), the temporal behavior of concentrations is given by the balance equation

$$\frac{\mathrm{d}S_i}{\mathrm{d}t} = \sum_{j=1}^{r} n_{ij} v_j \tag{2.7a}$$

(cf. Glansdorff and Prigogine, 1971; Horn and Jackson, 1972). This equation is a consequence of the definition (2.6) and the conservation of mass, so that the contributions of all reactions can be summed. Equation (2.7a) can be written in matrix notation as

$$\frac{\mathrm{d}\boldsymbol{S}}{\mathrm{d}t} = \mathbf{N}\boldsymbol{v}, \tag{2.7b}$$

where $\boldsymbol{v}$ and $\boldsymbol{S}$ denote the vectors of reaction rates and concentrations, respectively.

Because we wish to exclude diffusion and convection, we can apply Eq. (2.7) to the transport of substances from one compartment to another, both of which are spatially homogeneous (for example, proton transport from mitochondria into the cytosol). The transported substance then has to be indicated by different subscripts for the two compartments. If the compartments have different volumes, one must either express S_i in moles rather than moles/volume or divide the stoichiometric coefficients by the compartment volume, so that Eq. (2.7) can still be applied.

When the system is autonomous, Eq. (2.7) becomes, more specifically,

$$\frac{\mathrm{d}\boldsymbol{S}(t)}{\mathrm{d}t} = \mathbf{N}\boldsymbol{v}(\boldsymbol{S}(t)) = \boldsymbol{f}(\boldsymbol{S}(t)), \tag{2.8}$$

where $\boldsymbol{f}(\cdot)$ is a vector function of the time-dependent concentrations.

In the frequent situation that a biochemical system subsists in a steady state (cf. Section 2.3), the balance equation (2.8) becomes

$$\mathbf{N}\boldsymbol{v}(\boldsymbol{S}) = \boldsymbol{0}. \tag{2.9}$$

This represents an algebraic equation system in the variables S_i. For reaction rates at steady state, we will frequently use the term flux. Note that this term will normally refer to a scalar quantity rather than to a flow in space.

2.2. RATE LAWS

2.2.1. Generalized Mass-Action Kinetics

The functions $v_j(\boldsymbol{S})$ entering Eq. (2.8) represent rate laws (also called kinetic functions). More exactly, they should be written $v_j(\boldsymbol{S},\boldsymbol{p})$ with $\boldsymbol{p}$ being a vector of parameters p_k. Certain classes (types) of rate laws can be discerned. A very well known and fundamental kinetic function is the mass action rate law suggested by Guldberg and Waage in the last century (cf. Smith and Missen, 1992). It is derived from the idea that the reaction velocity is proportional to the probability of collision of reactants, which in turn is proportional to each concentration raised to the power of the respective molecularity, because this is the number of molecules that have to meet to initiate the reaction. This gives

$$v_j(\boldsymbol{S},\boldsymbol{p}) = k_{+j} \prod_i S_i^{n_{ij}^-} - k_{-j} \prod_i S_i^{n_{ij}^+} \tag{2.10}$$

(cf. Moore, 1972; Horn and Jackson, 1972), where k_{+j} and k_{-j} denote the forward and reverse rate constants, respectively, of reaction j. n_{ij}^- and n_{ij}^+ stand for the stoichiometric coefficients of reactants and products, respectively, that is,

$$n_{ij}^- = \begin{cases} -n_{ij} & \text{if } n_{ij} < 0 \\ 0 & \text{otherwise,} \end{cases} \tag{2.11a}$$

$$n_{ij}^+ = \begin{cases} n_{ij} & \text{if } n_{ij} > 0 \\ 0 & \text{otherwise.} \end{cases} \tag{2.11b}$$

This implies $n_{ij} = n_{ij}^+ - n_{ij}^-$. In the case of the rate law (2.10), k_{+j}, k_{-j} and the n_{ij}^-, n_{ij}^+ form the parameter vector $\boldsymbol{p}$. The rate constants depend on temperature and pressure.

At equilibrium (i.e. when $v_j = 0$), Eq. (2.10) implies

$$\frac{\prod_i S_i^{n_{ij}^+}}{\prod_i S_i^{n_{ij}^-}} = \prod_i S_i^{n_{ij}} = \frac{k_{+j}}{k_{-j}} = q_j, \tag{2.12}$$

which is the well-known law of mass action, with $q_j = k_{+j}/k_{-j}$ denoting the equilibrium constant of reaction j.

If external metabolites, P_i, participate in the reaction, their concentrations may be included in the mass-action term

$$v_j(\boldsymbol{S},\boldsymbol{p}) = k_{+j} \prod_i S_i^{n_{ij}^-} \prod_k P_k^{m_{kj}^-} - k_{-j} \prod_i S_i^{n_{ij}^+} \prod_k P_k^{m_{kj}^+}, \tag{2.13}$$

where the products over i and k run over all internal and external metabolites, respectively. The stoichiometric coefficients, m_{kj}^+ and m_{kj}^-, are defined similarly as in Eqs. (2.11a) and (2.11b). Alternatively, the external concentrations may be incorporated into the rate constants. An apparent equilibrium constant can then be defined as

$$\tilde{q}_j = q_j \prod_k P_k^{-m_{kj}}. \tag{2.14}$$

Any (bio)chemical rate law must satisfy the condition that upon insertion into the balance equation (2.8), the concentrations always remain non-negative. This condition is actually met by the kinetics (2.10) and (2.13), as can be seen by the following. The initial concentration values are clearly non-negative. Assume that at some point in time, some concentration S_i becomes zero. For each reaction, one can distinguish the three following cases. If n_{ij} is positive, the second term on the right-hand sides of Eqs. (2.10) and (2.13) equals zero, because of the assumptions $S_i = 0$ and $n_{ij}^+ > 0$. The velocity therefore remains non-negative, which ensures, owing to the balance equation (2.8) and $n_{ij} > 0$, that S_i cannot decrease below zero. If n_{ij} is negative, the first term on the right-hand sides of Eqs. (2.10) and (2.13) is zero, due to $S_i = 0$ and $n_{ij}^- > 0$. The rate is then zero or negative, so that the balance equation (2.8) implies that S_i cannot decrease further. If $n_{ij} = 0$, the rate has no effect on S_i owing to Eq. (2.8).

For reactions in nonideal solutions, it is sensible to use the rate law (2.10) in a more general way, by allowing the exponents to differ from the stoichiometric coefficients and even to be noninteger (Othmer, 1981).

Enzyme-catalyzed reactions can be described at least at two different levels. First, all elementary steps of enzyme-substrate binding, isomerization, and dissociation of enzyme intermediates may be taken into account. For these steps, the mass-action rate law is normally well suited. Kinetic modeling of enzymic systems can, however, be simplified considerably if overall enzymic reactions rather than all the elementary steps are treated as basic units, because the order of the governing differential equation system and the number of parameters are reduced. Rate laws of these enzyme-catalyzed reactions can then be derived, in which only the concentrations of nonenzymic substrates and products but not the concentrations of enzyme intermediates occur (cf. Cornish-Bowden and Wharton, 1988; Kuby, 1991). To derive enzymatic rate laws, usually certain approximations are employed; for example, the assumption that enzyme-containing intermediates are at equilibrium with the substrates (equilibrium models) or the quasi-steady-state hypothesis, which says that enzyme intermediates attain a quasi-steady-state even when the concentrations of the nonenzymic substances still change in time (see Section 4.2).

At the level of overall enzymic reactions, the enzyme-kinetic rate laws exhibit features that cannot immediately be described by mass-action kinetics (e.g., saturation, cooperativity, inhibition or activation by effectors). The phenomenon of saturation, for example, arises from the fact that at high substrate concentration, nearly all enzyme molecules are bound to the substrate, so that a further increase in substrate concentration has almost no effect on reaction rate.

To cope with the various specific phenomena in enzyme kinetics in a general way, Schauer and Heinrich (1983) proposed a generalized mass-action rate law of the form

$$v_j(\boldsymbol{S},\boldsymbol{p}) = F_j(\boldsymbol{S},\boldsymbol{p})\left[k_{+j}\prod_i S_i^{n_{ij}^-} - k_{-j}\prod_i S_i^{n_{ij}^+}\right]. \tag{2.15}$$

The $F_j(\boldsymbol{S},\boldsymbol{p})$ are positive functions which describe the above-mentioned specific, nonlinear effects. The parameter vector $\boldsymbol{p}$ in this notation contains all parameters apart from the rate constants and the stoichiometric coefficients, n_{ij}^- and n_{ij}^+. Note that the requirement that concentrations remain always non-negative is again satisfied and that this kinetics is consistent with the law of mass action, because at equilibrium, where $v_j = 0$, Eq. (2.15) entails Eq. (2.12). Moreover, the usual mass-action kinetics (2.10) is comprised in Eq. (2.15) as a special case with $F_j(\boldsymbol{S},\boldsymbol{p}) \equiv 1$.

The mass-action kinetics can be written in terms of the reaction affinity, which is defined as

$$A_j = RT\ln\left(\bar{q}_j\prod_i S_i^{-n_{ij}}\right). \tag{2.16}$$

From Eq. (2.15), one obtains

$$v_j = G_j(\boldsymbol{S},\boldsymbol{p})\left[\exp\left(\frac{A_j}{RT}\right) - 1\right], \tag{2.17}$$

where

$$G_j(\boldsymbol{S},\boldsymbol{p}) = F_j(\boldsymbol{S},\boldsymbol{p})k_{-j}\prod_i S_i^{n_{ij}^+}. \tag{2.18}$$

From Eq. (2.17), it can be seen that for nonzero concentrations, a velocity v_j is zero if and only if the reaction affinity A_j is zero, because for positive concentrations, the function $G_j(\boldsymbol{S},\boldsymbol{p})$ is unequal to zero.

A drawback of Eq. (2.17) is that it is only applicable to reversible reactions, whereas Eq. (2.15) also describes irreversible reactions, when $k_{+j} = 0$ or $k_{-j} = 0$.

2.2.2. Various Enzyme-Kinetic Rate Laws

A fundamental rate law for enzymic reactions is the *Michaelis–Menten kinetics,* which applies to enzymes following the *uni-uni mechanism* shown in Scheme 1 (Henri, 1902; Michaelis and Menten, 1913). While it was first derived for irreversible reactions, it was later generalized for the case of reversible uni-uni reactions (Haldane, 1930). The temporal changes in the concentrations of the enzyme-substrate complex and free enzyme are determined by

$$\frac{\mathrm{d}ES}{\mathrm{d}t} = -\frac{\mathrm{d}E}{\mathrm{d}t} = k_1 S_1 \cdot E - (k_{-1} + k_2)ES + k_{-2}S_2 \cdot E. \tag{2.19}$$

This equation is consistent with the fact that the total enzyme concentration is constant, $E + ES = E_T = const.$ Using the quasi-steady-state assumption $\mathrm{d}ES/\mathrm{d}t \cong 0$ (cf. Section 4.2), one obtains the rate equation

$$v(S_1,S_2) = \frac{V_m^+ S_1/K_{m1} - V_m^- S_2/K_{m2}}{1 + S_1/K_{m1} + S_2/K_{m2}}, \tag{2.20}$$

which contains several phenomenological constants. V_m^+ and V_m^- denote the maximal activities of the forward and reverse reactions, respectively. K_{m1} and K_{m2} are the Michaelis constants of S_1 and S_2, respectively.

$$\mathrm{E} + \mathrm{S}_1 \underset{k_{-1}}{\overset{k_1}{\rightleftharpoons}} \mathrm{ES} \underset{k_{-2}}{\overset{k_2}{\rightleftharpoons}} \mathrm{E} + \mathrm{S}_2 \qquad \text{Scheme 1}$$

In the case of Scheme 1, the phenomenological constants are linked with the elementary rate constants by

$$K_{m1} = \frac{k_{-1} + k_2}{k_1}, \qquad K_{m2} = \frac{k_{-1} + k_2}{k_{-2}}, \tag{2.21a,b}$$

$$V_m^+ = k_2 E_T, \qquad V_m^- = k_{-1} E_T. \tag{2.22a,b}$$

In the case that the concentrations are small compared to the respective Michaelis constant, the kinetics (2.20) simplifies to the linear mass-action kinetics

$$v(S_1,S_2) = \frac{V_m^+}{K_{m1}} S_1 - \frac{V_m^-}{K_{m2}} S_2. \tag{2.23}$$

The ratios of maximal activities and Michaelis constants then play the role of first-order rate constants.

Setting $S_2 = 0$ in Eq. (2.20), one obtains the irreversible Michaelis–Menten kinetics

$$v = \frac{V_m S}{K_{mS} + S}, \tag{2.24}$$

with the simplified notation $S = S_1$, $V_m = V_m^+$, and $K_{mS} = K_{m1}$.

The reversible Michaelis–Menten kinetics (2.20) fits into the generalized mass-action kinetics (2.15), with

$$F(S_1,S_2) = \frac{1}{1 + S_1/K_{m1} + S_2/K_{m2}} \tag{2.25}$$

and

$$q = \frac{V_m^+ K_{m2}}{V_m^- K_{m1}}. \tag{2.26}$$

Relation (2.26) interrelating the phenomenological coefficients with the equilibrium constant was derived by Haldane (1930).

The rate law (2.20) is a minimal model in that a minimal catalytic scheme was assumed, in which only one form of the free enzyme and one form of the enzyme-substrate complex exist. There is experimental evidence that for many enzymes, the catalytic mechanism is more complex; for example, involving distinct enzyme-substrate and enzyme-product complexes. The three-step mechanism depicted in Scheme 2 is more realistic than the two-step mechanism shown in Scheme 1, because it involves separate steps of binding and catalytic conversion.

$$E + S_1 \underset{k_{-1}}{\overset{k_1}{\rightleftharpoons}} ES_1 \underset{k_{-2}}{\overset{k_2}{\rightleftharpoons}} ES_2 \underset{k_{-3}}{\overset{k_3}{\rightleftharpoons}} E + S_2$$

Scheme 2

The three-step mechanism can also be described by the reversible Michaelis–Menten kinetics (2.20), but the relations between the phenomenological constants and the elementary rate constants read now

$$V_m^+ = \frac{k_2 k_3 E_T}{k_2 + k_3 + k_{-2}}, \tag{2.27a}$$

$$V_m^- = \frac{k_{-1}k_{-2}E_T}{k_2 + k_{-1} + k_{-2}}, \quad (2.27b)$$

$$K_{m1} = \frac{k_2k_3 + k_{-1}k_3 + k_{-1}k_{-2}}{k_1(k_2 + k_3 + k_{-2})}, \quad (2.27c)$$

$$K_{m2} = \frac{k_2k_3 + k_{-1}k_3 + k_{-1}k_{-2}}{k_{-3}(k_2 + k_{-1} + k_{-2})}. \quad (2.27d)$$

A generalized rate law can be derived for unbranched catalytic schemes with any number, r, of elementary steps

$$v = (\tilde{q} - 1)E_T\left(\prod_{i=1}^{r}\frac{1}{k_{-i}} + \prod_{i=1}^{r}\prod_{j=1}^{r-1}\frac{1}{k_{-(i-1)}}\prod_{h=i}^{i+j-1}\frac{k_h}{k_{-h}}\right)^{-1}, \quad (2.28)$$

where by definition $k_{\pm(n+i)} = k_{\pm i}$ and $k_{-0} = k_{-n}$ (Wilhelm *et al.*, 1994). This type of scheme covers all ordered mechanisms with one or more substrates and products. This means, in particular, that bimolecular reactions are also comprised. The concentrations of substrates and products are incorporated into the rate constants of the respective steps where they enter the scheme, giving rise to apparent rate constants. The apparent equilibrium constant is linked with the (apparent) rate constants by

$$\tilde{q} = \prod_{i=1}^{r}\frac{k_i}{k_{-i}}. \quad (2.29)$$

The kinetics (2.28) fits in the generalized mass-action kinetics (2.15), with $F_j(\boldsymbol{S},\boldsymbol{p})$ being the reciprocal of a polynomial in the concentrations. Special cases of Eq. (2.28) for monomolecular reactions (uni-uni reactions) had been derived by Peller and Alberty (1959), and for systems involving two reactants and two products (bi-bi reactions) by Bloomfield *et al.* (1962). It is easy to see that for uni-uni reactions with the substrate binding in the first step and the product being released in the final step, Eq. (2.28) leads, irrespective of the number of elementary steps, to the phenomenological rate law (2.20).

Those reactions of molecularity higher than one (e.g., bi-uni and bi-bi) following a catalytic scheme other than the ordered mechanism (e.g., random or ping-pong mechanisms) do not obey the rate law (2.28), but are comprised in the generalized mass-action kinetics (2.15) with $F_j(\boldsymbol{S},\boldsymbol{p})$ being a quotient of two polynomials. Specific rate laws of this type are given, for example, by Kuby (1991).

As was seen above, the phenomenological constants are linked with the elementary rate constants in a different way and, hence, have different physical meanings for different reaction mechanisms. Although experimental determination of the values of elementary rate constants has advanced appreciably (Baykov *et al.,* 1990, 1993; Christensen *et al.,* 1990; Patel *et al.,* 1991), it is generally impossible to assign values to all of these constants in a mechanism which involves even a reasonable number of intermediates. On the other hand, this knowledge is often not necessary to write down phenomenological rate laws.

For deriving phenomenological rate laws, the catalytic mechanism can be considered, to some extent, as a black box. The question arises about which extensions to a given enzyme scheme would change the rate law and which would not. An example of a uni-uni mechanism where the phenomenological rate law has a structure different from Eq. (2.20) is provided by a three-step mechanism in which the free enzyme isomerizes between two different forms. The rate law can be derived from Eq. (2.28). Setting the number of steps $r = 3$, assigning the index 3 to the isomerization step of the enzyme and replacing k_1 by k_1S_1, and k_{-2} by $k_{-2}S_2$, we obtain the formula

$$\begin{aligned} v &= \frac{E_T}{D}(k_1k_2k_3S_1 - k_{-1}k_{-2}k_{-3}S_2) \\ D &= k_{-1}k_{-3} + k_2k_{-3} + k_2k_3 + k_{-1}k_3 + k_1(k_2 + k_3)S_1 \\ &\quad + k_{-2}(k_{-1} + k_{-3})S_2 + k_1k_{-2}S_1S_2, \end{aligned} \tag{2.30}$$

which has also been derived by Cornish-Bowden (1994). It can be seen that this rate law is not comprised in the Michaelis–Menten kinetics (2.20), because the denominator includes a term proportional to S_1S_2.

An important aspect in enzyme kinetics is the *effect of inhibitors and activators.* There are numerous catalytic mechanisms including the action of effectors. For illustration, we discuss an example of a mixed-type inhibitor of a reversible, uni-uni reaction. Consider a two-step catalytic mechanism with an effector binding to the free enzyme and to the enzyme-substrate complex with the dissociation constants

$$K_{\mathrm{Ia}} = \frac{E \cdot I}{EI}, \tag{2.31}$$

$$K_{\mathrm{Ib}} = \frac{ES \cdot I}{ESI}. \tag{2.32}$$

EI and ESI are dead-end complexes which are not transformed into product (see Scheme 3).

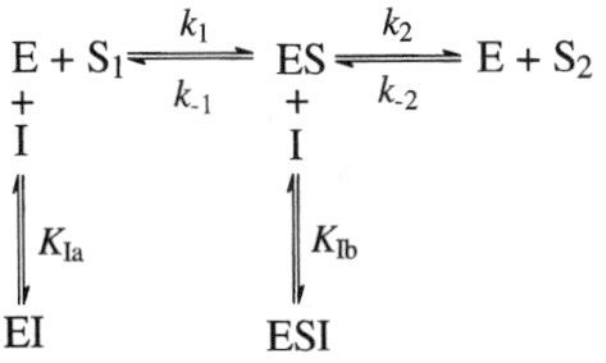

Scheme 3

The enzyme species are subject to the relation

$$E + EI + ES + ESI = E_T = \text{const.} \tag{2.33}$$

Applying the quasi-steady-state approximation to ES (cf. Section 4.2), dES/dt is assumed to equal zero. This gives

$$ES = \frac{E_T(k_1S_1 + k_{-2}S_2)}{(k_1S_1 + k_{-2}S_2)(1 + I/K_{\mathrm{Ib}}) + (k_{-1} + k_2)(1 + I/K_{\mathrm{Ia}})}. \tag{2.34}$$

With $v = k_2ES - k_{-2}E \cdot S_2$ and the definitions (2.21) and (2.22), one obtains for the quasi-steady-state reaction rate

$$v(S_1,S_2) = \frac{V_{\mathrm{m}}^{+}S_1/K_{\mathrm{m1}} - V_{\mathrm{m}}^{-}S_2/K_{\mathrm{m2}}}{1 + I/K_{\mathrm{Ia}} + (S_1/K_{\mathrm{m1}} + S_2/K_{\mathrm{m2}})(1 + I/K_{\mathrm{Ib}})}. \tag{2.35}$$

This inhibition kinetics also fits in the general form (2.15). It coincides with the reversible Michaelis–Menten rate law (2.20) when $I = 0$, as should be expected.

Three special cases of the inhibition kinetics (2.35) are of particular interest:

(a) *Competitive inhibition* ($0 < K_{\mathrm{Ia}} < \infty$, $K_{\mathrm{Ib}} \to \infty$). In this case, the inhibitor only binds to the free enzyme. Equation (2.35) then specifies to

$$v(S_1,S_2) = \frac{V_{\mathrm{m}}^{+}S_1/K_{\mathrm{m1}} - V_{\mathrm{m}}^{-}S_2/K_{\mathrm{m2}}}{1 + I/K_{\mathrm{Ia}} + S_1/K_{\mathrm{m1}} + S_2/K_{\mathrm{m2}}}. \tag{2.36}$$

(b) *Noncompetitive inhibition* ($0 < K_{\mathrm{Ia}} = K_{\mathrm{Ib}} < \infty$). In this case, binding of the inhibitor to E and to ES is equally tight. Equation (2.35) then simplifies to

$$v(S_1,S_2) = \frac{V_{\mathrm{m}}^{+}S_1/K_{\mathrm{m1}} - V_{\mathrm{m}}^{-}S_2/K_{\mathrm{m2}}}{(1 + S_1/K_{\mathrm{m1}} + S_2/K_{\mathrm{m2}})(1 + I/K_{\mathrm{Ia}})}. \tag{2.37}$$

This type of inhibition amounts to an effective diminution of total enzyme.

(c) *Uncompetitive inhibition* ($K_{Ia} \to \infty$, $0 < K_{Ib} < \infty$). Equation (2.35) specifies to

$$v(S_1,S_2) = \frac{V_m^+ S_1/K_{m1} - V_m^- S_2/K_{m2}}{1 + (S_1/K_{m1} + S_2/K_{m2})(1 + I/K_{Ib})}. \tag{2.38}$$

Although in the above kinetic equations describing enzyme inhibition, I denotes, strictly speaking, the free-inhibitor concentration, it is often approximately identified with its total concentration, I_T, because this quantity is normally better known, for example when the inhibitor is added in experiment. This approximation is justified whenever total inhibitor concentration is much higher than total enzyme concentration. In the case of high dissociation constants K_{Ia} and K_{Ib} (poor binding), a weaker condition for validity of the approximation $I \cong I_T$ results in that the total enzyme concentration may be of the same order of magnitude as, or even lower than, the total inhibitor concentration, because most of the inhibitor then subsists in the free form. When the contribution of the bound inhibitor to the mass balance is not neglected, the quasi-steady-state equation gives rise to a quadratic equation in ES (cf. Gellerich *et al.*, 1990). Things become simpler in the case of irreversible inhibition; that is, when either or both of the equilibrium constants K_{Ia} and K_{Ib} are extremely low. If $I_T < E_T$ almost all of the inhibitor then subsists in the form bound to the enzyme. If $I_T > E_T$, all of the enzyme is bound to the inhibitor and, hence, inactive, whereas the excess inhibitor remains in the free form. The reversible Michaelis–Menten rate law (2.20) can then be modified by the consideration that the total enzyme concentration is approximately diminished by the inhibitor concentration to give

$$v_j(S_1,S_2) \cong \begin{cases} \dfrac{V_m^+ S_1/K_{m1} - V_m^- S_2/K_{m2}}{1 + S_1/K_{m1} + S_2/K_{m2}}\left(1 - \dfrac{I_T}{E_T}\right) & \text{if } I_T < E_T \\ 0 & \text{if } I_T > E_T \end{cases}. \tag{2.39}$$

Wang and Tsou (1987) stressed the fact that the enzyme-kinetic literature on enzyme inhibition mainly concerns reversible inhibition, although irreversible inhibition is also very important. Many chemotherapeutic agents as well as pesticides are irreversible enzyme inhibitors by alkylating, phosphorylating, or acylating at the active sites.

The above rate laws for inhibition mechanisms are based on the assumption that EI and ESI are dead-end complexes (see Scheme 3). In the more general case that an interconversion of EI and S_1 to ESI and perhaps a slow reaction from ESI to EI and S_2 occur, the quasi-steady-state rate laws are more complex and generally involve terms proportional to S_1^2, I^2, S_1^2I, and S_1I^2 even under the simplifying assumption that the final steps of product formation are irreversible (cf. I.H. Segel and Martin, 1988). Rate laws for enzymes involving two substrates in the presence of modifiers can be found, for example, in the works of Wang and Tsou (1987) and Kuby (1991, Chap. 5). A special case of inhibition studied in recent years is the so-called suicide inactivation, where the enzyme-substrate complex converts into enzyme and product(s) and in parallel into an inactive complex (cf. Casas *et al.,* 1993).

Derivation of quasi-steady-state rate laws generally involves solving a system of linear algebraic equations for the enzyme-containing species (such as $\mathrm{d}ES/\mathrm{d}t = 0$ in the mechanism shown in Scheme 3). The number of concentration variables can be diminished by consideration of the fact that total enzyme concentration is conserved [such as expressed in Eq. (2.33)]. The solution of the resulting inhomogeneous linear equation system can be found by standard methods of linear algebra. The solving procedure can be simplified by a graph-theoretical method developed by King and Altman (1956). This method is explained in detail in a number of textbooks (e.g., Cornish-Bowden, 1976b; T.L. Hill, 1977; Kuby, 1991). Due to the ever increasing facilities of symbolic computation software such as MATHEMATICA or MAPLE, enzymatic rate laws can nowadays also be derived with the aid of these programs.

Plots of enzymatic activity versus substrate concentration often exhibit *sigmoidal kinetics,* that is, the rate increases more than linearly for low substrate concentrations. One of the first to detect such behavior was A.V. Hill (1910) when studying the binding of oxygen to hemoglobin. He used an empirical equation to describe this binding mathematically, which can be translated into the following rate equation:

$$v = \frac{V_{\mathrm{m}}(S/K_{\mathrm{S}})^{n_{\mathrm{H}}}}{1 + (S/K_{\mathrm{S}})^{n_{\mathrm{H}}}} \tag{2.40}$$

(n_{H}: Hill coefficient, K_{S}: half-saturation constant). A mechanistic explanation of this equation would be that n_{H} molecules bind simultaneously to the enzyme E before transformation into the product occurs. In the more realistic situation that the substrate molecules bind sequentially, Eq. (2.40) has to be replaced by a quotient of polynomials involving powers of concentrations with exponents from 0 to n_{H} (cf. Ricard and Noat, 1986).

For example, consider an enzyme that can bind two molecules of substrate, so that the ternary complex $E(S)_2$ is able to irreversibly generate the product, whereas the binary complex ES is not. The corresponding quasi-steady-state rate law fits into the general form

$$v(S) = \frac{aS + \beta S^2}{\gamma + \delta S + \varepsilon S^2} \tag{2.41}$$

with $a = 0$ (cf. Kuby, 1991, Chap. 7). In case that both an active and an inactive form of the ES complex exist, an equation of the form (2.41) with $a \neq 0$ results (Frieden, 1964). Rate laws of this type also apply to other mechanisms, for example, in the situation that the substrate contains an impurity that forms an inactive complex with the enzyme (Kuby, 1991), or when the enzyme can exist in two forms (e.g., conformers) with two different activities (Ricard, 1978). Note that Eq. (2.41) does not necessarily produce sigmoidal kinetics. Under the condition

$$\beta\gamma > \alpha\delta, \tag{2.42}$$

one obtains $\partial^2 v/\partial S^2 > 0$ for $S = 0$ and, therefore, sigmoidal kinetics.

A biphasic, but not sigmoidal, rate law is obtained in the case that one and the same reaction is catalyzed by two enzymes, or by an enzyme with two catalytic sites per molecule. Indeed, the sum of two Michaelis–Menten rate laws of the type given in Eq. (2.24) gives, when written over a common denominator, an equation of the type (2.41). A similar procedure can be applied to multiphase saturation curves. Recently, Holzhütter *et al.* (1994) fitted experimental data of the transport rate of the oxoglutarate-malate exchanger of rat-heart mitochondria to a sum of five Hill equations.

Sigmoidal kinetics is also obtained for enzymes composed of several subunits that interact with each other. A classical model was established by Monod *et al.* (1965). It starts from the assumptions that the enzyme can exist in two conformations with different catalytic activity and that binding of substrate, activator, and/or inhibitor leads to transitions between these states. We now give the Monod-Wyman-Changeux equation for the case that one conformation is completely inactive and the reaction is irreversible:

$$v(S) = \frac{V_m(S/K_S)(1 + S/K_S)^{n-1}}{(1 + S/K_S)^n + L'}, \tag{2.43a}$$

$$L' = L\left(\frac{1 + I/K_I}{1 + A/K_A}\right)^n \tag{2.43b}$$

(n: number of subunits of the allosteric enzyme, L: allosteric constant of the transition from active to inactive state, K_S: intrinsic dissociation constants of the enzyme-substrate complex, I: inhibitor concentration, A: activator concentration, K_I: inhibition constant, K_A: activation constant). Equation (2.43) takes into account both homotropic and heterotropic effects (i.e., the interactions between identical ligands and the interaction between different ligands, respectively). The Monod-Wyman-Changeux model is based on the assumption that all subunits switch simultaneously from one conformation to the other. When the corresponding equilibrium constant L becomes negligibly small, Eq. (2.43) simplifies to the

Michaelis–Menten equation (2.24). A more general case where the subunits can exist in different conformations simultaneously is described by a model of Koshland *et al.* (1966).

The presentation of enzyme kinetics in this section focuses on time-independent rate laws, which indeed constitute the main objective of research in this field. In addition, the temporal behavior of enzymes has been treated since the very beginning of investigations in enzyme kinetics (Henri, 1902). Recent developments in this direction are, among others, the description of non-steady-state enzyme kinetics (Schauer and Heinrich, 1979; Segel, 1988; Frenzen and Maini, 1988; Chou, 1993) and the kinetics of enzyme-catalyzed reactions in the presence of an unstable modifier (which is assumed to decay exponentially) (Topham, 1990).

In recent years, the putative importance of heterologous enzyme-enzyme complexes and of the direct transfer of metabolic intermediates between these have been discussed intensively (Srivastava and Bernhard, 1986; Srere, 1987; Cheung *et al.,* 1989; Ovádi, 1991; Anderson *et al.,* 1991). This phenomenon is usually referred to as *metabolic channeling.* Traditional equations of enzyme catalysis, which take into account complexes of enzymes with metabolites or with modifiers, but not with other enzymes, are insufficient in the situation of metabolic channeling.

We will here derive an enzymatic rate law for a very simple channel mechanism, which is depicted in Figure 2.1 and studied in more detail in Sections 5.6.4 and 5.15. The symbol w is used for the net rates of elementary steps. For the present derivation, this reaction system is further simplified. First, the channel is to be considered perfect; that is, the route of catalysis by the separate enzymes E_1 and E_2 is excluded (i.e., $w_{1b} = w_{2a} = 0$). Second, the two dissociation steps

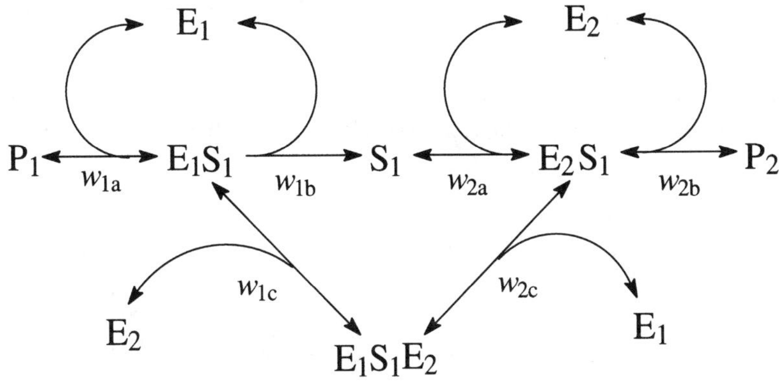

Figure 2.1 Reaction scheme with two sequential reactions including a parallel branch of dynamic channeling.

2b and 2c are assumed to be irreversible. A further idealization is that the association steps for the two enzymes are so fast as to be at quasi-equilibrium. This gives

$$\frac{P_1 \cdot E_1}{E_1S_1} = K_1, \tag{2.44a}$$

$$\frac{E_1S_1 \cdot E_2}{E_1S_1E_2} = K_2. \tag{2.44b}$$

The two enzymes obey the conservation relations

$$E_1 + E_1S_1 + E_1S_1E_2 = E_{T,1}, \tag{2.45a}$$

$$E_2 + E_2S_1 + E_1S_1E_2 = E_{T,2}. \tag{2.45b}$$

The quasi-steady-state assumption for E_2S_1 yields

$$E_2S_1 = \frac{k_{2c}}{k_{2b}} E_1S_1E_2. \tag{2.46}$$

Substituting E_1 and E_2 in Eqs. (2.44a) and (2.44b) by Eqs. (2.45a) and (2.45b), respectively, gives, under consideration of Eq. (2.46),

$$P_1(E_{T,1} - E_1S_1 - E_1S_1E_2) = K_1 \cdot E_1S_1, \tag{2.47a}$$

$$E_1S_1\left(E_{T,2} - E_1S_1E_2 - \frac{k_{2c}}{k_{2b}} E_1S_1E_2\right) = K_2 \cdot E_1S_1E_2. \tag{2.47b}$$

These two equations entail a quadratic equation for the concentration of the complex $E_1S_1E_2$. For the steady-state reaction velocity of the channel, $v = k_{2c} \cdot E_1S_1E_2$, it follows

$$v = \frac{k_{2c}}{2}\left[E_{T,1} + aE_{T,2} + K_2a\left(1 + \frac{K_1}{P_1}\right)\right] \left\{1 - \left(1 - \frac{4aE_{T,1}E_{T,2}}{(E_{T,1} + aE_{T,2} + K_2a(1 + K_1/P_1))^2}\right)^{1/2}\right\} \tag{2.48}$$

with

$$a = \frac{k_{2b}}{k_{2b} + k_{2c}}, \qquad 0 \leq a \leq 1. \tag{2.49}$$

The limiting cases for a are characterized by high k_{2b} and high k_{2c}, which imply that one of the dissociation steps is extremely fast. From among the two solutions to the quadratic equation, the solution with the plus sign is irrelevant, because for $E_{T,1} \to 0$ or $E_{T,2} \to 0$, the rate must vanish. For very large substrate concentrations, the rate v tends to a finite value.

An interesting situation is when the second term in the square root in Eq. (2.48) is small compared with unity. This occurs, for example, in the case of high dissociation constant K_2 compared with both enzyme concentrations, or if one enzyme concentration is much lower than the other one, or if the substrate concentration is very low compared with K_1. The square root in Eq. (2.48) can then be expanded into a Taylor series. Neglecting terms of order greater than unity yields the approximation formula

$$v = \frac{k_{2c} a E_{T,1} E_{T,2} P_1}{(E_{T,1} + aE_{T,2})P_1 + aK_2(P_1 + K_1)}. \tag{2.50}$$

For vanishing substrate concentration, the rate v tends to zero, as should be expected. In contrast to the usual Michaelis–Menten kinetics, where the rate is proportional to the total enzyme concentrations, now saturation occurs in the dependence of v on enzyme concentrations.

Equation (2.50) can be written in the form of the irreversible Michaelis–Menten equation (2.24) with $P_1 = S$,

$$V_m = \frac{k_{2c} a E_{T,1} E_{T,2}}{E_{T,1} + aE_{T,2} + aK_2}, \tag{2.51a}$$

$$K_{m1} = \frac{aK_1 K_2}{E_{T,1} + aE_{T,2} + aK_2}. \tag{2.51b}$$

The Michaelis constant now depends on enzyme concentrations, except for the case that K_2 is very high. In this case, we have

$$V_m = \frac{k_{2c} E_{T,1} E_{T,2}}{K_2}, \tag{2.52a}$$

$$K_{m1} = K_1. \tag{2.52b}$$

The rate of the channel is then a bilinear function of the two total enzyme concentrations.

We see that in contrast to the equations encountered in traditional enzyme kinetics, enzyme-enzyme interactions lead, for obvious reasons, to equations quadratic in the concentration of an enzyme-containing species. Therefore, classical methods such as the King–Altman procedure are no longer applicable.

Rate laws for overall enzymatic reactions are an instructive example for the situation that a scientific description at some level can be derived, under certain assumptions, from a description at a lower level (the elementary steps in this case). There are other situations in science where such derivation is not or only incompletely feasible, so that the description at a higher level must again start from first principles (axioms). An example is provided by the interrelation between classical mechanics and thermodynamics. Interestingly, in enzyme kinetics, the type of the relevant equations changes depending on the level of description. Whereas the simple mass-action kinetics relevant for the elementary reactions is linear in substrate and product concentrations, the overall enzyme-kinetic equations are nonlinear in these variables.

2.2.3. Thermodynamic Flow–Force Relationships

The rate laws given in the preceding sections are usually referred to as kinetic. They express reaction velocities as functions of concentrations. In an alternative approach based on principles of nonequilibrium thermodynamics, velocities are expressed in terms of thermodynamic forces. Besides (bio)chemical reactions driven by reaction affinities, standard situations are the heat flow as driven by temperature gradients and the electric current as driven by electric potential gradients. The fundamentals of irreversible thermodynamics and its application to chemical and biological processes can be found, for example, in the textbooks of Katchalsky and Curran (1965), Nicolis and Prigogine (1977), and Jou *et al.* (1993).

Substance fluxes across biological membranes are also flows in the thermodynamic sense, with the electrochemical potential difference being the corresponding force. It can, such as any reaction affinity, be defined as the negative change in Gibbs free energy accompanying the membrane flux or reaction.

A basic postulate of irreversible thermodynamics is that the flows in a given system can be expressed as functions of all thermodynamic forces acting in or on the system,

$$J_i = J_i(X_1, X_2, \ldots, X_\rho), \quad i = 1, \ldots, \rho. \tag{2.53}$$

When reaction affinities are the only macroscopically relevant forces in the system, this equation can be specified to a "thermodynamic rate law,"

$$v_i = v_i(A_1, A_2, \ldots, A_r), \quad i = 1, \ldots, r. \tag{2.54}$$

What may be important are electric flows and forces, in particular, in the case of membrane transport processes. They can, however, be included in the variables v_i and A_i, respectively, by considering the electric current as a superposition of ion fluxes and by using the concept of electrochemical potential gradient (cf. Guggenheim, 1967).

In thermodynamic equilibrium, all affinities and all reaction rates are zero. The function (2.54) must therefore satisfy the condition

$$v_i(0, 0, \ldots, 0) = 0. \qquad (2.55)$$

The thermodynamic approach is particularly important for systems the internal mechanisms of which are incompletely known. Lacking detailed knowledge of the functions (2.53) and (2.54), one often uses a linear approximation in the vicinity of equilibrium,

$$J_i = \sum_{k=1}^{p} L_{ik} X_k, \qquad (2.56a)$$

$$v_i = \sum_{k=1}^{r} L_{ik} A_k. \qquad (2.56b)$$

The L_{ik} are called phenomenological coefficients or Onsager coefficients. In Eq. (2.56a), a coefficient L_{ii} expresses the influence of the force X_i on its conjugate flow J_i (e.g., the effect of the temperature gradient in driving heat flow), whereas the coefficients L_{ki} reflect the cross-effects on other flows (e.g., the effect of the temperature gradient on diffusion). As far as chemical reactions are concerned [Eq. (2.56b)], cross-effects can usually be excluded ($L_{ki} = 0$ for $i \neq k$), provided that the reactions are described at a sufficiently detailed level. Equation (2.56b) can then be written as

$$\boldsymbol{v} = (\mathrm{dg}\,\boldsymbol{L})\boldsymbol{A} \qquad (2.57)$$

with A and (dg $\boldsymbol{L}$) denoting the vector of affinities of the elementary reactions and the (here diagonal) matrix of Onsager coefficients, respectively.

This assumption concerning the absence of cross-effects is no longer fulfilled when two or more distinct reactions share common intermediary species. It depends on the level of description whether such (possibly short-living) intermediates are included explicitly in the model. In biochemical modeling, this is of particular interest for enzymes coupling exergonic to endergonic processes. An example is provided by the mitochondrial H^+-transporting ATP synthase (H^+-ATPase, EC 3.6.1.34). A scheme representing the basic steps involved in this enzyme are shown in Figure 2.2 (cf. also Pietrobon and Caplan, 1985). As was stressed by T.L. Hill (1977), energy-transducing enzymes operate along at least two interconnected reaction cycles. In the example shown in Figure 2.2, these are the cycle of proton influx (reactions 1, 7, 5, and 6) and the cycle of ATP—ADP interconversion (reactions 2, 3, 4, and 7).

When the condition is imposed that the system subsists in a steady state, the particular reactions do become interdependent due to Eq. (2.9). In the example shown in Figure 2.2, steady-state conditions applied to the enzyme-containing species yield equations of the form $v_2 = v_3$, $v_2 + v_7 = v_1$, and so on. The coupling by the steady-state condition brings about that the number of independent fluxes and forces is decreased. One can therefore introduce a reduced flow vector, $\boldsymbol{J}'$, and a reduced affinity vector, $\boldsymbol{A}'$, for which

$$\boldsymbol{J}' = \mathbf{L}'\boldsymbol{A}' \tag{2.58}$$

holds true. $\mathbf{L}'$ is a reduced matrix of Onsager coefficients, which is not normally diagonal. In the considered example, $\boldsymbol{A}'$ could encompass the proton-motive force and phosphate potential, and $\boldsymbol{J}'$ would comprise the proton influx rate and phosphorylation rate (cf. Westerhoff and Van Dam, 1987).

According to the fundamental Onsager reciprocity relations (Onsager, 1931; cf. Guggenheim, 1967), $\mathbf{L}'$ is a symmetric matrix,

$$L'_{ik} = L'_{ki}. \tag{2.59}$$

For the considered case of coupled enzymic processes, this relation will be analyzed in detail in Section 3.3.3.

The original idea underlying linear irreversible thermodynamics was to linearize the function (2.53) around an equilibrium point. Later, it was found both experimentally and theoretically that this function often exhibits a multidimensional inflection point, that is, a point where some or all second derivatives $\partial^2 J_i/\partial X_k^2$ vanish (Rottenberg, 1973b; cf. Caplan, 1981). In the neighborhood of such a point, linearity between flows and forces (possibly with an additive constant), but not necessarily proportionality, approximately holds.

Thermodynamic approaches were intensely used in biochemical modeling several years ago (Kedem and Caplan, 1965; Kedem, 1972; Rottenberg, 1973b; Stucki *et al.,* 1983; Caplan, 1981; Westerhoff *et al.,* 1983; Westerhoff and Van Dam, 1987; Westerhoff, 1989; Groen *et al.,* 1990). In a number of models, both kinetic and thermodynamic rate laws were used (Bohnensack, 1981, 1985; Holzhütter *et al.,* 1985a; Pietrobon and Caplan, 1985; Pietrobon *et al.,* 1986; Stoner, 1992). It is worth noting that these approaches were employed to describe exclusively processes of biological energy transduction (oxidative phosphorylation in mitochondria and bacteria, photosynthesis, etc.). This is certainly due to the fact

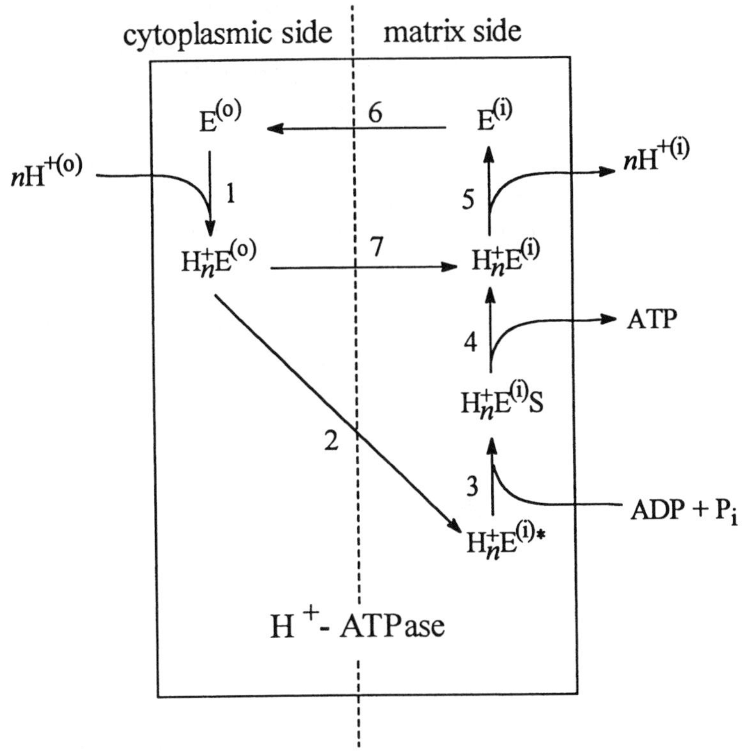

Figure 2.2 Reaction scheme of H^+-ATPase including elementary reaction steps. Symbols: E, ATPase; the supercripts (o) and (i) refer to the outside and inside of the membrane, respectively; Step 7 represents a slip reaction.

that energy is a central notion in thermodynamics. It is not clear, however, why the description by irreversible thermodynamics should be more appropriate for biological energy transduction than for other biochemical systems such as amino acid synthesis or the tricarboxylic acid cycle, which have usually been modeled by kinetic approaches. Even most models of glycolysis, which is part of energy metabolism, are kinetic (Selkov, 1975b; Ataullakhanov *et al.,* 1981; Werner and Heinrich, 1985; Markus and Hess, 1986; R. Schuster *et al.,* 1988; Joshi and Palsson, 1989a). This argument can be put also the other way around. Given that kinetic modeling has turned out to be very powerful in the modeling of many biochemical networks of anabolism and catabolism, why not apply it to biological energy transduction? Several advantages and drawbacks of thermodynamic and kinetic modeling were outlined in an interesting dispute between Westerhoff (1982) and Wilson (1982).

It is our impression that nowadays, for the construction of simulation models, kinetic rate laws are more and more favored compared to flow-force relationships (Heinrich *et al.,* 1977; Goldbeter *et al.,* 1990; Majewski and Domach, 1990b; Novak and Tyson, 1993; Ko *et al.,* 1994, as well as the papers on glycolysis cited earlier). The kinetic paradigm gives a deeper insight into biochemical systems, as it allows one to make use of more detailed knowledge about the enzymatic processes involved (cf. Korzeniewski and Froncisz, 1991).

A comparison of the (linear or nonlinear) thermodynamic flow-force relationships (2.54) with the general kinetic rate law (2.15) shows that the affinities are insufficient in number to reflect all degrees of freedom that determine the reactions. Consider, for example, a single monomolecular reaction interconverting S_1 and S_2. In the thermodynamic description, the rate would be written as a function of the affinity

$$v = v(A), \tag{2.60a}$$

$$A = RT\ln\left(\frac{S_1 q}{S_2}\right), \tag{2.60b}$$

whereas in the kinetic description, it reads

$$v = F(S_1,S_2)(k_+S_1 - k_-S_2). \tag{2.61}$$

Imagine now the situation that both the concentrations S_1 and S_2 are multiplied by some factor λ, $S_1' = \lambda S_1$, $S_2' = \lambda S_2$. The reaction velocity given by Eq. (2.60a) would not change in this situation, not even if $v(A)$ were a nonlinear function; but the rate given by Eq. (2.61) generally would, which reflects reality more adequately. To elucidate this discrepancy, we rewrite Eq. (2.61) as

$$v = k_-(S_2^{\mathrm{eq}} + \Delta S_2)\left[\exp\left(\frac{A}{RT}\right) - 1\right], \tag{2.62}$$

where the factor $F(S_1,S_2)$ has been dropped for the sake of simplicity and S_2^{eq} denotes the concentration of S_2 at some equilibrium point. v can now be regarded as a function of ΔS_2 and A, that is, of the deviation of the concentration S_2 from its value at the chosen equilibrium point and of the reaction affinity. The usual way to derive flow-force relationships to describe reaction kinetics is to expand Eq. (2.62) into a Taylor series. This gives, because $A = 0$ at equilibrium,

$$v = k_-S_2^{\mathrm{eq}}\frac{A}{RT} + k_-\Delta S_2\frac{A}{RT} + \frac{k_-S_2^{\mathrm{eq}}}{2}\left(\frac{A}{RT}\right)^2 + \ldots \tag{2.63}$$

which contains only one linear term and two second-order terms. Only in the linear approximation can v be written as a function of the only argument A [cf. (2.60a)]. To be able to describe the kinetically relevant situation that all concentrations are increased by (approximately) the same factor (see above), one has to include at least the second-order terms. The reduction of two variables (S_1 and S_2) to only one degree of freedom (affinity) is therefore an oversimplification.

Reaction rates depend not only on concentration ratios but also on their absolute values because the probability of collision of molecules becomes higher with increasing concentrations. For example, the dependence of ATP synthesis on internal and external pH in chloroplasts cannot be subsumed under a dependence on the proton-motive force, as was shown by experiment (Possmeyer and Gräber, 1994). Furthermore, the fluxes across biological membranes are not normally a unique function of the overall force given by the electrochemical potential difference, but depend on the electric potential difference in a different way than on the concentration ratio (Pietrobon and Caplan, 1985; Skulachev, 1988).

It has sometimes been argued that the rates of many biochemical reactions depend on the concentration ratio ATP/ADP, so that this ratio is a global signal variable. One should not, however, forget that changes in the absolute values of ADP and ATP generally also have an effect. Indeed, such changes are excluded when the sum of ADP and ATP is constant, but there are many instances where this is not the case, for example when adenylate kinase is operative, so that AMP must be included in the balance of adenine nucleotides. In some cases, it is convenient to work with concentration ratios, namely when they can be measured more easily than the particular concentrations (e.g., pH differences by the distribution of special substances, cf. Skulachev, 1988), but in many cases they cannot.

Alternatively to the ATP/ADP ratio, the energy charge, $(ATP + \frac{1}{2}ADP)/(ATP + ADP + AMP)$ was suggested to be a key variable by Atkinson (1968) (see Reich and Selkov, 1981). The aim in defining such global variables is to compress the body of data into a tractable number of variables easy to survey. Indeed, the energy charge obtains as a pool variable under some conditions concerning the separation of time constants of the reactions of interconversion of adenine nucleotides (Heinrich *et al.*, 1977). However, also in this case, the signal effect on other enzymes by the particular adenine nucleotides cannot necessarily be expressed by the energy charge only. Therefore, such compression of variables into the ATP/ADP ratio or energy charge usually eliminates relevant variables.

We have seen above for the case of a monomolecular reaction that thermodynamic rate laws are more restrictive than kinetic functions. Let us now consider the linear flow-force relationship resulting in the more general case of reactions of any stoichiometry. We start from the generalized mass-action rate law (2.15) for a single reaction with the stoichiometric coefficients n_i. It can be split up into the rate laws for the forward and reverse reactions:

$$v^+(S) = F(S)k_+ \prod_{i=1}^{n} S_i^{n_i^-}, \tag{2.64a}$$

$$v^-(S) = F(S)k_- \prod_{i=1}^{n} S_i^{n_i^+}. \tag{2.64b}$$

respectively. Dividing these two equations by each other, we find

$$\frac{v^+}{v^-} = q \prod_{i=1}^{n} S_i^{-n_i} = \exp\left(\frac{A}{RT}\right). \tag{2.65}$$

This equation can be written in logarithmic form as

$$\ln\left(1 + \frac{v^+ - \bar{v}}{\bar{v}}\right) - \ln\left(1 + \frac{v^- - \bar{v}}{\bar{v}}\right) = \frac{A}{RT}, \tag{2.66}$$

where $\bar{v}$ denotes the arithmetic mean of v^+ and v^-. In the neighborhood of thermodynamic equilibrium states, the unidirectional rates are much higher than net velocity ($v^+, v^- \gg |v^+ - v^-|$) and, hence, much higher than the quantities $|v^+ - \bar{v}|$ and $|v^- - \bar{v}|$. Accordingly, we can approximate Eq. (2.66) by using the Taylor expansion $\ln(1 + x) \cong x$ at a thermodynamic equilibrium state, where $v^+ = v^- = \bar{v}$. This gives

$$v = \frac{\bar{v}}{RT}A. \tag{2.67}$$

The net rate depends on the average unidirectional velocity, $\bar{v}$, and hence, on the equilibrium point chosen for the approximation. For a single reaction, this point depends on the conservation quantity of the concentrations involved (in the case of a monomolecular reaction, $S_1 + S_2$). In the analysis of multienzyme systems, the question of how to choose an appropriate reference state is even more problematic. For example, when the system involves conservation relations, the equilibrium state depends on all conservation quantities and is, therefore, a systematic property. However, a rate law should be defined in a very general way to be applicable wherever the considered reaction occurs. This requirement is fulfilled by kinetic rate laws such as the Michaelis–Menten kinetics (2.20).

This kinetics can be rewritten as a function of the affinity and the sum of substrate and product concentrations, $\Omega = S_1 + S_2$,

$$v = \left[\exp\left(\frac{A}{RT}\right) - 1\right]\left[\frac{1}{V_m^-}\left(1 + \frac{K_{m2}}{\Omega}\right) + \frac{1}{V_m^+}\exp\left(\frac{A}{RT}\right)\left(1 + \frac{K_{m1}}{\Omega}\right)\right]^{-1} \tag{2.68}$$

(Rottenberg, 1973b; Westerhoff and Van Dam, 1987). When the reaction is studied in isolation, Ω is constant. The reaction rate can then be regarded as a function of the affinity only, that is, as a (nonlinear) flow-force relationship (2.60a), which contains parameters that are only available by a kinetic characterization of the reaction. When the reaction is embedded in a metabolic system, the dependence on Ω becomes essential. Accordingly, rewriting the Michaelis–Menten kinetics in the form (2.68) is not of particular use.

A way to circumvent some difficulties in applying flow-force relationships to chemical reactions is to define separate affinities for the forward and reverse reactions (Lengyel, 1989), but this amounts to rewriting kinetic equations in other variables.

Due to the treatment of systems as black boxes, the thermodynamic approach only necessitates a small number of parameters. On the other hand, by a moderate increase in the number of parameters and using some knowledge of the internal mechanisms of the system, an enormous gain in modeling power can be achieved. For example, replacing the linear version of the thermodynamic rate law (2.60a) for an enzymic uni-uni reaction by the reversible Michaelis–Menten rate law (2.20) increases the number of parameters to be estimated from one to four. Due to the Haldane relation (2.26), only three of these four parameters are independent. On the other hand, the Michaelis–Menten rate law, which is based on the knowledge that the reaction is catalyzed by an enzyme, allows one to describe a number of additional phenomena, such as saturation, the limit case of irreversibility ($q \to 0$ or $q \to \infty$), and the fact that the rate depends on the absolute values of concentrations. Apart from situations where the knowledge about the system is really very limited, there is no reason why not to prefer a kinetic description in the case of biochemical reactions. For several other processes such as diffusion, electrodiffusion, and heat flow, thermodynamic flow-force relationships are very helpful.

An attempt to marry the simplicity of thermodynamics with the adequacy of kinetics is the Mosaic Non-Equilibrium Thermodynamics (MNET) developed by Westerhoff and Van Dam (1987). Flow-force relationships are combined with enzyme-kinetic rate laws like in a mosaic. The requirement of proportionality between fluxes and forces is relaxed so as to only require linearity, possibly including additive terms. This also serves to extend the validity of the treatment to the vicinity of multidimensional inflection points. Onsager symmetry is no longer invoked. The number of applications of this approach is, however, limited until now.

In this book, we focus on kinetic approaches. Nevertheless, reference to thermodynamics is made whenever appropriate [e.g., in the determination of the direction of fluxes (as will be done in optimization, see Section 6.2) or of detailed balanced subnetworks (see Section 3.3)]. Thermodynamics plays the role of an "accountant" deciding what is feasible in terms of energy balances and what is not. Its tools are not, however, sufficient to predict steady states and time-dependent behavior of biochemical systems far from equilibrium.

2.2.4. Power-Law Approximation

Rate laws of enzymic reactions in terms of the concentrations of substrates and products are usually nonlinear. Therefore, analytical treatment of models of larger reaction systems is very difficult or even impossible. For example, the flux through a chain of enzymic reactions endowed with reversible Michaelis–Menten kinetics cannot be expressed in terms of parameters in explicit form (cf. Section 5.4.3.1).

In many fields of science, intricate mathematical functions are often approximated by more concise expressions to simplify mathematical treatment. Care must be taken, however, that the essential properties are still reflected in this simplification. The most common approximation method is linearization. Accordingly, biochemical rate laws have often been linearized in terms of concentrations (Heinrich and Rapoport, 1974a; Palsson *et al.,* 1985; Liao and Lightfoot., 1987; S. Schuster and Heinrich, 1987; Cornish-Bowden, 1991) or in terms of reaction affinities (Kedem and Caplan, 1965; Rottenberg, 1973a; Westerhoff and Van Dam, 1987). By this method, however, many biochemically relevant effects such as saturation and sigmoidicity cannot be described.

Another possibility is to use power-law approximations (Savageau, 1969, 1976; Savageau *et al.,* 1987a; Peschel and Mende, 1986; Cascante *et al.,* 1989a, 1989b). Development of this method was inspired by the mathematical structure of usual mass-action kinetics, which involves products of concentrations raised to some power each. Any reaction rate v_j can be written as the difference of a forward and a reverse reaction rate, which are functions of concentrations [cf. Eq. (2.15)],

$$v_j = v_j^+(\mathbf{S}) - v_j^-(\mathbf{S}). \tag{2.69}$$

It can also be written in terms of logarithmic concentrations,

$$v_j = v_j^+(\ln \mathbf{S}) - v_j^-(\ln \mathbf{S}). \tag{2.70}$$

An essential prerequisite of the power-law approach is that some operating point must be chosen, which is, in most cases, some stationary state. Thereafter, the logarithms of the forward and reverse rates are linearized around the operating point,

$$\ln v_j^{\pm}(\ln \mathbf{S}) = \ln v_j^{\pm}(\ln \mathbf{S}^0) + \sum_{i=1}^{n} \left.\frac{\partial \ln v_j^{\pm}}{\partial \ln S_i}\right|_{S^0} (\ln S_i - \ln S_i^0), \tag{2.71}$$

where S_i^0 denotes the concentrations at the operating point. With the abbreviations

$$\ln v_j^+(\ln \mathbf{S}^0) - \sum_i \frac{\partial \ln v_j^+}{\partial \ln S_i} \ln S_i^0 = \ln \alpha_j, \tag{2.72}$$

$$\ln v_j^-(\ln \mathbf{S}^0) - \sum_i \frac{\partial \ln v_j^-}{\partial \ln S_i} \ln S_i^0 = \ln \beta_j, \tag{2.73}$$

$$\left.\frac{\partial \ln v_j^+}{\partial \ln S_i}\right|_{\mathbf{S}^0} = g_{ji}, \tag{2.74a}$$

$$\left.\frac{\partial \ln v_j^-}{\partial \ln S_i}\right|_{\mathbf{S}^0} = h_{ji}, \tag{2.74b}$$

one can write

$$\ln v_j^+(\ln \mathbf{S}) = \ln \alpha_j + \sum_i g_{ji} \ln S_i, \tag{2.75}$$

$$\ln v_j^-(\ln \mathbf{S}) = \ln \beta_j + \sum_i h_{ji} \ln S_i. \tag{2.76}$$

Transforming back into Cartesian coordinates, one obtains

$$v_j = \alpha_j \prod_i S_i^{g_{ji}} - \beta_j \prod_i S_i^{h_{ji}}. \tag{2.77}$$

It can be seen that the power-law approximation is equivalent to a linearization in logarithmic space. A directly linearized rate law,

$$v_j = a_j + \sum_{i=1}^{n} b_{ji} S_i, \tag{2.78}$$

has $n + 1$ parameters (for a given reaction j), whereas the power law (2.77) involves $2(n + 1)$ parameters. The latter function can therefore be expected to exhibit a larger richness in different curve shapes.

In the considered approach, the parameters α_j and β_j in Eq. (2.77) play the role of rate constants. The concentrations of external metabolites can either be incorporated into these rate constants or be written in the same way as the concentrations of internal substances. The exponents g_{ji} and h_{ji} are referred to as kinetic orders. Note that if the approximation procedure outlined above [Eqs. (2.70)–(2.76)] is applied to the mass-action rate law (2.10), the latter remains unchanged. In that rate law, the kinetic orders are given by the stoichiometric coefficients. In the general power-law function (2.77), the kinetic orders are phenomenological parameters which may or may not be integer.

An important aspect in enzyme kinetics is the effect of inhibitors. It is obvious that in the power-law equation (2.77), the concentration of an inhibitor, I, must be raised to a negative power. This gives rise to the singularity $v \to \infty$ at $I \to 0$, which would paradoxically imply an activation for very small inhibitor concentrations. Accordingly, several models using the power-law approach lead to huge concentration values of many orders of magnitude higher than realistic values (e.g. Torres *et al.*, 1993). In contrast, usual inhibition kinetics such as in Eq. (2.35) ensures that the reaction rate remains finite and nonzero as I tends to zero.

Replacing enzyme-kinetic rate laws such as Michaelis–Menten kinetics by power laws means that certain knowledge about the mechanism of enzyme catalysis and, accordingly, about kinetic properties such as saturation is deliberately sacrificed for the sake of the simplicity of mathematical treatment. As computational resources are nowadays no longer limiting in biochemical modeling, usage of power-law approaches only seems to be acceptable for processes for which a detailed kinetic description is not yet available.

2.3. STEADY STATES OF BIOCHEMICAL NETWORKS

2.3.1. General Considerations

To restrict modeling analysis to essential features, one often investigates the asymptotic time behavior of dynamical systems only, (i.e., the behavior after a sufficiently long time span). The asymptotic behavior may be oscillatory or even chaotic, but in many important situations, the systems will reach steady states.

The concept of steady state plays an outstanding role in kinetic modeling. A metabolic or any other macroscopic system is said to subsist in a steady state (also called stationary or time-invariant state) if the macroscopic variables (in the case of biochemical pathways, these are usually concentrations and fluxes) do not change within a tolerable accuracy over a certain time span of interest. As a matter of course, the concept of steady state is a mathematical idealization that can describe real situations only in an approximative way, due to fluctuations of different nature. Steady states comprise, as special cases, thermodynamic equilibrium states, in which all net flows as well as entropy production are zero. In general, however, net flows in steady states are not equal to zero (but constant), so that entropy production is positive.

Static situations are widespread in biology. Well-known examples are the fairly constant body temperature of homeothermic animals, the glucose concentration in the blood, and the pH in a great variety of living cells. Biochemical examples of virtually time-invariant states are erythrocyte glycolysis (cf. Ataullakhanov *et al.*, 1981; Werner and Heinrich, 1985; Joshi and Palsson, 1989a) and amino acid synthesis (cf. Fell and Snell, 1988). On the other hand, many biological events such as growth, nerve excitation, heart activity, and, on a longer time scale, biological evolution are clearly nonstationary. Nevertheless, examination of steady states and their neighborhood often helps to better understand the behavior of biological processes (cf. Edelstein-Keshet, 1988).

A more detailed analysis shows that the frequent occurrence of stationary behavior in biochemical networks results from the phenomenon of separation of time constants. In living cells, fast and slow processes are coupled with each other. The fast processes attain, under some stability conditions, a quasi-steady-state after an initial transient period (cf. Chapter 5). Every steady state can, in fact, be considered as a quasi-steady-state of a subsystem embedded in a larger, nonstationary system.

The usefulness of analyzing time-invariant states of biochemical systems becomes clearer by consideration of the general approach of classical thermodynamics. This theory starts with the study of equilibrium states, because a number of physical quantities such as temperature and entropy can be much more easily defined for these states. Moreover, several extremality principles derived from the Second Law of Thermodynamics are related to equilibrium states. Nevertheless, assertions about nonequilibrium systems can be made, as long as transition processes between equilibrium states are analyzed.

Because biological organisms are characterized by a throughput of energy and matter, it has to be acknowledged that time-invariant regimes in biology must usually be nonequilibrium phenomena. The German word *Fliessgleichgewicht* (equilibrium of flows) coined by von Bertalanffy (1953) properly expresses the fact that in steady states input flows balance output flows.

The modeling of nonequilibrium systems is one step up the ladder compared to equilibrium thermodynamics. Here, the analysis of steady states is the first step, because of the widespread occurrence of these states and the favorable property that differential equations containing time as an independent variable simplify to algebraic equations (similar to the equilibrium states studied in thermodynamics). Also for these states, extremality principles such as the principle of minimum entropy production (cf. Glansdorff and Prigogine, 1971) can be derived. As for biochemical reaction systems, the analysis of steady states implies that the differential equation system (2.8) is replaced by the algebraic equation system (2.9).

In analogy to equilibrium thermodynamics, transitions between steady states can be studied also. Relaxation processes (i.e., attainment of a steady state after a perturbation) and oscillatory regimes with small amplitudes can be approximately described by linearizing the equations in the neighborhood of the stationary state, which allows analytical solution of these equations. Moreover, linearization is a well-suited prerequisite for stability analysis.

2.3.2. Stable and Unstable Steady States

A steady state, $\bar{S}$, of a metabolic system described by the autonomous differential equation system (2.8) is considered to be *stable* if after an initial perturbation

$$\Delta S_i(0) = S_i(0) - \bar{S}_i \tag{2.79}$$

the concentrations $S_i(t)$ remain within a close neighborhood of the original steady state. Otherwise, the steady state is unstable. It is said to be *asymptotically stable* if

$$\Delta S_i(t) \underset{t\to\infty}{=} 0. \tag{2.80}$$

[For a more rigorous definition of stability, compare textbooks on differential equations (e.g., Andronov *et al.*, 1966; Guckenheimer and Holmes, 1983)]. Very often it is sufficient to analyze stability with respect to infinitesimally small perturbations $\delta S_i^0 = \delta S_i(0)$ (local stability). In this case, stability analysis can be performed on the basis of a Taylor expansion of the system equations (2.8), that is,

$$\frac{\mathrm{d}(\delta S_i)}{\mathrm{d}t} = \sum_{j=1}^{n} \frac{\partial f_i}{\partial S_j}\delta S_j + \frac{1}{2}\sum_{j,k=1}^{n} \frac{\partial^2 f_i}{\partial S_j\,\partial S_k}\delta S_j \delta S_k + \ldots, \tag{2.81}$$

where the derivatives of the functions $f_i(S_1, \ldots, S_n)$ defined in Eq. (2.8) have to be calculated at the reference steady state $\bar{S}$.

If infinitesimally small perturbations are considered, the quadratic terms on the right-hand side of Eq. (2.81) can often be neglected and one arrives at a linear differential equation system for $\delta S_i(t)$. In matrix notation this linear equation system may be written as

$$\frac{\mathrm{d}(\delta \mathbf{S})}{\mathrm{d}t} = \mathbf{M}\delta \mathbf{S}, \tag{2.82}$$

where the matrix **M** with the elements $m_{ij} = \partial f_i/\partial S_j$ denotes the Jacobian of the original differential equation system. Taking into account Eq. (2.8), **M** may be expressed in terms of the stoichiometry matrix and the rate laws,

$$m_{ij} = \sum_{k=1}^{r} n_{ik}\frac{\partial v_k}{\partial S_j} \quad \text{or} \quad \mathbf{M} = \mathbf{N}\frac{\partial \mathbf{v}}{\partial \mathbf{S}}. \tag{2.83}$$

(The derivatives $\partial v_k/\partial S_j$ which enter the matrix **M** also play a basic role in metabolic control analysis where they are called elasticity coefficients; see Chapter 5).

The solutions of the linear differential equation system (2.82) may be expressed in different ways. A formal solution may be given as

$$\delta \boldsymbol{S}(t) = \exp(\mathbf{M}t)\delta \boldsymbol{S}^0, \tag{2.84}$$

where the exponential function is a matrix which transforms the initial perturbation $\delta \boldsymbol{S}^0$ into $\delta \boldsymbol{S}(t)$. It is defined as the following expansion of the matrix $\mathbf{M}t$:

$$\exp(\mathbf{M}t) = \mathbf{I} + \mathbf{M}t + \frac{1}{2!}(\mathbf{M}t)^2 + \frac{1}{3!}(\mathbf{M}t)^3 + \ldots, \tag{2.85}$$

where $\mathbf{I}$ denotes the identity matrix. For more practical purposes, the solutions of Eq. (2.82) may be written as

$$\delta \boldsymbol{S}(t) = \sum_{i=1}^{n} c_i \boldsymbol{b}_i \exp(\lambda_i t), \tag{2.86}$$

where $\boldsymbol{b}_i$ denote the eigenvectors and λ_i the eigenvalues of the Jacobian. The n unknown constants c_i are determined by the initial perturbations. The eigenvalues may be calculated by solving the characteristic equation

$$\mathrm{Det}(\mathbf{M} - \lambda \mathbf{I}) = a_n\lambda^n + a_{n-1}\lambda^{n-1} + \ldots + a_1\lambda + a_0 = 0, \tag{2.87}$$

which is a polynomial equation of order n.

From the solution (2.86) of the linearized differential equation system (2.82), it can immediately be seen that a steady state of this system is asymptotically stable if, and only if, all the eigenvalues of the Jacobian for $S_i = \bar{S}_i$ have negative real parts. If at least one eigenvalue has a positive real part, then the steady state is unstable. Strictly speaking, Eq. (2.86) can only be applied if all the solutions of the characteristic equation are distinct. Otherwise, some functions $\delta S_i(t)$ involve polynomial functions, which do not, however, have any effect on stability (cf. Section 4.4).

Upon reduction of the complete equation (2.8) to the linearized equation (2.82), nonlinear terms have been neglected. Concerning the problem of whether the stability behavior is affected by these terms, one may prove the following theorem (cf. Hahn, 1967):

Theorem 2A. *If the steady state of the linearized system* (2.82) *is asymptotically stable then the steady state of the complete system* (2.8) *is also asymptotically stable. If the steady state of the linearized system is unstable* (*at least one eigenvalue of the Jacobian has a positive real part*), *then the steady state of the complete system is also unstable.*

In the framework of the linear theory, conclusions on the behavior of the trajectories for the case that some eigenvalues have negative real parts and the remaining eigenvalues have zero real parts can hardly be drawn. In that case the stability behavior can only be determined when the nonlinear terms in the expansion (2.81) are taken into account.

From Theorem 2A, it follows that for stability analysis it is often sufficient to test the signs of the roots of the characteristic equation (2.87). To this end, the so-called Hurwitz criterion may be applied (cf. Hahn, 1967). The coefficients of the characteristic equation are used to form the Hurwitz matrix

$$\mathbf{H} = \begin{pmatrix} a_{n-1} & a_n & 0 & 0 & \dots & 0 & 0 \\ a_{n-3} & a_{n-2} & a_{n-1} & a_n & \dots & 0 & 0 \\ a_{n-5} & a_{n-4} & a_{n-3} & a_{n-2} & \dots & 0 & 0 \\ \dots & \dots & \dots & \dots & \dots & \dots & \dots \\ 0 & 0 & 0 & 0 & \dots & a_1 & a_2 \\ 0 & 0 & 0 & 0 & \dots & 0 & a_0 \end{pmatrix}. \tag{2.88}$$

The elements H_{ij} of $\mathbf{H}$ are defined as

$$H_{ij} = \begin{cases} a_{n-2i+j} & \text{for } 0 < 2i - j \leq n \\ 0 & \text{otherwise.} \end{cases} \tag{2.89}$$

The Hurwitz determinants D_i are defined by the following sequence of principal subdeterminants

$$\begin{aligned} D_1 &= a_{n-1}, \qquad D_2 = a_{n-1}a_{n-2} - a_n a_{n-3}, \dots, \\ D_{n-1} &= a_1 D_{n-2}, \qquad D_n = a_0 D_{n-1} = \mathrm{Det}(\mathbf{H}). \end{aligned} \tag{2.90}$$

The following theorem is due to Hurwitz (1895).

Theorem 2B. *The characteristic equation* (2.87) *has only roots with negative real parts if, and only if, the inequalities*

$$\frac{a_{n-1}}{a_n} > 0, \qquad \frac{a_{n-2}}{a_n} > 0, \dots, \frac{a_0}{a_n} > 0 \tag{2.91a}$$

and

$$D_1 > 0, \qquad D_2 > 0, \dots, D_n > 0 \tag{2.91b}$$

hold true.

[For a proof of this theorem, cf. Hahn (1967).]

It follows immediately that for one-component systems ($n = 1$), a steady state is asymptotically stable if

$$\left.\frac{\partial f_1}{\partial S_1}\right|_{S_1=\bar{S}_1} < 0. \tag{2.92}$$

Let us discuss in more detail the case $n = 2$. Despite the fact that real metabolic systems contain a huge number of variable concentrations, many important models are formulated as two-component systems (cf. Sections 4.2 and 4.3 for the reduction of the number of variables using quasi-steady-state and rapid-equilibrium approximations). Moreover, the behavior of two-component systems is mathematically very well understood (Andronov *et al.*, 1966; Guckenheimer and Holmes, 1983). The time-dependent properties of a two-component system are governed by the differential equations

$$\frac{\mathrm{d}S_1}{\mathrm{d}t} = f_1(S_1,S_2), \tag{2.93a}$$

$$\frac{\mathrm{d}S_2}{\mathrm{d}t} = f_2(S_1,S_2), \tag{2.93b}$$

where the functions f_1 and f_2 include dependencies on parameters. The linear approximation in the neighborhood of a steady state gives

$$\frac{\mathrm{d}(\delta S_1)}{\mathrm{d}t} = \frac{\partial f_1}{\partial S_1}\delta S_1 + \frac{\partial f_1}{\partial S_2}\delta S_2, \tag{2.94a}$$

$$\frac{\mathrm{d}(\delta S_2)}{\mathrm{d}t} = \frac{\partial f_2}{\partial S_1}\delta S_1 + \frac{\partial f_2}{\partial S_2}\delta S_2. \tag{2.94b}$$

The characteristic equation reads

$$\lambda^2 - tr \cdot \lambda + \Delta = 0, \tag{2.95}$$

where

$$tr = \frac{\partial f_1}{\partial S_1} + \frac{\partial f_2}{\partial S_2}, \qquad \Delta = \frac{\partial f_1}{\partial S_1}\frac{\partial f_2}{\partial S_2} - \frac{\partial f_1}{\partial S_2}\frac{\partial f_2}{\partial S_1} \tag{2.96}$$

denote the trace (tr) and the determinant (Δ), respectively, of the Jacobian. Because $a_2 = 1$ the Hurwitz criterion leads to the stability conditions

$$tr < 0, \tag{2.97a}$$

$$\Delta > 0. \tag{2.97b}$$

For $n = 2$ the stability conditions follow directly from the explicit solutions of the characteristic equation (2.87)

$$\lambda_{1/2} = \frac{1}{2}\left(tr \pm \sqrt{tr^2 - 4\Delta}\right). \tag{2.98}$$

The dynamical properties of a metabolic system with only two variable concentrations may be studied by consideration of the motion within a plane which is spanned by the two concentration variables. There, the solutions of the differential equation system (2.93) may be represented by the solution of the differential equation

$$\frac{dS_2}{dS_1} = \frac{f_2(S_1,S_2)}{f_1(S_1,S_2)} = F(S_1,S_2) \tag{2.99}$$

with given initial conditions $S_i(0)$ for $i = 1, 2$, where the time variable has been eliminated. The curves determined by Eq. (2.99) are called *trajectories* of the system.

After small perturbations, a dynamical system may approach an asymptotically stable steady state in different ways. Accordingly, a stable steady state may be of different type. For $n = 2$ the following two cases are possible.

1. The steady state is a *stable node* if both eigenvalues λ_1 and λ_2 of the Jacobian are real and negative, that is, if in addition to relations (2.97a) and (2.97b), the following condition is fulfilled:

$$tr^2 > 4\Delta. \tag{2.100}$$

2. The steady state is a *stable focus* if the Jacobian has a pair of complex eigenvalues λ_1 and λ_2 with a negative real part. This is the case if in addition to relations (2.97a) and (2.97b), the following condition is fulfilled

$$tr^2 < 4\Delta. \tag{2.101}$$

Conditions (2.97), (2.100), and (2.101) follow directly from the solution (2.98) of the characteristic equation and by consideration of the general solution (2.86).

Unstable steady states may be classified in a similar way. For an *unstable node* (λ_1 and λ_2 are real and positive) or an *unstable focus* (complex eigenvalues λ_1 and λ_2 with a positive real part) condition (2.97a) is replaced by $tr > 0$.

There exists a third type of unstable steady states, called a *saddle point,* which corresponds to the occurrence of two real eigenvalues with opposite sign. According to Eq. (2.98), this is the case for $\Delta < 0$ irrespective of the sign of the trace of the Jacobian.

The stability of stationary states depends on the system parameters. Upon changes of the parameters, the eigenvalues of the Jacobian alter as well. At certain critical points called bifurcations, this will lead to a change in the character of the trajectories in the neighborhood of the steady state. For two-component systems, the following situations deserve special interest. For $\Delta > 0$ and $tr = 0$, the eigenvalues of the Jacobian are pure imaginary, and solution (2.86) of the linearized system (2.82) predicts closed trajectories in the phase space around the steady-state point, which is called a *center.* This corresponds to solutions periodic in time. The frequency of this oscillation reads, in the linear approximation,

$$f = \frac{\sqrt{\Delta}}{2\pi}. \tag{2.102}$$

The transition from a parameter region with $tr < 0$ to a region with $tr > 0$ is called Hopf bifurcation (Hopf, 1942; cf. Guckenheimer and Holmes, 1983). Transitions from a parameter domain with $\Delta < 0$ to a domain with $\Delta > 0$ generally results in a change of the number of stationary states (cf. Section 2.3.3).

For three-variable systems ($n = 3$), the characteristic equation (2.87) reads with, $a_3 = 1$,

$$\lambda^3 + a_2\lambda^2 + a_1\lambda + a_0 = 0, \tag{2.103}$$

where $-a_0$ and $-a_2$ denote the determinant and trace of the Jacobian, respectively. For the coefficient a_1, one derives

$$a_1 = \frac{\partial f_1}{\partial S_1}\frac{\partial f_2}{\partial S_2} + \frac{\partial f_1}{\partial S_1}\frac{\partial f_3}{\partial S_3} + \frac{\partial f_2}{\partial S_2}\frac{\partial f_3}{\partial S_3} - \frac{\partial f_1}{\partial S_2}\frac{\partial f_2}{\partial S_1} - \frac{\partial f_1}{\partial S_3}\frac{\partial f_3}{\partial S_1} - \frac{\partial f_2}{\partial S_3}\frac{\partial f_3}{\partial S_2}. \tag{2.104}$$

From the Hurwitz criterion, it follows that the stationary state of a three-variable system is asymptotically stable if the conditions

$$a_0 > 0, \quad a_1 > 0, \quad a_2 > 0, \quad a_1 a_2 - a_0 > 0 \tag{2.105}$$

are fulfilled.

It is worth mentioning that Theorem 2A makes assertions about stability but not about the shape of the trajectories of the complete system (2.93) compared to that of the trajectories of the linearized system (2.94) in the neighborhood of steady states. It may occur that the complete system predicts an unstable focus, whereas the solution of the linearized system represents an unstable node.

If one is interested in the asymptotic stability of a steady state after finite perturbations (global stability) the analysis may be performed by using so-called *Lyapunov functions.* These functions, $V_L = V_L(\xi_1, \ldots, \xi_n)$ with $\xi_i(t) = \Delta S_i(t) = S_i(t) - \bar{S}_i$, are constructed in such a way that they have the following properties:

1. $V_L(0, \ldots, 0) = 0,$ (2.106a)

2. $V_L > 0$ for $\xi \neq \mathbf{0}$ in a certain region D around the stationary state, (2.106b)

3. $\dfrac{dV_L}{dt} < 0$ for all $\xi \neq \mathbf{0}$ in D. (2.106c)

Theorem 2C (Lyapunov's Second Stability Theorem). *If there is a Lyapunov function in a region D, then the steady state $S_i = \bar{S}_i$ is globally asymptotically stable in D.*

The proof of this theorem can be found, for example, in the books of Hahn (1967) and Guckenheimer and Holmes (1983). Note that if relation (2.106c) does not hold as a strict inequality, a weaker conclusion concerning stability can be drawn (Lyapunov's First Stability Theorem).

The time derivative of V_L along the solution curves reads

$$\frac{dV_L}{dt} = \sum_{i=1}^{n} \frac{\partial V_L}{\partial \xi_i}\frac{d\xi_i}{dt} = \sum_{i=1}^{n} \frac{\partial V_L}{\partial \xi_i} f_i. \tag{2.107}$$

In practical applications of Theorem 2C difficulties may arise because there are no general methods for finding suitable Lyapunov functions. In mechanical systems often the energy may play the role of V_L. For the analysis of steady states of chemical systems, it may be useful to consider the entropy production.

2.3.3. Multiple Steady States

When the system equations (2.8) are nonlinear with respect to the system variables S_i, the steady-state solutions are not always unique. This means that for a given parameter vector $\mathbf{p}$ more than one vector $\bar{\mathbf{S}}$ fulfills Eq. (2.9). Furthermore, when the kinetic parameters vary, the number of possible steady states may also change at critical values of the parameters. The phenomenon of multistationarity may be important for explaining switching processes between different branches of metabolic systems.

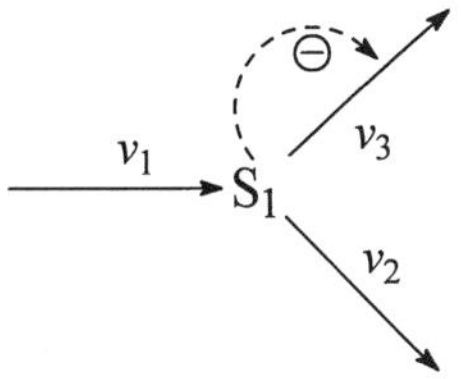

Scheme 4

Multiple steady states may occur even for one-component systems ($n = 1$). Let us consider the reaction system depicted in Scheme 4 where the kinetics of reaction 3 is characterized by cooperative substrate inhibition. All three reactions are assumed to be irreversible. The following kinetic equations are used

$$v_1 = \text{const.}, \tag{2.108a}$$

$$v_2 = k_2 S_1, \tag{2.108b}$$

$$v_3 = \frac{k_3 S_1}{1 + (S_1/K_I)^{n_H}} \tag{2.108c}$$

(k_2 and k_3: rate constants; K_I: inhibition constant of S_1; n_H: Hill coefficient). The kinetic equation (2.108c) describes, for example, essential characteristics of the phosphofructokinase reaction in glycolysis (EC 2.7.1.11) as a function of the concentration of its substrate ATP.

Figure 2.3 shows the reaction rates given in Eq. (2.108) as functions of the concentration S_1. The three different straight lines for $v_1 - v_2$ correspond to different values of the rate constant k_2. In the present case, steady states are determined by $v_1 - v_2 = v_3$, that is, by the intersection points of the curves for $v_1 - v_2$ and v_3 in Figure 2.3. Three cases are possible. For low and high values of k_2, there is only one intersection point corresponding to high or low concentrations S_1, respectively. In these cases, the steady states are unique. They are stable, as can be shown by Eq. (2.92). For intermediate values of k_2 three steady states are possible for one and the same set of kinetic parameters. The steady state in the middle is unstable. It can easily be seen in Figure 2.3 that in the stable steady state with low concentration of S_1, this intermediate is mainly metabolized via reaction 3 (low inhibition), whereas in the stable state with high S_1, reaction 2 is more active than reaction 3.

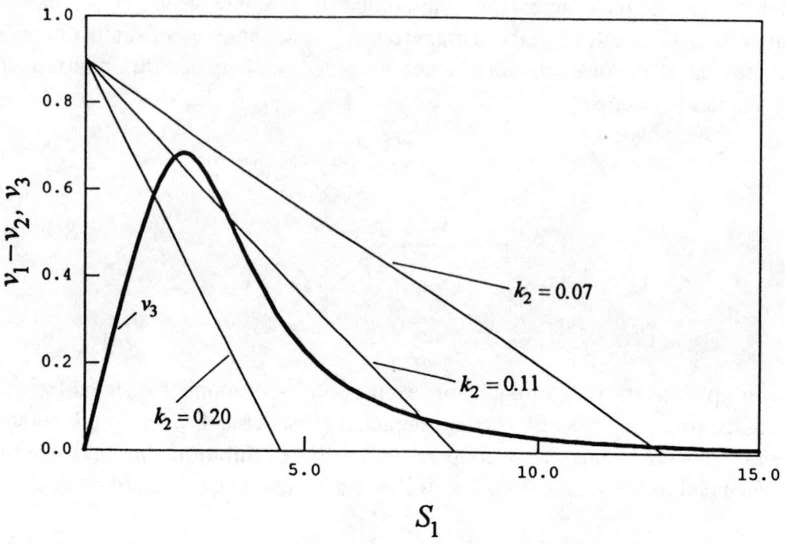

Figure 2.3 Plot of the difference of reaction rates v_1 and v_2 (straight lines) and of the reaction rate v_3 for the system shown in Scheme 4 as functions of S_1. Parameter values: $v_1 = 0.9$, $n_H = 4$, $k_3 = 0.4$, and $K_I = 3$.

A bifurcation occurs if at an intersection point both curves have the same tangent, so that the following equations are fulfilled simultaneously:

$$v_1 - v_2 = v_3, \tag{2.109a}$$

$$\frac{\partial(v_1 - v_2)}{\partial S_1} = \frac{\partial v_3}{\partial S_1}. \tag{2.109b}$$

These equations allow one to determine all critical parameter values where the number of steady-state solutions changes. Due to the nonlinearities, it may be difficult to solve one of the equations ([(2.109a) or (2.109b)] to obtain a parameter-dependent function for S_1 which may be used in the other equation to derive an expression for the critical parameter values. However, one easily derives a "parametric representation" of the bifurcation line within the (k_2,v_1)-plane. From Eq. (2.109b) one obtains

$$k_2 = \frac{k_3[(n_H - 1)(S_1/K_I)^{n_H} - 1]}{[1 + (S_1/K_I)^{n_H}]^2}, \tag{2.110a}$$

and from Eq. (2.109a)

$$v_1 = k_2 S_1 + \frac{k_3 S_1}{1 + (S_1/K_I)^{n_H}}. \tag{2.110b}$$

Equation (2.110a) may be used to calculate k_2 for varying S_1 values. Introducing the resulting function $k_2(S_1)$ into Eq. (2.110b), one obtains a curve $v_1(k_2(S_1),S_1)$, where bifurcations occur. Figure 2.4 shows the bifurcation line of the reaction system depicted in Scheme 4 within the (k_2,v_1)-plane for fixed values of the kinetic parameters of reaction 3. For parameters taken from region B, the steady states are unique, whereas for parameters of region A, three steady states are obtained.

The steady-state concentration $\bar{S}_1$ as a function of the first-order rate constant k_2 for various values of v_1 is depicted in Figure 2.5. Solid and broken lines correspond to stable and unstable states, respectively. It is seen that in the vicinity of bifurcation points, a switching between two stable steady states may be brought about by very small parameter changes. Furthermore, slow variations of the parameter k_2 from low values to high values and backward may lead to a hysteretic cycle. Crossing the region of multiple stationary states, it depends on the "history" of the system which stable steady state will be reached.

It follows from the theory of implicit functions that for all one-component systems described by the parameter-dependent differential equation $dS_1/dt = f_1(S_1,p_1, \ldots, p_m)$, bifurcation points for multiple steady states are determined by the conditions

$$f_1 = 0, \tag{2.111a}$$

$$\left.\frac{\partial f_1}{\partial S_1}\right|_{S_1 = \bar{S}_1} = 0, \tag{2.111b}$$

which for the example given above are equivalent to Eqs. (2.109a) and (2.109b), respectively.

For *two-component systems* the location of steady-state points within the phase plane is determined by intersection points of nullclines, that is, the curves defined in an implicit manner by the equations

$$f_1(S_1,S_2) = 0, \tag{2.112a}$$

$$f_2(S_1,S_2) = 0. \tag{2.112b}$$

A bifurcation implying a change in the number of steady-state solutions, occurs if in addition to Eqs. (2.112a) and (2.112b) the condition that the nullclines are tangential to each other at the intersection point is fulfilled. By use of implicit differentiation of Eqs. (2.112a) and (2.112b), this condition can be written as

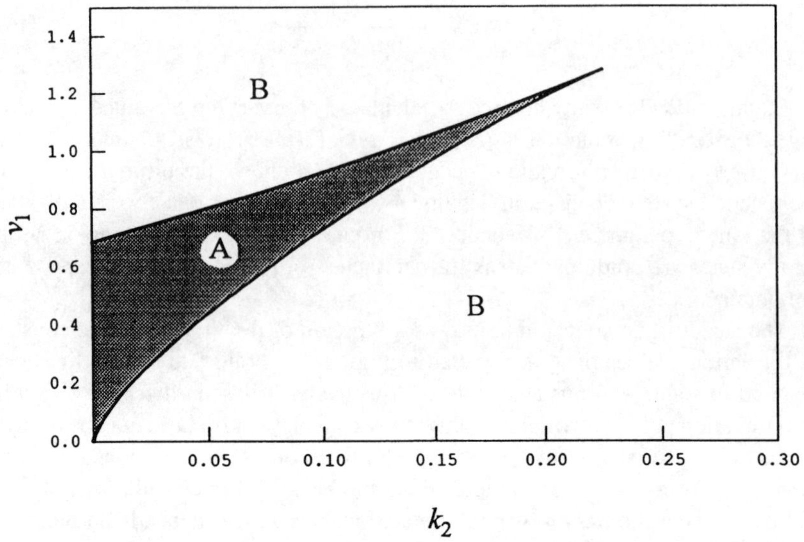

Figure 2.4 Bifurcation diagram of the reaction system depicted in Scheme 4 in the parameter space (k_2, v_1). Parameter values: $n_H = 4$, $k_3 = 0.4$, $K_I = 3$. Region A, three steady states; region B, one steady state.

$$\left.\frac{\partial f_1/\partial S_1}{\partial f_1/\partial S_2}\right|_{\bar{S}} = \left.\frac{\partial f_2/\partial S_1}{\partial f_2/\partial S_2}\right|_{\bar{S}}. \tag{2.113}$$

This equation is equivalent to

$$\Delta = \left.\left(\frac{\partial f_1}{\partial S_1}\frac{\partial f_2}{\partial S_2} - \frac{\partial f_1}{\partial S_2}\frac{\partial f_2}{\partial S_1}\right)\right|_{\bar{S}} = 0. \tag{2.114}$$

As follows from the theory of implicit functions the condition $\Delta = \mathrm{Det}(\mathbf{M}) = 0$ is necessary for a change of the number of steady states at varying parameter values also for the general case of n-component systems with $n > 2$.

The phenomenon of bistability has been extensively studied for the glycolytic system. For example, in a mathematical model of erythrocyte glycolysis, three stationary states are obtained if the rate constant of ATP-consuming processes is below a critical value (cf. Section 5.4.4.3, in particular Fig. 5.7). Bistability in glycolysis results mainly from the special kinetic properties of phosphofructokinase; in particular, the substrate inhibition by ATP and the activation by AMP. This has been demonstrated experimentally in open reconstituted enzyme systems using a stirred flow-through reactor (Eschrich *et al.*, 1980; Schellenberger *et al.*, 1988). The system contains the enzymes phosphofructokinase (EC 2.7.1.11), py-

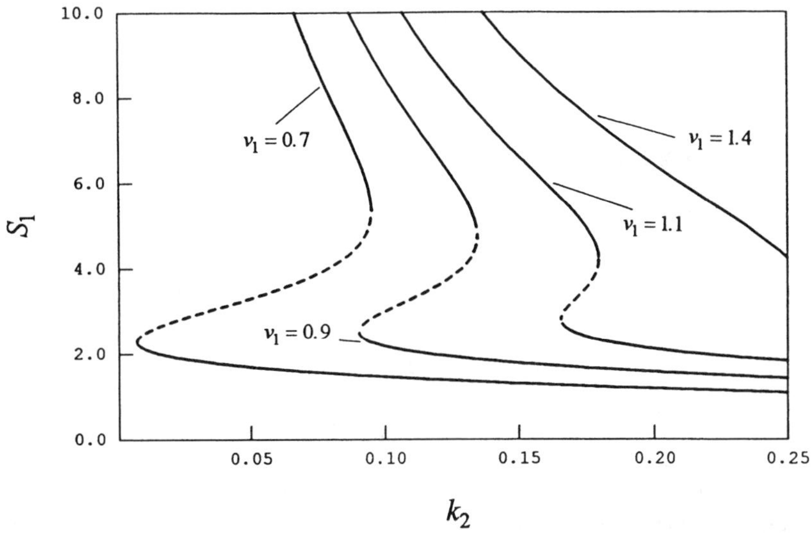

Figure 2.5 Stationary intermediate concentration S_1 for the system shown in Scheme 4 as a function of the rate constant k_2. Solid lines: stable steady states; broken lines: unstable steady states. Parameter values: $n_H = 4$, $k_3 = 0.4$, $K_m = 1$, $K_I = 3$.

ruvate kinase (EC 2.7.1.40), adenylate kinase (EC 2.7.4.3), and phosphogluco-isomerase (EC 5.3.1.9). Depending on the enzyme concentrations which are adjustable parameters, transitions between alternate stable stationary states characterized by high or low ATP concentration were observed. Furthermore, variations of the maximum activity of phosphofructokinase from low to high values and *vice versa* gave rise to a hysteretic cycle. Eschrich *et al.* (1990) studied experimentally as well as theoretically the dynamics of part of the glycolytic reaction sequence in cell-free yeast extracts. In addition to the enzymes converting glucose-6-phosphate to triose phosphates, it contains the enzyme fructose-1,6-bisphosphatase (EC 3.1.3.11). It has been shown that in a certain domain of parameter values, two different stable stationary states may coexist. While, at small perturbations, the system relaxes to the original steady state, suprathreshold perturbations may drive the system to the alternative steady state. In these experiments, *reversible* hysteretic effects were observed by varying the influx rate of glucose-6-phosphate. Schellenberger and Hervagault (1991) analyzed mathematically the reaction cycle formed by phosphofructokinase and fructose-1,6-bisphosphatase and drew attention to the fact that the occurrence of bistability may give rise to *irreversible* transitions. If one of the two bifurcation points, which separate regions of unique steady states and multiple steady states, are located outside the range of accessible parameter values, the formation of a full hysteretic loop (i.e., restoring the original state after large parameter perturbations) may become impossible.

2.4. METABOLIC OSCILLATIONS

2.4.1. Background

Metabolic systems may exhibit self-sustained oscillations, that is, the concentrations and fluxes may be periodic functions of time. In dynamical systems theory the occurrence of stable periodic solutions of nonlinear differential equation systems is a well-known phenomenon. Its discovery dates back to the work of Poincaré (1880–1890, 1899) in celestial mechanics, followed mainly by the work of Andronov and coworkers (Andronov *et al.,* 1966; cf. Minorsky, 1962; Guckenheimer and Holmes, 1983). In the space spanned by the system variables (state space), oscillations represent closed curves. In contrast to oscillations in linear systems (such as the harmonic pendulum), which are known for a much longer time, stable oscillations in nonlinear systems may have the property that after fluctuations of variables, the trajectory returns to the original orbit. Closed trajectories with this feature are called *limit cycles.* From the thermodynamic point of view, self-sustained oscillations are possible only in open systems far from equilibrium (Glansdorff and Prigogine, 1971; Feistel and Ebeling, 1989). Oscillations in biological systems were first analyzed mathematically by Lotka (1910) and Volterra (1931). Interestingly, the differential equation system nowadays referred to as the Lotka–Volterra system is conservative (i.e., the solutions represent a continuum of oscillations rather than limit cycles). However, Volterra (1931) also considered dissipative (nonconservative) systems (cf. also May, 1974). The possibility of oscillations on the genetic level, that is, periodic enzyme synthesis has been analyzed in early theoretical studies by Goodwin (1963, 1965).

Experimentally, autonomous biochemical oscillations have first been observed in glycolysis in intact yeast cells with typical periods in the order of minutes (Ghosh and Chance, 1964; Chance *et al.,* 1964a, 1964b). Thereafter, glycolytic oscillations have been studied in cell-free extracts (Hess and Boiteux, 1968; Pye, 1969). Furthermore, oscillations have been studied in open reconstituted enzyme systems using a stirred flow-through reactor (Schellenberger *et al.,* 1988). First models of glycolytic oscillations are due to Higgins (1964), Selkov (1968, 1975b), and Goldbeter and Lefever (1972).

Oscillations have been detected also in other enzymic systems, in particular, the periodic synthesis of cyclic AMP in the cellular slime mold *Dictyostelium discoideum* (Gerisch and Hess, 1974). More recently, the existence of hormone-induced oscillations in intracellular calcium concentrations has attracted much attention of experimentalists (Berridge, 1989) as well as of theoreticians (Goldbeter *et al.,* 1990; Dupont *et al.,* 1991; Somogyi and Stucki, 1991; cf. Section 2.4.4).

For reviews and textbooks on the modeling of self-sustained oscillations in biochemical and chemical systems, see the works of Higgins (1967), Goldbeter (1990), Winfree (1990), and Gray and Scott (1994).

2.4.2. Mathematical Conditions for Oscillations

Systems of autonomous first-order differential equations may exhibit oscillations only if they involve more than one variable ($n > 1$). For $n = 1$, periodic solutions are excluded because $dS/dt = f(S)$ is a unique function of S and, hence, cannot have opposite signs for a given S at different times t. For *two-dimensional systems* ($n = 2$), there are a number of theorems giving conditions for the occurrence of oscillations. The proof to the following two theorems can be found in Guckenheimer and Holmes (1983).

Theorem 2D (Theorem of Poincaré and Bendixson). *If, and only if, a trajectory remains for* $t_0 \leqslant t < \infty$ *within a finite region D of the phase plane without approaching a stationary state this trajectory is a periodic trajectory (closed cycle) or tends to such a trajectory for* $t \rightarrow \infty$.

Theorem 2E (Criterion of Bendixson). *If the expression* $\partial f_1/\partial S_1 + \partial f_2/\partial S_2$ *does not change sign in a region D of the phase plane then D contains no periodic trajectory.*

Whereas Theorems 2D and 2E apply to any two-dimensional dynamical system, the following two theorems concern systems of chemical reactions.

Theorem 2F. *Chemical systems with two variable compounds cannot exhibit limit cycles if only monomolecular and bimolecular reactions are involved.*

The proof was given by Hanusse (1972) on the basis of the mass-action kinetics (2.10). It follows that for chemical systems with mass-action kinetics to exhibit stable oscillations (limit cycles), it is necessary that they contain at least trimolecular reactions or involve more than two variable substances (cf. Section 2.4.5). Note that oscillations of the Lotka-Volterra type (which are not limit cycles) are not excluded by this theorem.

The following statement is valid for chemical systems involving any number of reacting substances.

Theorem 2G. *For systems composed only of first-order reactions, the eigenvalues of the Jacobian are always real and negative, which excludes both damped oscillations and limit cycles.*

In the proof of this theorem (Hearon, 1953; Bak, 1959) the fact was used that in any reaction cycle occurring in the system, Wegscheider's condition must be fulfilled (cf. Section 3.3.1).

For enzymic systems, a large number of possible regulatory mechanisms exist (activation or inhibition by internal metabolites) which may provide the necessary nonlinearities in the rate equations for generating oscillations. For the special case $n = 2$, Higgins (1967) was able to derive several necessary conditions on the form of the rate equations for the existence of a Hopf bifurcation.

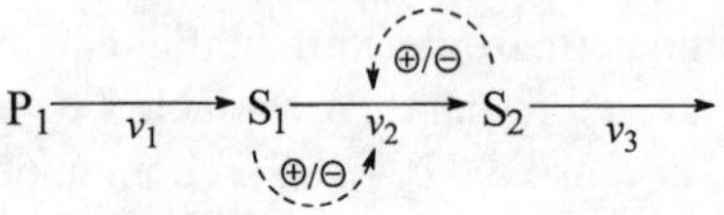

Scheme 5

Let us consider, for example, an unbranched pathway involving two intermediate compounds and three irreversible reactions (Scheme 5). Such a scheme can be used, for example, for analyzing glycolytic oscillations (cf. Section 2.4.3). We include the possibility that the activity of the enzyme catalyzing reaction 2 may be regulated allosterically by S_1 and/or S_2. Effector actions are designated by dashed lines, with activations and inhibitions symbolized by plus and minus signs, respectively. For the kinetic properties of reactions 1 and 3, it is assumed that

$$v_1 = \text{const.}, \tag{2.115a}$$

$$\frac{\partial v_3}{\partial S_2} > 0, \tag{2.115b}$$

$$\frac{\partial v_3}{\partial S_1} = 0. \tag{2.115c}$$

Relation (2.115b) is fulfilled, for example, by a Michaelis–Menten equation for reaction 3. Equation (2.115c) means that S_1 is no effector of reaction 3. The trace and determinant of the Jacobian of the system depicted in Scheme 5 may then be expressed as follows:

$$tr = -\frac{\partial v_2}{\partial S_1} + \frac{\partial v_2}{\partial S_2} - \frac{\partial v_3}{\partial S_2}, \tag{2.116a}$$

$$\Delta = \frac{\partial v_2}{\partial S_1}\frac{\partial v_3}{\partial S_2} \tag{2.116b}$$

For a Hopf bifurcation it is necessary that $\Delta > 0$ and that tr may become positive upon a change of the kinetic parameters (see Section 2.3.2). With relation (2.115b), one immediately obtains from Eqs. (2.116a) and (2.116b) the following necessary conditions:

$$\frac{\partial v_2}{\partial S_1} > 0, \tag{2.117a}$$

$$\frac{\partial v_2}{\partial S_2} > 0. \tag{2.117b}$$

Relation (2.117a) is generally fulfilled without effector regulation. From relation (2.117b) it follows that under the given assumptions, a positive feedback of the metabolite S_2 on its own production is a potential regulatory mechanism for the generation of oscillations ("back activation oscillator"; cf. Section 2.4.3).

Likewise, relations (2.115) and (2.116) imply that a "forward inhibition oscillator" is impossible under the considered assumptions, as $\partial v_2/\partial S_1 < 0$ implies $\Delta < 0$ which excludes a Hopf bifurcation. The qualitative effect of other regulatory loops (for example, of S_1 on reaction 3 or of S_2 on reaction 1) can be predicted in a similar way (Higgins, 1967)

2.4.3. Glycolytic Oscillations

Experimental studies on glycolytic oscillations (Betz and Selkov, 1969) have clearly shown that there is a phase angle shift of about 180° in the changes of the concentrations of fructose-6-phosphate (F6P) and fructose-bisphosphate (FP_2). According to the *crossover theorem* (Holmes 1959; Higgins, 1965) (see Section 5.10) this indicates that the phosphofructokinase (PFK) reaction may play an important role in the generation of the oscillations. From that, it was concluded that a back activation of PFK by FP_2 provides an explanation for the observed periodic behavior. Accordingly, Higgins (1964, 1967) proposed the two-component model depicted in Scheme 5 with only one regulatory loop, namely an activation of reaction 2 by S_2. It is governed by the differential equations

$$\frac{dS_1}{dt} = v_1 - v_2(S_1,S_2), \tag{2.118a}$$

$$\frac{dS_2}{dt} = v_2(S_1,S_2) - v_3(S_2), \tag{2.118b}$$

where $S_1 = F6P$ and $S_2 = FP_2$. v_1 and v_2 represent the rates of the hexokinase and phosphofructokinase reactions, respectively, whereas v_3 denotes the rate of a reaction degrading FP_2. The concentrations of glucose as well as of the cofactors ADP and ATP are considered to be constant which implies $v_1 = const.$ With the conditions $\partial v_2/\partial S_1 > 0$ and $\partial v_3/\partial S_2 > 0$, which are normally fulfilled for Michaelis–Menten kinetics, and the above-mentioned activation of PFK by FP_2

($\partial v_2/\partial S_2 > 0$), this model meets the necessary conditions for generation of oscillations mentioned in Section 2.4.2 [Eqs. (2.115) and (2.117)].

First models of glycolytic oscillations were based on kinetic equations taking into account saturation of v_2 and of v_3 by S_2. Later, the following more simple model had been proposed:

$$\frac{\mathrm{d}S_1}{\mathrm{d}t} = v_1 - k_2 S_1 S_2^{\gamma}, \tag{2.119a}$$

$$\frac{\mathrm{d}S_2}{\mathrm{d}t} = k_2 S_1 S_2^{\gamma} - k_3 S_2 \tag{2.119b}$$

with $\gamma > 1$ (Selkov, 1968). It has also been shown that back activation of PFK by FP_2 is not operative under *in vivo* conditions and that it is more appropriate to consider alternative models including cooperative back activation of PFK by ADP or the even more effective AMP activation. Concentration changes of AMP are linked to those of ADP due the very fast adenylate kinase reaction (Betz and Selkov, 1969, cf. Goldbeter and Caplan, 1976).

Let us consider the model (2.119) in more detail. Despite its nonlinearity, the steady-state concentrations may be expressed analytically as functions of the kinetic parameters:

$$\bar{S}_1 = \frac{k_3^{\gamma}}{k_2 v_1^{\gamma-1}}, \qquad \bar{S}_2 = \frac{v_1}{k_3}. \tag{2.120}$$

The trace and determinant of the Jacobian read

$$tr = (\gamma - 1)k_3 - k_2\left(\frac{v_1}{k_3}\right)^{\gamma}, \tag{2.121a}$$

$$\Delta = k_2 k_3 \left(\frac{v_1}{k_3}\right)^{\gamma}. \tag{2.121b}$$

It is seen that the determinant is always positive, whereas for $\gamma > 1$, the trace may change its sign depending on the kinetic parameters. According to the stability criteria for two-component systems given in Eq. (2.97), the steady state (2.120) is unstable for

$$k_2 < \frac{(\gamma - 1)k_3^{\gamma+1}}{v_1^{\gamma}} = k_2^{\mathrm{crit}}; \tag{2.122}$$

that is, the systems exhibits a Hopf bifurcation at $k_2 = k_2^{crit}$. Furthermore, one easily derives that within the parameter region defined by

$$k_2^- = \frac{k_3^{1+\gamma}}{v_1^\gamma}(1 - \sqrt{\gamma})^2 < k_2 < \frac{k_3^{1+\gamma}}{v_1^\gamma}(1 + \sqrt{\gamma})^2 = k_2^+, \tag{2.123}$$

the periodicity condition $tr^2 - 4\Delta < 0$ is fulfilled, that is, the eigenvalues have an imaginary part. Consequently, for kinetic parameters which fulfill relations (2.122) and (2.123), the steady state is an unstable focus.

Figure 2.6 shows, within the (v_1, k_2)-plane, the curve $k_2 = k_2^{crit}$ which separates regions of stable and unstable steady states as well the curves $k_2 = k_2^-$ and $k_2 = k_2^+$ which separate regions of periodic and aperiodic behavior. Limit cycles are obtained for parameter values taken from the region of instability near the bifurcation curve $k_2 = k_2^{crit}$. Figures 2.7A and 2.7B show self-sustained oscillations of the concentrations S_1 and S_2 as functions of time and in the phase plane, respectively. It is seen that in accordance with the experimental facts, the oscillations of the concentrations of F6P (S_1) and FP_2 (S_2) are out of phase. Increasing the distance from the bifurcation line (e.g. lowering of k_2) results in an increase of the size of the limit cycle. Eventually, the system becomes globally unstable at very low k_2 values.

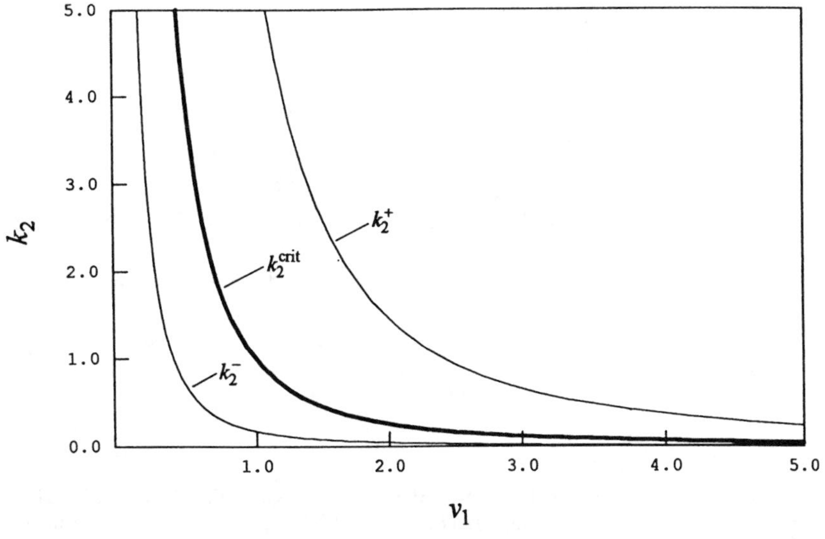

Figure 2.6 Stability diagram of the Higgins-Selkov oscillator. Across the curves, the following transitions between different types of steady states occur: $k_2^-(v_1)$, transitions from unstable nodes to unstable foci; $k_2^{crit}(v_1)$, Hopf bifurcations; $k_2^+(v_1)$, transitions from stable foci to stable nodes. Parameter values: $\gamma = 2$, $k_3 = 1$.

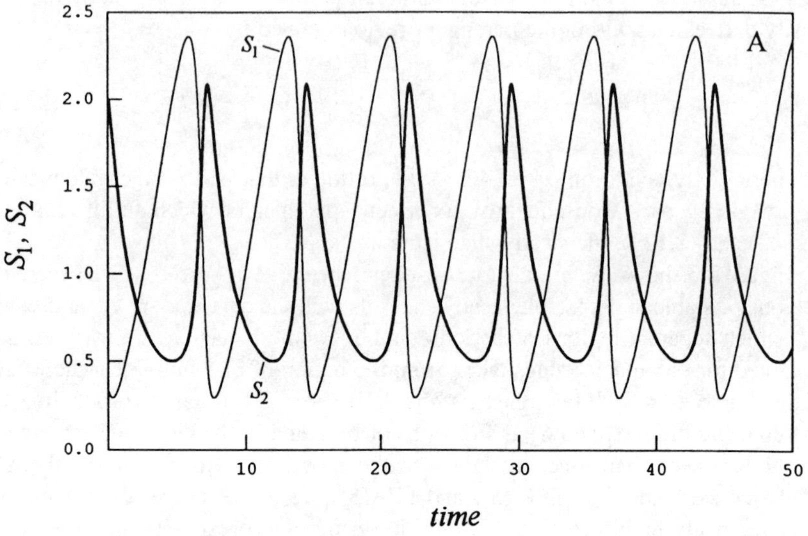

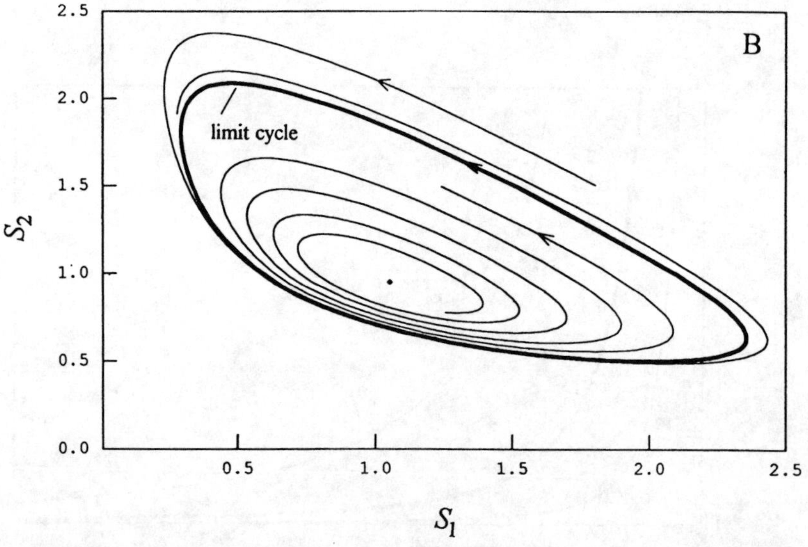

Figure 2.7 Self-sustained oscillations in the Higgins-Selkov system [Eqs. (2.119a) and (2.119b)]. (A) Concentrations S_1 and S_2 versus time. (B) Limit cycle (thick line) and neighboring trajectories in the phase plane. Parameter values: $v_1 = 0.95$, $k_2 = k_3 = 1$.

The basic models of Higgins (1964) and Selkov (1968) have been modified in different ways. For example, Goldbeter and Lefever (1972) extended the model by describing the phosphofructokinase reaction by a rate equation resulting from the allosteric model of Monod *et al.* (1965) [Eq. (2.43)]. In fact, it describes the kinetic properties of this enzyme rather well in a number of cells (Blangy *et al.,* 1968; Otto *et al.,* 1974). Furthermore, Goldbeter and Lefever (1972) considered the coupling of the glycolytic oscillator to diffusion of the metabolites, but neglecting the diffusion of enzymes. In this way, they could show that also in the case of glycolysis, reaction-diffusion processes may result in chemical waves (i.e., to dissipative structures in space and time).

Selkov (1975b) was able to demonstrate that allosteric regulation of phosphofructokinase is not the only possible explanation of glycolytic oscillations. He considered a model of cellular energy metabolism which includes only stoichiometric couplings but no regulatory interactions by internal modifiers of the enzymes. The corresponding reaction scheme is similar to that shown in Scheme 12 in Section 5.4.4, where its details will be explained in the framework of control analysis of glycolysis. In addition to glycolysis, the stoichiometric oscillation model involves an alternative source for ATP (e.g., oxidative phosphorylation). The occurrence of oscillations has been demonstrated only for the case that some of the reaction rates are described by saturation functions. The corresponding nonlinearities in the rate equations make the stability analysis rather cumbersome despite the fact that other nonlinearities resulting from allosteric interactions have not been included (cf. also Heinrich *et al.,* 1977).

2.4.4. Models of Intracellular Calcium Oscillations

A wide variety of cells exhibit oscillations of intracellular calcium (Ca^{2+}) in the form of repetitive spikes. For example, calcium oscillations may be stimulated by hormones or neurotransmitters in hepatocytes, where the oscillation period ranges from 0.5 to 10 min (Woods *et al.,* 1986, 1987). Shorter and longer periods of calcium oscillations have been observed depending on the cell type. Generally, the period of calcium oscillations decreases with increasing agonist concentration. It is generally assumed that the PI signaling pathway [i.e., the receptor-stimulated hydrolysis of phosphatidylinositol 4,5-bisphosphate (PIP_2) to inositol 1,4,5-trisphosphate (IP_3) and diacylglycerol catalyzed by phospholipase C (PLC, EC 3.1.4.3)] plays a crucial role in the generation of the calcium oscillations. PIP_2 is regenerated from IP_3 by inositol-1,4,5-trisphosphate 5-phosphatase (EC 3.1.3.56) and several subsequent enzymes. Concerning the oscillatory mechanism it is most likely that the "oscillator" is located within the cytoplasm and that its mechanism is closely related to the mobilization of calcium from intracellular stores [e.g., endoplasmic or sarcoplasmic reticulum (ER, SR) (Berridge, 1989)]. In a model proposed by Meyer and Stryer (1988), oscillations result from a positive feedback loop between cytosolic calcium and the formation of IP_3 (cf. Fig. 2.8). In particular, it is assumed that IP_3 triggers the release of calcium from intracellular stores (Ca^{2+}_{store}) into the cytosol (Ca^{2+}_{cyt}). The cytosolic calcium, in turn, activates IP_3 synthesis. The oscillating variables are the concentration of IP_3 and the concentrations of cytosolic and stored calcium [for a more elaborate version of this model, cf. Meyer and Stryer (1991)].

Other models of hormone-induced calcium oscillations are based on the phenomenon of *calcium-induced calcium release* (CICR mechanism) as first described by Endo *et al.* (1970) for skeletal muscle cells and later on by Fabiato and Fabiato (1975) for the sarcoplasmic reticulum in cardiac cells. Let us here consider the minimal two-variable model as proposed by Dupont and Goldbeter (1989) (cf. also Goldbeter *et al.,* 1990; Somogyi and Stucki, 1991). This model includes the following processes (see Fig. 2.9):

- Inward and outward transport of cytosolic calcium through the plasma membrane (rates v_1 and v_2, respectively)
- IP_3-activated release of calcium from an IP_3-sensitive intracellular store (rate v_3)
- Active transport of cytosolic calcium into an IP_3-insensitive store (rate v_4)
- Release of calcium from the IP_3-insensitive store which is activated by the cytosolic calcium (CICR mechanism, rate v_5)
- Leak flux of calcium from the latter store into the cytosol (rate v_6)

The model is based on the following differential equations:

$$\frac{dS_1}{dt} = v_1 - v_2 + v_3 - v_4 + v_5 + v_6, \tag{2.124a}$$

$$\frac{dS_2}{dt} = v_4 - v_5 - v_6, \tag{2.124b}$$

where S_1 and S_2 denote the concentrations of the calcium in the cytosol and the IP_3-insensitive stores, respectively. The influx rates v_1 and v_3 are considered to be constant. To justify a constant rate v_3, Dupont and Goldbeter (1989) use the contestable assumption that the IP_3-sensitive store is very fast replenished. For v_3, the expression $v_3 = \tilde{v}_3\beta$ is employed, where the factor β represents a saturation function of the release of calcium from the IP_3-sensitive store with respect to IP_3. From the mathematical point of view, the constant rates v_1 and v_3 may be subsumed into one parameter $v_0 = v_1 + v_3$.

It is reasonable to assume that the rates v_2 and v_4 of the two ATP-dependent calcium pumps are activated by cytosolic calcium, whereas they are independent of the concentration of stored calcium, that is,

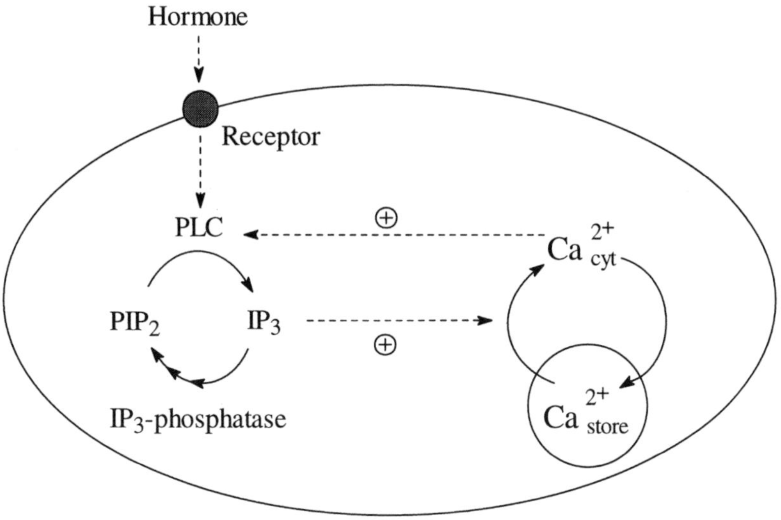

Figure 2.8 Model for generation of calcium oscillations by Meyer and Stryer (1988). For symbols see text.

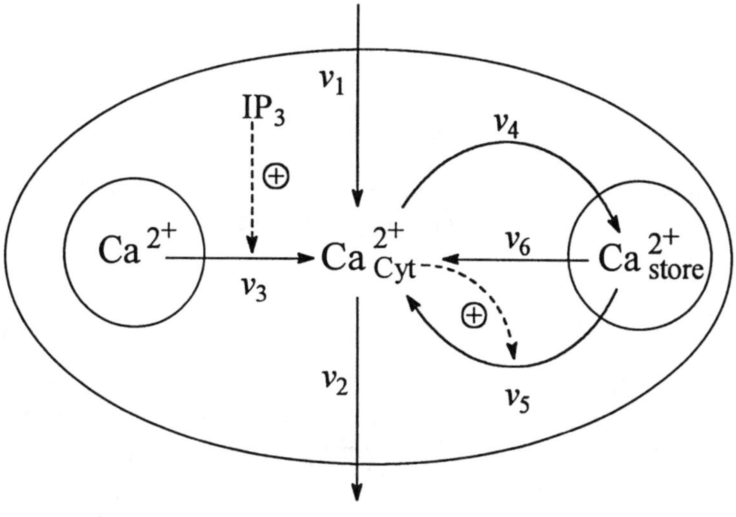

Figure 2.9 Schematic picture of the processes responsible for receptor-induced intracellular calcium oscillations [adapted from Dupont and Goldbeter (1989)]. For an explanation of the various processes, see the text.

$$\frac{\partial v_2}{\partial S_1} > 0, \tag{2.125a}$$

$$\frac{\partial v_4}{\partial S_1} > 0, \tag{2.125b}$$

$$\frac{\partial v_2}{\partial S_2} = 0, \tag{2.125c}$$

$$\frac{\partial v_4}{\partial S_2} = 0. \tag{2.125d}$$

For the leak rate v_6, one may assume

$$\frac{\partial v_6}{\partial S_1} = 0, \tag{2.126a}$$

$$\frac{\partial v_6}{\partial S_2} > 0. \tag{2.126b}$$

Indeed, the leak is likely to be reversible but the reverse reaction can formally be included into the rate v_4. Using relations (2.125) and (2.126) one derives for the trace, *tr*, and the determinant, Δ, of the Jacobian the expressions

$$tr = -\left(\frac{\partial v_2}{\partial S_1} + \frac{\partial v_4}{\partial S_1} + \frac{\partial v_5}{\partial S_2} + \frac{\partial v_6}{\partial S_2}\right) + \frac{\partial v_5}{\partial S_1}, \tag{2.127a}$$

$$\Delta = \frac{\partial v_2}{\partial S_1}\left(\frac{\partial v_5}{\partial S_2} + \frac{\partial v_6}{\partial S_2}\right). \tag{2.127b}$$

We are now looking for a Hopf bifurcation, for which $\Delta > 0$ is a necessary condition. Because we assume inequality (2.125a) to hold, it follows from Eq. (2.127b) that

$$\frac{\partial v_5}{\partial S_2} + \frac{\partial v_6}{\partial S_2} > 0 \tag{2.128}$$

at any Hopf bifurcation. Using this relation and inequalities (2.125a) and (2.125b), it is immediately seen from Eq. (2.127a) that the trace may only become positive if

$$\frac{\partial v_5}{\partial S_1} > 0. \tag{2.129}$$

This means that under the given assumptions, activation of the calcium release from the IP_3-insensitive store by cytosolic calcium is a necessary condition for a Hopf bifurcation. For the concentration dependent rates we use, in a more detailed analysis, the expressions proposed by Somogyi and Stucki (1991),

$$v_2 = k_2 S_1, \tag{2.130a}$$

$$v_4 = k_4 S_1, \tag{2.130b}$$

$$v_5 = \frac{k_5 S_2 S_1^{n\mathrm{H}}}{K_{0.5}^{n\mathrm{H}} + S_1^{n\mathrm{H}}}, \tag{2.130c}$$

$$v_6 = k_6 S_2, \tag{2.130d}$$

which fulfill conditions (2.125), (2.126), and (2.128), and (2.129). For the activation of v_5 by S_1, a Hill equation is used, where $K_{0.5}$ and n_{H} denote the half-saturation constants and the Hill coefficient, respectively.

The steady-state solution of equation system (2.124) reads

$$\bar{S}_1 = \frac{v_1 + v_3}{k_2} = \frac{v_0}{k_2}, \tag{2.131a}$$

$$\bar{S}_2 = \frac{v_0 k_4}{k_2}\left(k_6 + \frac{k_5 v_0^{n\mathrm{H}}}{v_0^{n\mathrm{H}} + (k_2 K_{0.5})^{n\mathrm{H}}}\right)^{-1}. \tag{2.131b}$$

Using these equations and expressions (2.127), one derives for the determinant and trace of the Jacobian at the steady state

$$\Delta = k_2\left(k_6 + \frac{k_5 v_0^{n\mathrm{H}}}{v_0^{n\mathrm{H}} + (k_2 K_{0.5})^{n\mathrm{H}}}\right), \tag{2.132}$$

$$tr = -\left(k_2 + k_4 + k_6 + \frac{k_5 v_0^{n\mathrm{H}}}{v_0^{n\mathrm{H}} + (k_2 K_{0.5})^{n\mathrm{H}}}\right) + \frac{k_4 k_5 n_{\mathrm{H}} (v_0 k_2 K_{0.5})^{n\mathrm{H}}}{[v_0^{n\mathrm{H}} + (k_2 K_{0.5})^{n\mathrm{H}}]\,[(k_5 + k_6) v_0^{n\mathrm{H}} + k_6 (k_2 K_{0.5})^{n\mathrm{H}}]}. \tag{2.133}$$

Whereas the determinant is always positive, the trace may change its sign depending on the parameter values. The boundary of the region of stability where Hopf bifurcations occur can be calculated by putting $tr = 0$. Figure 2.10 shows regions of stable and unstable behavior within a two-dimensional section of the parameter space defined by variable values of v_0 (combined calcium input) and k_5 (rate constant of CICR) and fixed values of the other parameters.

The existence of the oscillations is confirmed by numerical integration of the

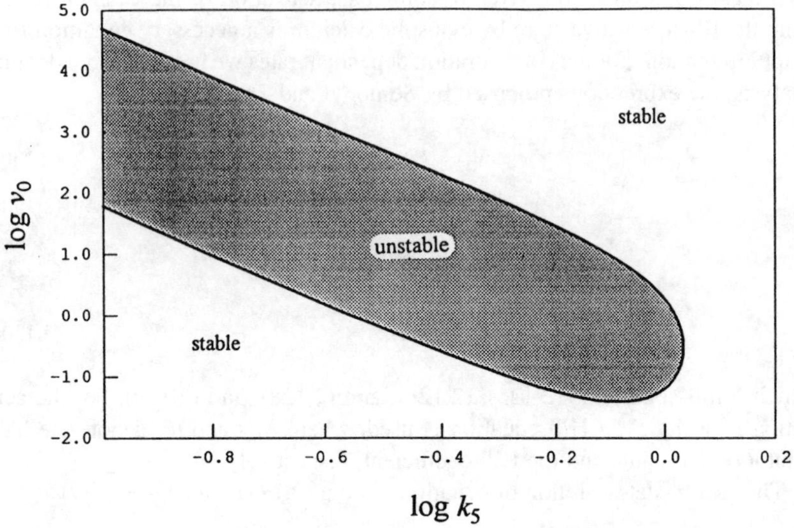

Figure 2.10 Stability diagram for the model of calcium oscillations given by the differential equations (2.124a) and (2.124b) and rate laws (2.130a)–(2.130d). Parameter values: $k_2 = 1$, $k_4 = 2$, $k_6 = 0.01$, $K_{0.5} = 3.1$, $n_H = 4$.

differential equations (2.124) by using parameter values from the region of instability. The oscillations shown in Figures 2.11A and 2.11B correspond to low and high values of v_0, respectively. It is seen that in both cases, the oscillations of the cytosolic calcium concentration have the form of repetitive spikes in accordance with experimental data (cf. Cuthbertson, 1989). In contrast, the oscillations of stored calcium have a sawtooth appearance [not shown, see Dupont and Goldbeter (1989)]. For high v_0 values, the oscillation frequency is much higher than for low values of this parameter. On the other hand, the amplitude of the cytosolic calcium oscillations is not as much affected by a change in v_0 as the frequency. Taking into account that an increase in v_0 may be brought about by an increase in the parameter β, these results may explain the experimental fact that the frequency of calcium oscillations increases with the concentration of IP_3 and in this way by the extent of receptor stimulation. The IP_3 concentration plays in the considered model the role of a parameter, which can be set to different values and thus act as a switch. Oscillatory behavior occurs even at constant IP_3 concentration, provided it has appropriate values. In contrast, in the above-mentioned model of Meyer and Stryer (1988) the IP_3 concentration is a system variable. The effect of the parameter β on the period and the amplitude of the oscillations has been systematically studied by Goldbeter *et al.* (1990). The results support the hypothesis that the physiological effect of calcium messenger oscillations is

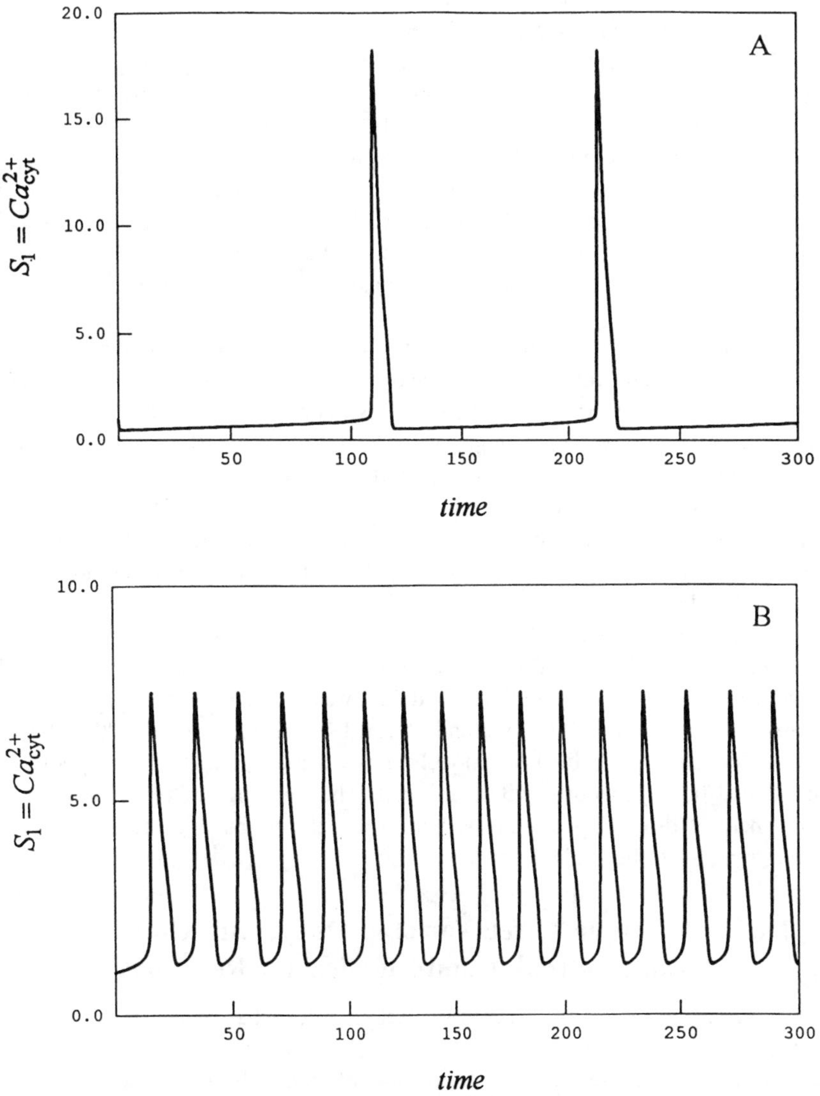

Figure 2.11 Time course of the cytosolic calcium concentration, $S_1 = Ca^{2+}_{cyt}$ as obtained by numerical integration of Eqs. (2.124a), (2.124b), and (2.130a)–(2.130d). Parameter values: (A) $k_5 = 1$, $v_0 = 1.4$; (B) $k_5 = 1$, $v_0 = 3.0$. The other parameter values are the same as in Figure 2.10.

brought about by a frequency-encoded mechanism rather than by an amplitude-dependent mechanism (cf. Section 2.4.6).

The fact that the rates v_1 and v_3 may be lumped into one quantity, $v_0 = v_1 + v_3$, indicates that the effect of Ca^{2+} efflux (v_3) from the IP_3-sensitive store can be mimicked by a change of the Ca^{2+} entry from the extracellular medium into the cytosol. It seems, therefore, that the assumption of two different calcium stores, one sensitive to IP_3 and one insensitive to IP_3 is not a necessary prerequisite of calcium oscillations. In fact, Dupont and Goldbeter (1993) have shown that calcium oscillations may also be explained on the basis of a one-pool model where the same Ca^{2+} channel is assumed to be sensitive to both IP_3 and Ca^{2+} behaving as coagonists.

The process of Ca^{2+} oscillations is often accompanied by a spatial propagation of Ca^{2+} waves. The velocity of the waves is of the order of 10 μm/s in oocytes (Jaffe, 1991) and of 30 μm/s in hepatocytes (Thomas *et al.,* 1991). In isolated cardiomyocytes the velocity ranges from 30 to 125 μm/s (Engel *et al.,* 1994). According to Meyer (1991) and Dupont and Goldbeter (1992), the calcium waves may be classified into two main types. For type 1, repetitive Ca^{2+} spikes move through the cytoplasm, whereas for type 2, the calcium concentration increases along the entire cell before it returns to its basal level in a nearly homogeneous manner ("tide" waves).

Mathematically, calcium waves may be described by adding to the differential equation system (2.124) a diffusion term for the concentration of cytosolic calcium. Models for the propagation of calcium waves in one or two spatial dimensions have been proposed (Meyer and Stryer, 1991; Thomas *et al.,* 1991; Dupont and Goldbeter, 1992). In these models it is assumed that the Ca^{2+} pools are distributed homogeneously within the cell. The influence of the geometric arrangement of discrete pools on the period and the propagation rate of calcium waves remains to be studied.

2.4.5. A Simple Three-Variable Model with Only Monomolecular and Bimolecular Reactions

According to Theorem 2F given in Section 2.4.2, chemical systems with two variable compounds cannot exhibit limit cycles if only monomolecular and bimolecular reactions are involved. As a matter of fact, well-known oscillatory two-component systems as the *Brusselator* (cf. Nicolis and Prigogine, 1977) or the system given by Selkov (1968) [cf. Eq. (2.119)] involve trimolecular reactions or contain nonlinearities higher than second order. Elementary chemical reactions are, in general, monomolecular or bimolecular because simultaneous collisions of more than two molecules are extremely improbable. Accordingly, the question arises of how complex a reaction system analyzed at the level of elementary reactions must be in order to allow limit cycle behavior. The problem of finding the smallest system showing Hopf bifurcations on the basis of the mass-action kinetics (2.10) was addressed by Wilhelm and Heinrich (1995).

Obviously, for the solution of this problem it is necessary to give a precise meaning to the term "smallest." The following characterization is proposed: (1) lowest number of variable reactants; (2) lowest number of quadratic terms in the differential equations; (3) minimal number of reactions, and (4) minimal number of bimolecular reactions. These four features are listed with descending importance.

Because it has been demonstrated that there exist three-variable mass-action systems showing Hopf bifurcations (Hanusse, 1973) and due to Theorem 2F, the smallest system must be searched for within the group of three-component systems (cf. point 1). For $n = 3$ local stability analysis may be performed on the basis of the characteristic equation (2.103).

A Hopf bifurcation takes place across the surface

$$a_1 a_2 - a_0 = 0, \quad a_0, a_1, a_2 > 0, \tag{2.134}$$

where a_0, a_1 and a_2 denote the coefficients of the characteristic polynomial.

Under condition (2.134) the characteristic equation (2.103) has one negative real eigenvalue ($\lambda_1 < 0$) and a pair of pure imaginary eigenvalues ($\lambda_2 = i\omega$, $\lambda_3 = -i\omega$).

By analyzing the stability of all three-component systems giving rise to only one nonlinear term in the differential equations (cf. point 2) it has been shown that there is exactly one system which follows from the given characterization of the smallest system with Hopf bifurcation (Wilhelm and Heinrich, 1995). Its mechanism involves five reactions with two of them being bimolecular, as is depicted in Figure 2.12. Because the scheme encompasses an autocatalytic step, it might be of importance in population kinetics also. The dynamic properties of this system are governed by the following differential equations:

$$\frac{\mathrm{d}S_1}{\mathrm{d}t} = (k_1 P_1 - k_4) S_1 - k_2 S_1 S_2, \tag{2.135a}$$

$$\frac{\mathrm{d}S_2}{\mathrm{d}t} = k_5 S_3 - k_3 S_2, \tag{2.135b}$$

$$\frac{\mathrm{d}S_3}{\mathrm{d}t} = k_4 S_1 - k_5 S_3, \tag{2.135c}$$

where P_1 denotes the fixed concentration of the outer reactant P_1 of the autocatalytic reaction.

The system has two steady states:

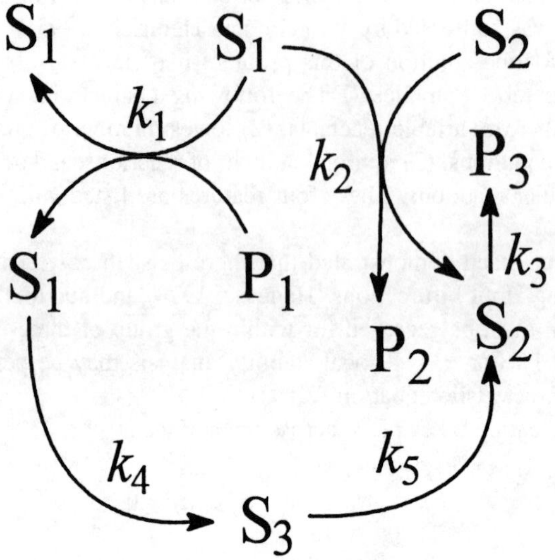

Figure 2.12 Scheme of a simple oscillating reaction system with three species, S_i, with variable concentrations and three external compounds, P_i.

$$\bar{S}_1 = \bar{S}_2 = \bar{S}_3 = 0, \tag{2.136a}$$

$$\bar{S}_1 = \frac{k_3(k_1P_1 - k_4)}{k_2k_4}, \qquad \bar{S}_2 = \frac{k_1P_1 - k_4}{k_2}, \qquad \bar{S}_3 = \frac{k_3(k_1P_1 - k_4)}{k_2k_5}. \tag{2.136b}$$

The coefficients of the characteristic polynomial for the first steady state read:

$$a_0 = -k_3k_5(k_1P_1 - k_4), \tag{2.137a}$$

$$a_1 = k_3k_5 - (k_1P_1 - k_4)(k_3 + k_5), \tag{2.137b}$$

$$a_2 = k_3 + k_4 + k_5 - k_1P_1. \tag{2.137c}$$

This gives

$$a_1a_2 - a_0 = (k_3 + k_5)(k_3 + k_4 - k_1P_1)(k_4 + k_5 - k_1P_1). \tag{2.138}$$

For the second steady state, one obtains

$$a_0 = (k_1P_1 - k_4)k_3k_5, \tag{2.139a}$$

$$a_1 = k_3 k_5, \tag{2.139b}$$

$$a_2 = k_3 + k_5, \tag{2.139c}$$

and from that

$$a_1 a_2 - a_0 = k_3 k_5 (k_3 + k_4 + k_5 - k_1 P_1). \tag{2.140}$$

Using Eqs. (2.137)–(2.140), it follows from the Hurwitz criterion (2.105) that the first steady state is stable within the range $0 \leq k_1 P_1 < k_4$ and the second one within the range $k_4 < k_1 P_1 < k_3 + k_4 + k_5$. Figure 2.13 shows the steady-state concentration $\bar{S}_1$ as a function of the parameter $k_1 P_1$ at fixed values of the other parameters. Stable and unstable steady states are characterized by solid and broken lines, respectively.

The system has two bifurcation points: a transcritical bifurcation at $k_1 P_1 = k_4$, and a Hopf bifurcation at $k_1 A = k_3 + k_4 + k_5$.

Figures 2.14A and 2.14B show numerical solutions of the differential equation system in the state space for parameter values $k_1 P_1$ from both sides of the Hopf bifurcation point.

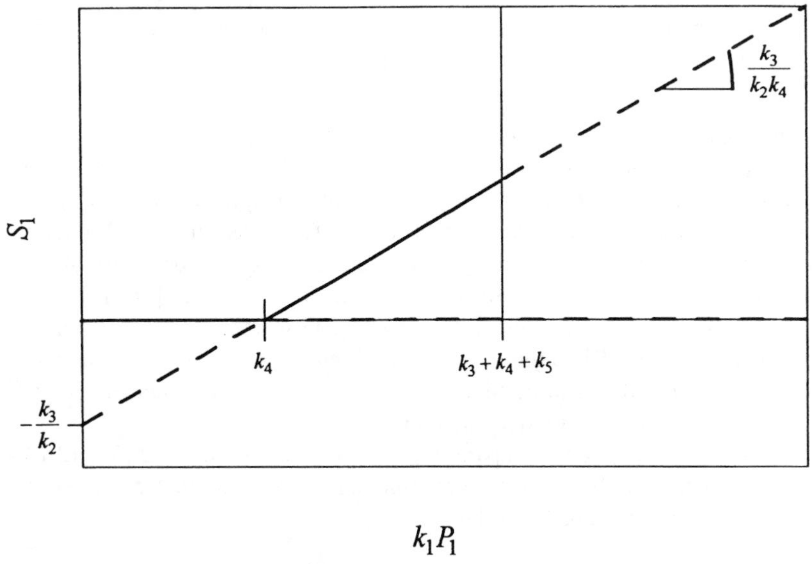

Figure 2.13 Bifurcation diagram of the steady-state concentration of S_1 for the system given by Eqs. (2.135a)–(2.135c). Solid and broken lines indicate stable and unstable states, respectively.

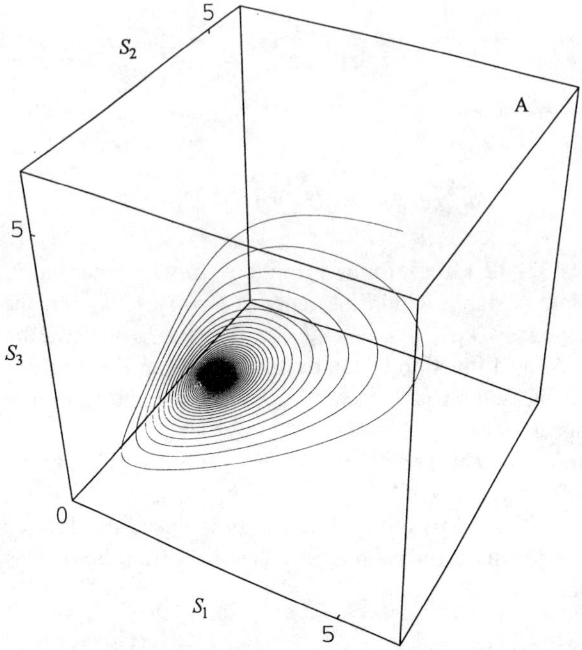

Figure 2.14 Numerical solutions of the differential equation system (2.135a)–(2.135c) in the state space in the neighborhood of the Hopf bifurcation point. Parameter values: (A) $k_1P_1 = 2.9$ (stable steady state). (*Figure continued on facing page*)

Obviously, the characterization of the "smallest system" given above is not the only one possible. As an alternative, one could give point 3 (i.e., *minimal number of reactions*) a higher priority than point 2 (i.e., *lowest number of quadratic terms*). Accordingly, other minimal chemical systems with Hopf bifurcations might exist. In particular, the analysis does not exclude the possibility of an oscillating three-component system with *two* quadratic terms but less than five reactions. Furthermore, it remains an open problem whether the irreversibility of all reactions depicted in Figure 2.12 is a crucial assumption. For example, if reaction 1 is considered to be reversible a second quadratic nonlinearity ($\propto S_1^2$) would appear in the differential equation.

It may be interesting to compare the system given by Eqs. (2.135a)–(2.135c) with other three-variable systems containing only one quadratic term. It has been demonstrated, for example, that the system

$$\frac{\mathrm{d}X}{\mathrm{d}t} = -Y - Z, \qquad (2.141\mathrm{a})$$

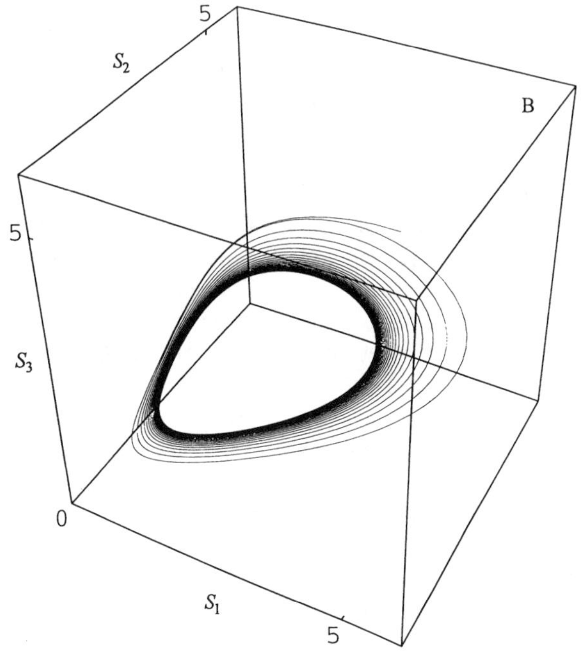

Figure 2.14 (*continued*) (B) $k_1P_1 = 3.1$ (stable limit cycle). Other parameter values: $k_2 = k_3 = k_4 = k_5 = 1$.

$$\frac{dY}{dt} = X + aY, \tag{2.141b}$$

$$\frac{dZ}{dt} = bX - cZ + XZ \tag{2.141c}$$

with positive parameters a, b and c may exhibit not only limit cycles but also chaotic behavior (Rössler, 1979). However, Eqs. (2.141a)–(2.141c) cannot describe a chemical system, because the trajectories are not confined to the non-negative orthant. This may be immediately seen from Eq. (2.141a) which predicts $dX/dt < 0$ for $X = 0$; $Y,Z > 0$. Note that for the generalized mass-action kinetics (2.15), the trajectories always remain in the non-negative orthant.

2.4.6. Possible Physiological Significance of Oscillations

There is ample evidence that oscillations are a ubiquitous phenomenon in biological systems. Periodic changes in different biological processes are observed in all types of organisms, from bacteria to the most complex multicellular organisms. The periods may range from seconds to years. Besides the glycolytic oscillations and calcium oscillations considered in Sections 2.4.3 and 2.4.4, these include oscillations in bacterial protein synthesis, periodic changes in photosynthesis, and periodicities in neural activity as well as in muscular contractions. Well known are the circadian and circannual rhythms in plants and the menstrual cycles in higher animals. For many biological processes, the physiological role of oscillations is obvious. For example, in cardiac cells, biochemical oscillations are transformed into periodic mechanical movements, whereas in neural cells, oscillations are used for the transmission of information. It is generally believed that oscillatory behavior is of functional advantage also for other processes. First, oscillations may, in contrast to steady states or transient states, play a role in the temporal coordination of various cellular processes. In this respect, the phenomenon of synchronization and related mechanisms for the entrainment of oscillations are of major importance. Taking into account the close relationship between oscillation and wave phenomena, metabolic oscillators are likely to be important for the spatial organization of cellular processes (cf. Rapp, 1987). Furthermore, it has been stated that oscillatory processes may be more efficient with respect to energy conversion in cells (Termonia and Ross, 1981).

It is widely accepted that the physiological effect of oscillations is frequency encoded. This view is supported by the fact that for receptor-stimulated oscillations, the frequency increases with the concentration of the agonist (see Section 2.4.4 for the effect of the hormone concentration on the frequency of calcium oscillations). The problem of the parameter dependence of oscillation frequencies is generally addressed in Section 5.8.5 using metabolic control analysis. There it is shown that for the glycolytic oscillator defined by Eq. (2.119), the frequency of the oscillations increases proportionally with increasing input rate of glucose, at least for parameter values near the Hopf bifurcation. Furthermore, there is strong evidence that frequency-encoded signal transduction is much more stable against noise than amplitude-dependent mechanisms. Frequency encoding may be even effective if the system enter regimes of chaotic behavior where after short times the information encoded in amplitudes is lost. [For more detailed considerations on frequency encoding, cf. the works of Rapp *et al.* (1981) and Goldbeter and Li (1989)].

For many biochemical systems, the functional significance of oscillations is still unknown. This concerns, for example, glycolysis, despite the fact that the glycolytic oscillator is most successfully investigated, experimentally as well as theoretically. Because all cells contain this pathway and glycolysis was historically the first way for the supply of energy in the form of ATP, one hardly believes that the regulatory couplings which make oscillations in this pathway possible are of only secondary importance. It has been argued by Selkov (1980) that, at an early stage of evolution, glycolysis has been responsible for oscillations on a circadian time scale. The period of glycolytic oscillations, which is normally in the order of minutes, may be drastically increased due the "deposition effect"; that is, the reversible transformation of the pool of glucose-6-phosphate and fructose-6-phosphate into polysaccharides. Using a refined model of anaerobic metabolism, Selkov (1980) could show that incorporation of the synthesis and breakdown of glycogen may increase the oscillation period from $T = 3$ min to $T = 25$ h.

However, it may well be that in many cases oscillations have no physiological meaning: that is, the occurrence of limit cycles is sometimes unavoidably concomitant to the fact that metabolic systems are highly nonlinear processes working under conditions far from equilibrium.

Generally, models of metabolic systems predict oscillations only for certain ranges of the parameter values and it is not always clear whether these parameter regions correspond to physiological states. It is also conceivable that cells need mechanisms to avoid oscillations in cases where they are not necessary for their function. One possible way is the optimization of kinetic parameters. It has been realized that great differences in the magnitudes of rate constants, which is a typical phenomenon in biochemical systems (cf. Chapter 4), often counteracts the generation of oscillations (cf. Savageau, 1975; Heinrich *et al.*, 1977; Dibrov *et al.*, 1982). This may be exemplified for unbranched reaction chains with feedback inhibition of the first reaction by the end product (cf. Scheme 6).

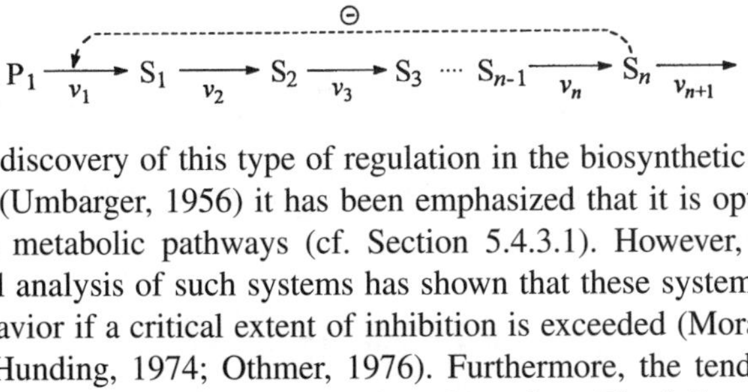

$$P_1 \xrightarrow[v_1]{} S_1 \xrightarrow[v_2]{} S_2 \xrightarrow[v_3]{} S_3 \cdots S_{n-1} \xrightarrow[v_n]{} S_n \xrightarrow[v_{n+1}]{} \qquad \text{Scheme 6}$$

Since the discovery of this type of regulation in the biosynthetic pathways of amino acids (Umbarger, 1956) it has been emphasized that it is optimal for homeostasis in metabolic pathways (cf. Section 5.4.3.1). However, the detailed mathematical analysis of such systems has shown that these systems exhibit oscillatory behavior if a critical extent of inhibition is exceeded (Morales and McKay, 1967; Hunding, 1974; Othmer, 1976). Furthermore, the tendency toward instability grows with the increasing number of reactions. The following condition for the emergence of unstable states has been derived:

$$n_{\mathrm{H}} > \frac{A}{\cos^n\left(\dfrac{\pi}{n}\right)} = n_{\mathrm{H}}^{\mathrm{crit}}(n), \qquad (2.142)$$

where n_{H} denotes the Hill coefficient characterizing the inhibition of the first reaction by the endproduct of the pathway. n stands for the number of intermediates (so that the number of reactions is $r = n + 1$) and A is a factor depending on the steady-state concentration of the end product (Viniegra-Gonzales and Martinez, 1969). n_H^{crit} decreases monotonically with increasing chain length. With $A = 1$, which is valid for high input rates (Hunding, 1974), one derives for the critical values of the inhibition constants $n_H^{crit} = 8$ when $n = 3$, and $n_H^{crit} = 1.6$ when $n = 10$. From these results one could conclude that long synthetic pathways of amino acids with feedback inhibition would generally be in the oscillatory regime. However, Eq. (2.142) has been derived for a chain where all the reactions are described by the same kinetic constants ($k_i = k$ for $i = 2, \ldots, n + 1$). If the kinetic constants k_i differ from each other, more complicated conditions for unstable steady states and oscillations arise. For unbranched pathways with three intermediates, the system becomes unstable under the condition

$$n_H > n_H^{crit} = \left(2 + \frac{k_1}{k_2} + \frac{k_2}{k_1} + \frac{k_2}{k_3} + \frac{k_3}{k_2} + \frac{k_1}{k_3} + \frac{k_3}{k_1}\right) \tag{2.143}$$

(Savageau, 1975). It is seen that n_H^{crit} defined in Eq. (2.143) may increase drastically as differences in the values of the rate constants increase. For example, $k_1 = 1$, $k_2 = 10$ and $k_3 = 100$ results in $n_H^{crit} = 122.21$ (i.e., a Hill coefficient much higher than those observed in enzyme kinetics).

If the rate constants do not differ very much from a common mean value $\langle k \rangle$ (i.e., $k_i = \langle k \rangle + \Delta k_i$ with $|\Delta k_i|/\langle k \rangle \ll 1$), one derives by a Taylor expansion of expression (2.143),

$$n_H^{crit} = 8 - \frac{2}{\langle k \rangle^2}(\Delta k_1 \Delta k_2 + \Delta k_1 \Delta k_3 + \Delta k_2 \Delta k_3) = 8 + \frac{1}{\langle k \rangle^2} \sum_{j=1}^{3} (\Delta k_j)^2, \tag{2.144}$$

where $\Delta k_1 + \Delta k_2 + \Delta k_3 = 0$ has been taken into account. Equation (2.144) may be rewritten as

$$p_{crit} = 8 + 3\left(\frac{\sigma}{\langle k \rangle}\right)^2, \tag{2.145}$$

where $\sigma^2 = \langle k^2 \rangle - \langle k \rangle^2$ denotes the variance of the kinetic parameters. Equations (2.143)–(2.145) substantiate the above assertion that separation of time constants may be instrumental to protect cells from oscillations if they are of no use for their functioning.

3

Stoichiometric Analysis

Stoichiometry concerns the proportions of changes in the concentrations of chemically reacting species. These proportions also indicate the topological structure of reaction networks, because they involve information about which substances are linked with each other by reactions. Stoichiometry does not primarily deal with the velocities of changes, which is the realm of kinetics. In contrast to kinetic properties, which can vary in biological systems quite rapidly due to inhibition and activation of enzymes, the stoichiometric properties are in a sense structural invariants, unless evolutionary time scales are considered (Aris, 1965; Clarke, 1988; Reder, 1988). Moreover, stoichiometric properties are often better known than kinetic parameters of reactions. Knowledge of the stoichiometric properties is a prerequisite for any simulation of biochemical reaction networks. Importantly, the stoichiometric properties of a model do not depend on whether the description is discrete, continuous, deterministic, or stochastic (cf. Érdi and Tóth, 1989).

One can distinguish two different approaches to stoichiometric analysis according to whether or not knowledge of the atomic composition of reacting substances is taken into account. The catalase reaction (2.2) and the hexokinase-phosphoglucomutase system (2.3) can be taken as examples of the two cases. (Although the molecular structure of glucose, ATP, and so forth is known, it is not essential for analyzing the kinetic properties of system (2.3).) In many instances, kinetic modeling in biochemistry does not require knowledge of the atomic composition of the substances involved. The system equations can be written down by just using information about the molecularities with which reactants and products enter the reactions.

As explained in Section 2.1, the proportions with which the substances in a reaction system are interconverted can be written in the form of a stoichiometry matrix, **N**. For illustration, let us consider the reaction scheme of the main processes of energy metabolism in erythrocytes (glycolysis and various membrane transport processes) depicted in Figure 3.1. This scheme will be studied in more detail in Section 5.4.4.1.

Using the numbering of reactions and substances as indicated in the legend to Figure 3.1, we can write the stoichiometry matrix for the internal metabolites and ions as given in Table 3.1. The external species are here chosen to comprise all

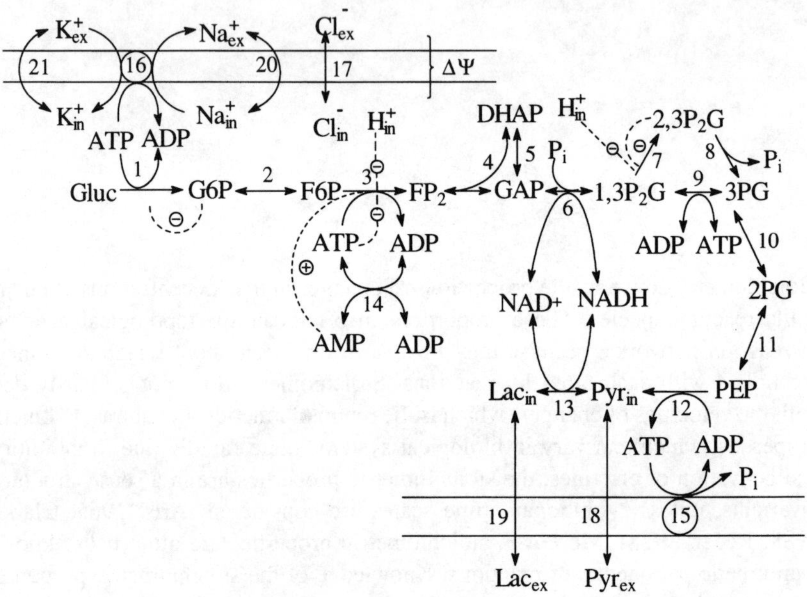

Figure 3.1 Glycolytic reactions and various membrane transport processes in erythrocytes. *Reactions:* (1) hexokinase (HK), (2) phosphoglucoisomerase (PGI), (3) phosphofructokinase (PFK), (4) aldolase (Ald), (5) triose-phosphate isomerase (TIM), (6) glyceraldehyde-phosphate dehydrogenase (GADP), (7) bisphosphoglycerate mutase (P_2GM), (8) 2,3-bisphosphoglycerate phosphatase (P_2Gase), (9) phosphoglycerate kinase (PGK), (10) phosphoglycerate mutase (PGAM), (11) enolase (Enol), (12) pyruvate kinase (PK), (13) lactate dehydrogenase (LDH), (14) adenylate kinase (AK), (15) ATP consumption by membrane phosphorylation, (16) Na/K-ATPase, (17) passive transport of chloride ions, (18 and 19) passive transport of pyruvate and lactate, respectively, (20 and 21) passive transport of sodium and potassium, respectively.

Metabolites and ions: (1) glucose-6-phosphate (G6P), (2) fructose-6-phosphate (F6P), (3) fructose-1,6-bisphosphate (FP_2), (4) glyceraldehyde-3-phosphate (GAP), (5) dihydroxyacetone phosphate (DHAP), (6) 1,3-bisphosphoglycerate ($1,3P_2G$), (7) 2,3-bisphosphoglycerate ($2,3P_2G$), (8) 3-phosphoglycerate (3PG), (9) 2-phosphoglycerate (2PG), (10) phosphoenolpyruvate (PEP), (11) intracellular pyruvate (Pyr_{in}), (12) intracellular lactate (Lac_{in}), (13) AMP, (14) ADP, (15) ATP, (16 and 17) NAD, NADH, (18, 19, and 20) intracellular potassium (K^+_{in}), sodium (Na^+_{in}) and chloride (Cl^-_{in}), respectively, (21) glucose (Gluc), (22 and 23) extracellular pyruvate (Pyr_{ex}) and lactate (Lac_{ex}), respectively, (24, 25, and 26) extracellular potassium (K^+_{ex}), sodium (Na^+_{ex}), and chloride (Cl^-_{ex}), respectively, (27) inorganic phosphate (P_i). Substances 21–27 are taken as external species.

Table 3.1 Stoichiometric Matrix **N** of Glycolysis for the Internal Metabolites According to the Numbering of Reactions and Metabolites Given in the Legend to Figure 3.1

1	−1	0	0	0	0	0	0	0	0	0	0	0	0	0	0	0	0	0	0	0
0	1	−1	0	0	0	0	0	0	0	0	0	0	0	0	0	0	0	0	0	0
0	0	1	−1	0	0	0	0	0	0	0	0	0	0	0	0	0	0	0	0	0
0	0	0	1	1	−1	0	0	0	0	0	0	0	0	0	0	0	0	0	0	0
0	0	0	1	−1	0	0	0	0	0	0	0	0	0	0	0	0	0	0	0	0
0	0	0	0	0	1	−1	0	−1	0	0	0	0	0	0	0	0	0	0	0	0
0	0	0	0	0	0	1	−1	0	0	0	0	0	0	0	0	0	0	0	0	0
0	0	0	0	0	0	0	1	1	−1	0	0	0	0	0	0	0	0	0	0	0
0	0	0	0	0	0	0	0	0	1	−1	0	0	0	0	0	0	0	0	0	0
0	0	0	0	0	0	0	0	0	0	1	−1	0	0	0	0	0	0	0	0	0
0	0	0	0	0	0	0	0	0	0	0	1	−1	0	0	0	0	−1	0	0	0
0	0	0	0	0	0	0	0	0	0	0	0	1	0	0	0	0	0	−1	0	0
0	0	0	0	0	0	0	0	0	0	0	0	0	−1	0	0	0	0	0	0	0
1	0	1	0	0	0	0	0	−1	0	0	−1	0	2	1	1	0	0	0	0	0
−1	0	−1	0	0	0	0	0	1	0	0	1	0	−1	−1	−1	0	0	0	0	0
0	0	0	0	0	−1	0	0	0	0	0	0	1	0	0	0	0	0	0	0	0
0	0	0	0	0	1	0	0	0	0	0	0	−1	0	0	0	0	0	0	0	0
0	0	0	0	0	0	0	0	0	0	0	0	0	0	0	2	0	0	0	0	−1
0	0	0	0	0	0	0	0	0	0	0	0	0	0	0	−3	0	0	0	−1	0
0	0	0	0	0	0	0	0	0	0	0	0	0	0	0	0	−1	0	0	0	0

substances outside the cell and glucose, because, in erythrocytes, hexokinase is nearly saturated with glucose.

In case the atomic composition is relevant for the modeling study, one can use the atomic matrix, **A**. Its elements, a_{ik}, give the number of atoms of the chemical element k involved in one molecule of substance S_i (cf. Aris, 1965; Érdi and Tóth, 1989). For example, the composition of hydrogen peroxide, water and molecular oxygen participating in the catalase reaction (2.2), can formally be written as

$$\begin{pmatrix} H_2O_2 \\ H_2O \\ O_2 \end{pmatrix} = \mathbf{A} \begin{pmatrix} H \\ O \end{pmatrix} = \begin{pmatrix} 2 & 2 \\ 2 & 1 \\ 0 & 2 \end{pmatrix} \begin{pmatrix} H \\ O \end{pmatrix}. \tag{3.1}$$

Without knowledge of the reaction mechanism, the columns of the stoichiometry matrix are indeterminate to the extent of multiplication by arbitrary factors. For instance, we could write $H_2O_2 \rightarrow H_2O + \frac{1}{2}O_2$ instead of reaction (2.2). This indeterminacy does not affect the stoichiometric analysis throughout this chapter. It does, however, matter for the kinetic equations because the stoichiometric coefficients enter, as exponents, the mass-action-type rate laws (cf. Section 2.2.1). One should then choose such a scaling of stoichiometric coefficients that reflects the number of molecules really colliding to initiate elementary reaction events.

From knowledge of the stoichiometric structure of a reaction system, interesting conclusions can be derived without using information about the kinetic properties. This concerns, in particular, conservation relations and dependencies between steady-state fluxes.

3.1. CONSERVATION RELATIONS

3.1.1. Linear Dependencies Between the Rows of the Stoichiometry Matrix

Frequently, the concentrations of several substances involved in biochemical reaction systems enter so-called conservation sums, such as

$$ATP + ADP + AMP = \text{const.}, \tag{3.2}$$

which holds, for example, in the system depicted in Figure 3.1. (Note: Italicized symbols of substances indicate their concentrations). Conservation relations can involve coefficients other than unity. For example, the conservation of phosphate gives, in the same system, a relation in which *ATP* occurs with the coefficient 3, whereas *ADP*, $1.3P_2G$ and $2.3P_2G$ enter it with the coefficient 2. Negative coefficients attached to some concentrations may also occur. For example, in the reaction $S_1 + S_2 \rightarrow S_3$, the conservation relation $S_1 - S_2 = \text{const.}$ holds. More complex, biochemically relevant examples are given in R. Schuster *et al.* (1988) and in Section 4.3.

In general terms, conservation relations can be written as

$$\boldsymbol{g}^{\mathrm{T}}\mathbf{S}(t) = T = \text{const.} \tag{3.3}$$

with $\boldsymbol{g}$ and T denoting a vector of constant coefficients and a conservation quantity, respectively. The Roman T stands for the transpose. Mathematically, conservation relations cause some rows of the stoichiometry matrix to be linearly dependent. This can be written as

$$\boldsymbol{g}^{\mathrm{T}}\mathbf{N} = \boldsymbol{0}^{\mathrm{T}}. \tag{3.4}$$

Postmultiplication of this equation by the rate vector $\mathbf{v}$ and integration yield Eq. (3.3), due to the system equation (2.7). Equation (3.3) means that a linear combination of metabolite concentrations is conserved in time.

It may occur that there are more than one (linearly independent) vectors $\boldsymbol{g}$ that fulfill Eq. (3.4). It follows from linear algebra that the number of independent conservation vectors $\boldsymbol{g}$ is given by $n - \text{rank}(\mathbf{N})$, with n denoting the number of rows of the stoichiometry matrix (see Groetsch and King, 1988). Accordingly, the number of independent conservation quantities, T_k, equals $n - \text{rank}(\mathbf{N})$. On simulating the dynamic behavior of the system, one can therefore eliminate this number of concentration variables. If **N** is full rank [i.e., if $\text{rank}(\mathbf{N}) = n$], the system has no conservation relations.

For any given reaction system, a complete set of linearly independent vectors $\boldsymbol{g}$ can be arranged into a matrix, **G**, which is called conservation matrix (cf. Park Jr., 1988) and fulfills the equation

$$\mathbf{GN} = \mathbf{0}. \tag{3.5}$$

This matrix is not uniquely determined because each matrix $\hat{\mathbf{G}} = \mathbf{PG}$ with **P** being any nonsingular square matrix of appropriate dimension is a conservation matrix as well (the rows of $\hat{\mathbf{G}}$ then are linear combinations of the rows of **G**). Accordingly, the set of all vectors $\boldsymbol{g}$ resulting from Eq. (3.4) for a given matrix **N** constitutes a vector space of dimension $n - \text{rank}(\mathbf{N})$. Independence of conservation relations can be defined as linear independence of the corresponding vectors $\boldsymbol{g}$. Note that the scaling indeterminacy of stoichiometric coefficients mentioned above does not affect the conservation relations.

An alternative way of representing the linear dependencies between the rows of the stoichiometry matrix was proposed by Reder (1988). By rearranging the rows of **N** so that the upper rank(**N**) rows are linearly independent, one can decompose **N** as

$$\mathbf{N} = \begin{pmatrix} \mathbf{N}^0 \\ \mathbf{N}' \end{pmatrix}, \tag{3.6}$$

where the submatrix $\mathbf{N}^0$ has rank (**N**) rows. As the rows of $\mathbf{N}'$ are linearly dependent on the rows of $\mathbf{N}^0$, we can write

$$\mathbf{N} = \mathbf{L}\mathbf{N}^0 = \begin{pmatrix} \mathbf{I} \\ \mathbf{L}' \end{pmatrix}\mathbf{N}^0. \tag{3.7}$$

L is called the link matrix. Therefore, the system equations of the reaction network assume the form

$$\frac{\mathrm{d}}{\mathrm{d}t}\begin{pmatrix} \boldsymbol{S}_{\mathrm{a}} \\ \boldsymbol{S}_{\mathrm{b}} \end{pmatrix} = \begin{pmatrix} \mathbf{I} \\ \mathbf{L}' \end{pmatrix}\mathbf{N}^0\boldsymbol{v}, \tag{3.8}$$

where the vector of concentrations is split into two vectors, $\boldsymbol{S}_{\mathrm{a}}$ and $\boldsymbol{S}_{\mathrm{b}}$, of dimensions rank (**N**) and $n - \text{rank}(\mathbf{N})$. The metabolites should be renumbered according to the rearrangement of the rows of **N**. From Eq. (3.8) one derives the relations

$$\frac{dS_b}{dt} = \mathbf{L}' \frac{dS_a}{dt}, \tag{3.9}$$

and from that

$$S_b = \mathbf{L}'S_a + \text{const.}, \tag{3.10}$$

that is, the metabolite concentrations S_i with $i >$ rank(**N**) may be expressed as linear functions of the concentrations S_i with $i \leq$ rank(**N**). Therefore, the system dynamics is completely described by the upper part of Eq. (3.8) (i.e., the reduced system $dS_a/dt = \mathbf{N}^0\boldsymbol{v}$).

As the two equations (3.5) and (3.7) express the same fact, it is quite natural that the matrix **G** can be written in terms of the matrix **L**. For example, we can choose

$$\mathbf{G} = (-\mathbf{L}' \ \mathbf{I}). \tag{3.11}$$

Together with the non-negativity condition for concentrations,

$$S(t) \geq \mathbf{0}, \tag{3.12}$$

the set of all conservation relations (3.3) for a given reaction system determines a region to which the concentration vector is confined. This region is a (possibly unbounded) convex polyhedron (cf. Rockafellar, 1970) and is called invariant manifold (Gavalas, 1968) or reaction simplex (Horn and Jackson, 1972). Denoting the difference of the concentration vectors for two different points in time, t_1 and t_2, by ΔS, we obtain, from Eq. (3.3),

$$\mathbf{G}\Delta S = \mathbf{0}. \tag{3.13}$$

This means that any vector lying in the concentration polyhedron is orthogonal to all conservation vectors.

3.1.2. Non-negative Conservation Relations

Conservation relations are frequently brought in relation to conservation of atoms or atom groups (Aris, 1965; Gavalas, 1968; Park, 1974; Cavallotti *et al.*, 1980; Hofmeyr *et al.*, 1986; Park Jr., 1988). Consider, for example, the pyruvate decarboxylase reaction (EC 4.1.1.1),

$$CH_3COCOO^- + H^+ \rightarrow CH_3CHO + CO_2. \qquad (3.14)$$

The stoichiometry matrix reads $\mathbf{N} = (-1\ -1\ 1\ 1)^T$. A conservation matrix $\mathbf{G}$ can be found by consideration of the conservation conditions for the atomic species carbon, oxygen, and hydrogen,

$$\mathbf{G} = \begin{pmatrix} 3 & 0 & 2 & 1 \\ 3 & 0 & 1 & 2 \\ 3 & 1 & 4 & 0 \end{pmatrix}. \qquad (3.15)$$

Another feasible conservation matrix is

$$\mathbf{G} = \begin{pmatrix} 1 & 0 & 1 & 0 \\ 0 & 1 & 1 & 0 \\ 1 & 0 & 0 & 1 \end{pmatrix}, \qquad (3.16)$$

whose rows correspond to conservation of the CH_3CO group, proton, and carboxyl group. There are also vectors fulfilling Eq. (3.4) which contain negative components [e.g., $\boldsymbol{g} = (1\ -1\ 0\ 0)^T$].

A necessary condition for a conservation relation to represent conservation of chemical units is that all coefficients be non-negative,

$$\boldsymbol{g} \geq \boldsymbol{0}, \qquad (3.17a)$$

$$\boldsymbol{g} \neq \boldsymbol{0}. \qquad (3.17b)$$

Relation (3.17b) excludes the trivial case that all coefficients are zero. The case with only one coefficient being positive occurs if some substance does not participate in any reaction, so that it can be canceled from the network. Therefore, condition (3.17) actually implies that at least two coefficients are positive. We shall call vectors satisfying relations (3.4) and (3.17) *non-negative conservation vectors* [in view of condition (3.17b), a more exact term is *semipositive conservation vectors* (cf. Schuster and Höfer, 1991)].

Attention has to be paid to systems containing electrically charged molecules or atoms (ions). If these systems are closed, not only some atom groups but also electric charge is conserved. An example is provided by the superoxide dismutase reaction (EC 1.15.1.1) proceeding in many living cells,

$$2\dot{O}_2^- + 2H^+ \rightarrow H_2O_2 + O_2. \qquad (3.18)$$

The stoichiometry matrix reads $\mathbf{N} = (-2\ -2\ 1\ 1)^T$. Because $n - \text{rank}(\mathbf{N}) = 3$ and the number of atomic species is 2, one has to include the conservation of electric charge to obtain a conservation matrix in which all rows are amenable to physico-chemical interpretation. An admissible conservation matrix is

$$\mathbf{G} = \begin{pmatrix} 2 & 0 & 2 & 2 \\ 0 & 1 & 2 & 0 \\ -1 & 1 & 0 & 0 \end{pmatrix}. \tag{3.19}$$

The elements of the third row of **G** represent electric charge. The corresponding conservation relation is not, however, non-negative. Nevertheless, one can always express charge conservation as a non-negative vector by considering all electrons involved. As to reaction (3.18), the vector relating to conservation of electrons is $\boldsymbol{g} = (17\ 0\ 18\ 16)^{\mathrm{T}}$, as neutral hydrogen and oxygen atoms have one and eight electrons, respectively.

There are different methods for calculating conservation matrices. One method is by determining a set of basic solutions to the homogeneous linear equation

$$\mathbf{N}^{\mathrm{T}}\boldsymbol{g} = \boldsymbol{0} \tag{3.20}$$

[which is equivalent to Eq. (3.4)], using the Gaussian elimination method (cf. Groetsch and King, 1988). The matrix **G** thus obtained has the form given in Eq. (3.11); that is, it contains an identity matrix as submatrix. A modified method for calculating **G** was given by Park Jr. (1988). It is based on the Gauss-Jordan inversion (cf. Groetsch and King, 1988). Both of these methods do not, however, guarantee that **G** be non-negative. Sauro and Fell (1991) proposed to determine conservation relations with only non-negative coefficients by computing the matrix **L** several times, with a different order of the rows of the stoichiometry matrix.

For detecting non-negative conservation matrices in a more systematic way, one can use methods of convex analysis. In that mathematical theory, it is shown that the solution sets to linear homogeneous equation systems subject to linear homogeneous inequality constraints, such as Eq. (3.4) and inequality (3.17), are unbounded *pointed convex polyhedral cones* (Rockafellar, 1970). Convex analysis further says that such cones can be represented as non-negative linear combination of *generating vectors,* which are unique up to scalar multiples; that is, the cone, $\mathcal{K}$, representing all non-negative conservation relations for a given reaction system can be written as

$$\mathcal{K} = \{\boldsymbol{g} \in \mathrm{R}^n | \boldsymbol{g} = \sum_{k=1}^{p} \eta_k \boldsymbol{e}_k, \eta_k \geq 0, k = 1, \ldots, p\}, \tag{3.21}$$

where $\boldsymbol{e}_k$ are the generating vectors and p is their number. By definition, a generating vector of a cone is a vector that belongs to this cone and cannot be represented as convex linear combination of two different vectors belonging to this cone as well. Figure 3.2 shows a schematic representation of a polyhedral cone. It can be seen that the generating vectors are located on the edges of the cone. Their number may be greater than the dimension of the cone. This dimension, in turn, is less than, or equal to, $n - \text{rank}(\mathbf{N})$.

The conservation relations corresponding to the generating vectors are to be called *extreme non-negative conservation relations.* In the following, we shall impose the additional condition that for each generating vector, the only common divisor of its components be unity, which can easily be fulfilled by appropriate reduction (rescaling). The vectors thus obtained shall be called *reduced generating vectors.*

A complete set of generating vectors can be found by an algorithm developed in convex analysis (Nožička *et al.,* 1974). A variant of this algorithm specified so as to be applicable to the problem dealt with in this section was given in S. Schuster and Höfer (1991). In this method, a sequence of matrices is consecutively computed, starting from the stoichiometry matrix augmented with the $n \times n$ identity matrix. Such sequences of matrices are usually called *tableaux.* For illustration, consider reaction (2.2). The stoichiometry matrix reads $\mathbf{N} = (-2\ 2\ 1)^{\mathrm{T}}$. The initial tableau reads

$$\mathbf{T}^{(0)} = \left(\begin{array}{c:ccc} -2 & 1 & 0 & 0 \\ 2 & 0 & 1 & 0 \\ 1 & 0 & 0 & 1 \end{array}\right). \tag{3.22}$$

One now calculates a new tableau, $\mathbf{T}^{(1)}$, by constructing all possible non-negative linear combinations of pairs of rows of $\mathbf{T}^{(0)}$ so that the elements of the first column

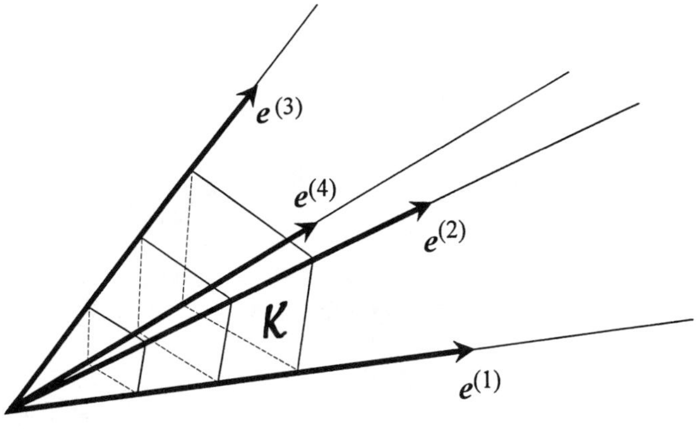

Figure 3.2 Convex polyhedral cone, $\mathcal{K}$, spanned by four generating vectors, $\boldsymbol{e}_1$ to $\boldsymbol{e}_4$.

become zeros. To this end, in Eq. (3.22), we sum up the first and second rows, and the first row with the double of the third row,

$$\mathbf{T}^{(1)} = \left(\begin{array}{c|ccc} 0 & 1 & 1 & 0 \\ 0 & 1 & 0 & 2 \end{array}\right). \tag{3.23}$$

Non-negative combination of the second and third rows does not lead to a zero. Therefore, $\mathbf{T}^{(1)}$ has here less rows than $\mathbf{T}^{(0)}$. As this special system has only one reaction, $\mathbf{T}^{(1)}$ is the final tableau. The right-hand submatrix contains the generating vectors $\boldsymbol{e}_1 = (1\ 1\ 0)^{\mathrm{T}}$ and $\boldsymbol{e}_2 = (1\ 0\ 2)^{\mathrm{T}}$. Owing to Eq. (3.3) and relation (3.21), this means that every non-negative conservation relation of the catalase reaction can be written as

$$\eta_1(H_2O_2 + H_2O) + \eta_2(H_2O_2 + 2O_2) = \text{const.}, \quad \eta_1, \eta_2 \geq 0. \tag{3.24}$$

A graphical representation of the corresponding cone is given in Figure 3.3. The vectors $(1\ 1\ 0)^{\mathrm{T}}$ and $(1\ 0\ 2)^{\mathrm{T}}$ span a two-dimensional cone. In the state space, any concentration vector lies in a manifold given by the intersection of the conservation relations $H_2O_2 + H_2O = T_1$ and $H_2O_2 + 2O_2 = T_2$. The concentration polyhedron is therefore a straight line. Its direction is given by the fact that it is orthogonal to the cone $\mathcal{K}$ [cf. Eq. (3.13)]. Its location depends on the values of the conservation quantities T_1 and T_2.

For systems with several reactions, further tableaux are successively calculated in the algorithm, so that not only the first column of $\mathbf{N}$ but also the others become null vectors. For constructing $\mathbf{T}^{(j+1)}$ from $\mathbf{T}^{(j)}$, one first determines, for each row of $\mathbf{T}^{(j)}$, a set $I(i)$ which contains the column indices, h, of all the elements $t_{ih}^{(j)}$ of the right-hand side part of $\mathbf{T}^{(j)}$ that are zero; that is,

$$I(i) = \{h |\ h > r,\ t_{ih}^{(j)} = 0\}. \tag{3.25}$$

Thereafter, one calculates the vectors

$$\boldsymbol{\vartheta} = \left|t_{i,j+1}^{(j)}\right| \cdot \boldsymbol{t}_k^{(j)} + \left|t_{k,j+1}^{(j)}\right| \cdot \boldsymbol{t}_i^{(j)}, \tag{3.26}$$

where $\boldsymbol{t}_i^{(j)}$ and $\boldsymbol{t}_k^{(j)}$ are the ith and kth rows of $\boldsymbol{T}^{(j)}$, respectively. Vectors $\boldsymbol{\vartheta}$ have to be computed, by Eq. (3.26), for all pairs of indices i and k that fulfill the conditions

$$t_{i,j+1}^{(j)} \cdot t_{k,j+1}^{(j)} < 0 \tag{3.27}$$

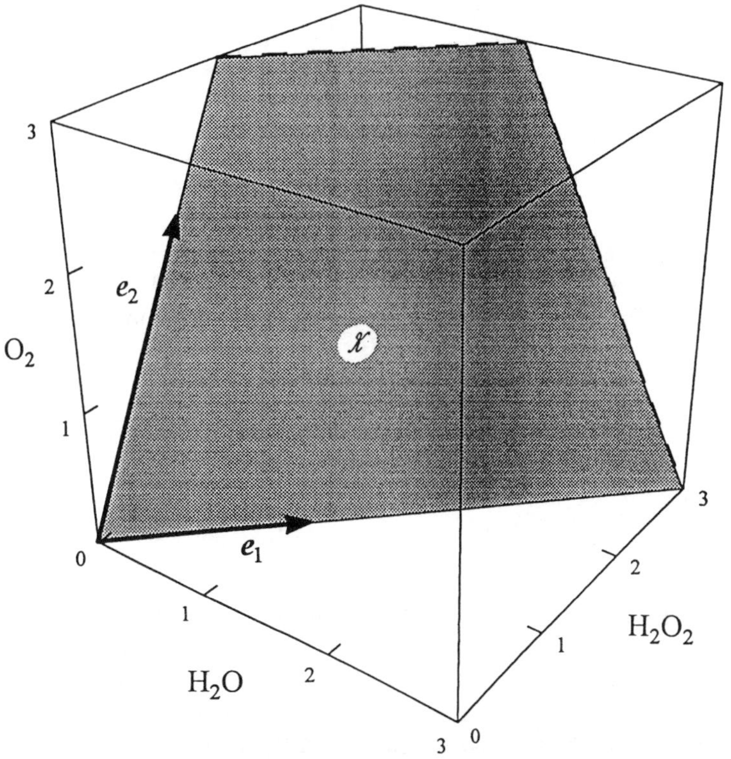

Figure 3.3 Cone, $\mathcal{K}$, of conservation vectors for the catalase reaction. $\boldsymbol{e}_1 = (1\ 1\ 0)^T$ and $\boldsymbol{e}_2 = (1\ 0\ 2)^T$ are the generating vectors of cone $\mathcal{K}$.

and

$$I(i) \cap I(k) \not\subseteq I(l) \tag{3.28}$$

for all row indices, l, of $\mathbf{T}^{(j)}$ with $l \neq i,k$. The tableau $\mathbf{T}^{(j+1)}$ is constructed by using, as rows, all the vectors $\boldsymbol{\vartheta}$ calculated by Eq. (3.26) as well as all rows of $\mathbf{T}^{(j)}$ with $t_{i,j+1}^{(j)} = 0$. Note that the number of tableau rows may change in this procedure. The row vectors of the right-hand submatrix of the final tableau, $\mathbf{T}^{(r)}$, which originate from the identity matrix in $\mathbf{T}^{(0)}$, are the generating vectors of cone $\mathcal{K}$.

We now illustrate the algorithm by a more complex stoichiometry matrix,

$$\mathbf{N} = \begin{pmatrix} -1 & 0 & 0 \\ 0 & 1 & -1 \\ 0 & 1 & 0 \\ 0 & -1 & 2 \\ 1 & -1 & 0 \\ 0 & 0 & -1 \end{pmatrix}. \tag{3.29}$$

which corresponds to a model of part of erythrocyte metabolism studied by Schauer and Heinrich (1983). After the second iteration, we obtain the tableau

$$\mathbf{T}^{(2)} = \left(\begin{array}{ccc|cccccc} 0 & 0 & -1 & 1 & 1 & 0 & 0 & 1 & 0 \\ 0 & 0 & 0 & 1 & 0 & 1 & 0 & 1 & 0 \\ 0 & 0 & 1 & 0 & 1 & 0 & 1 & 0 & 0 \\ 0 & 0 & 2 & 0 & 0 & 1 & 1 & 0 & 0 \\ 0 & 0 & -1 & 0 & 0 & 0 & 0 & 0 & 1 \end{array}\right). \tag{3.30}$$

In the third column, condition (3.27) is fulfilled for the index pairs (1,3), (1,4), (3,5) and (4,5). Rows 1 and 4 must not, however, be combined, because $I(1) = \{6,7,9\}$, $I(2) = \{5,7,9\}$, and $I(4) = \{4,5,8,9\}$, so that the intersection of the index sets indicating the location of zeros in the first and fourth rows is a subset of the index set, $I(2)$, for the second row. The final tableau reads

$$\mathbf{T}^{(3)} = \left(\begin{array}{ccc|cccccc} 0 & 0 & 0 & \boldsymbol{e}_1^{\mathrm{T}}: 1 & 2 & 0 & 1 & 1 & 0 \\ 0 & 0 & 0 & \boldsymbol{e}_2^{\mathrm{T}}: 1 & 0 & 1 & 0 & 1 & 0 \\ 0 & 0 & 0 & \boldsymbol{e}_3^{\mathrm{T}}: 0 & 1 & 0 & 1 & 0 & 1 \\ 0 & 0 & 0 & \boldsymbol{e}_4^{\mathrm{T}}: 0 & 0 & 1 & 1 & 0 & 2 \end{array}\right). \tag{3.31}$$

If we had combined the first and fourth rows of $\mathbf{T}^{(2)}$, we would have obtained the row (2 2 1 1 2 0), which is no extreme vector because it is the sum of $\boldsymbol{e}_1$ and $\boldsymbol{e}_2$.

For closed reaction systems, one can always find n − rank($\mathbf{N}$) linearly independent non-negative conservation relations, so that the cone $\mathcal{K}$ has dimension n − rank($\mathbf{N}$). The proof of this statement starts from the fact that these systems fulfill mass conservation, that is,

$$\sum_{i=1}^{n} \mu_i S_i(t) = \text{const.} \tag{3.32}$$

with the μ_i denoting the molar masses of the substances S_i. Equation (3.32) is a special case of the conservation relations (3.3). Any conservation relation with some coefficients being negative can be replaced by the sum of this relation and relation (3.32) with the latter multiplied by a sufficiently large positive number. This linear combination yields, due to $\mu_i > 0$ for any i, a non-negative conservation relation.

Open reaction systems may have less than $n - \text{rank}(\mathbf{N})$ independent non-negative conservation relations, as exemplified by the system

$$P_1 \rightarrow S_1 + S_2 \rightarrow P_2 + P_3, \tag{3.33}$$

for which we have $S_1 - S_2 = \text{const.}$, but no non-negative conservation relation.

3.1.3. Conserved Moieties

Chemical entities (atoms, ions, assemblies of atoms or ions) participating in a reaction system without loss of integrity and always remaining in the system (even if it is an open one) are called *conserved moieties.* Because any part of a conserved entity is also conserved, it is often of interest to find *maximal conserved moieties* (i.e., the largest molecular assemblies that are conserved in a given reaction system).

The moiety structure of closed reaction systems can be obtained by factorizing the atomic matrix (Park Jr., 1986). In many situations, however, this matrix is not available. For example, in the association reaction of the α and βγ subunits of the G-protein (cf. Alberts *et al.,* 1983), and in numerous other reactions involving macromolecules, the atomic composition and structure of some or all participating species are unknown. On the other hand, this information is unnecessary for detecting how many units of how many different moieties enter the particular reacting substances. For instance, to derive the Michaelis–Menten equation for enzyme kinetics, the atomic structure of the enzyme need not be known and, moreover, often is not known. Nevertheless, one employs a conservation relation stating that the sum of free enzyme and enzyme–substrate complexes is constant.

In the present section, we deal with the situation that a stoichiometry matrix is given at the outset and information about the conserved-moiety structure is sought. We first formalize the concept of conserved moiety from a purely stoichiometric viewpoint, generalizing the analysis of Park Jr. (1986). We regard conserved moieties as some physicochemical entities unspecified in their concrete structure. The constitution of a reaction system by these entities can be written in terms of vectors, $\boldsymbol{z}$, whose elements, z_i, indicate how many units of a moiety are contained in one molecule of the reacting species S_i. Obviously, every vector $\boldsymbol{z}$ is a special conservation vector, $\boldsymbol{g}$, which fulfills Eq. (3.4).

For a given reaction system, a vector $\boldsymbol{z}$ with n components is called *elementary conserved-moiety vector* if, and only if, it has the following four properties:

(P1) Conservation property:

$$z^{\mathrm{T}}\mathbf{N} = \boldsymbol{0}^{\mathrm{T}} \tag{3.34}$$

(P2) Integer-element property:

$$z_i \text{ are integers for all } i \tag{3.35}$$

(P3) Non-negativity:

$$z_i \geq 0 \text{ for any } i, \quad z \neq \boldsymbol{0} \tag{3.36}$$

(P4) Nondecomposability (maximal size of moiety): For any couple of vectors z' and z'' fulfilling conditions (P1)–(P3), z is no linear combination of these vectors with integer coefficients greater than, or equal to, 1:

$$z \neq \eta' z' + \eta'' z'', \quad \eta', \eta'' \geq 1, \text{ integer.} \tag{3.37}$$

To explain the meaning of condition (P4), we have to consider two cases according to whether or not z' and z'' are identical. If condition (P4) were not fulfilled in the former case, $\eta' + \eta''$ entities of some moiety could be combined into one moiety, which would then be larger than the one corresponding to z. In the catalase reaction given in Eq. (2.2), $z = (2\ 2\ 0)^{\mathrm{T}}$ would correspond to hydrogen atoms. This vector can, however, be written as two times the vector $z' = z'' = (1\ 1\ 0)^{\mathrm{T}}$, which corresponds to the larger H_2 moiety.

Now consider the case that z' and z'' are different. Consider, for example, the pyruvate decarboxylase reaction given in Eq. (3.14). The second row, (3 0 1 2), of the matrix **G** indicated in Eq. (3.15) corresponds to the conservation of oxygen atoms. It satisfies conditions (P1)–(P3), but not condition (P4), because it can be decomposed as (1 0 1 0) + 2(1 0 0 1). These two row vectors, which are involved in the matrix **G** given in Eq. (3.16), correspond to the oxygen in the keto group and the O_2 in the carboxyl group. They can be combined with other moieties represented by the same vectors, so that larger moieties obtain (the CH_3CO and carboxyl groups, respectively). A more detailed discussion of the four properties (P1)–(P4) can be found in the work of S. Schuster and Hilgetag (1995).

As the example of the pyruvate decarboxylase reaction shows, two moieties having the same conserved-moiety vector (e.g., O_2 and C) can be combined into one moiety, because they are "inherited" together in the reactions. Such combination might be questionable when the two moieties are located at different sites in the molecule. However, we wish to consider conservation in terms of the empirical formula (i.e., with no reference to structure). Therefore, there should be no two vectors z that are identical. Accordingly, we define a matrix **Z** with n

columns to be a *conserved-moiety matrix* if, and only if, it contains, as rows, all transposed vectors fulfilling properties (P1)–(P4) and has the property

(P5) Dissimilarity: All row vectors of **Z** are different,

$$z_i \neq z_k \text{ for all } i \text{ and } k. \tag{3.38}$$

The following example shows that the present analysis also applies to open systems:

$$\text{Gluc} + \text{ATP} \rightarrow \text{G6P} + \text{ADP}, \tag{3.39a}$$

$$\text{G6P} \rightarrow \text{F6P}, \tag{3.39b}$$

$$\text{PEP} + \text{ADP} \rightarrow \text{Pyr} + \text{ATP}, \tag{3.39c}$$

where an experimental setup is considered which guarantees that the concentrations of glucose (Gluc), fructose-6-phosphate (F6P), phosphoenolpyruvate (PEP), and pyruvate (Pyr) are fixed, so that these substances are external. The (here unique) moiety matrix contains only one row, (1 0 1) with the ones corresponding to ATP and ADP and the zero to glucose-6-phosphate (G6P). This row actually reflects conservation of the adenosine group contained in ATP and ADP. The example shows that open systems may involve substances, here G6P, which do not contain any conserved moiety, in contrast to closed systems. Note that many open systems, such as the unbranched reaction chains studied in Section 5.4.3.1, do not involve any conserved moiety at all.

It is worth noting that in certain reaction systems, the number of chemical elements is smaller than the number of independent conservation relations, $n - \text{rank}(\mathbf{N})$, as the following example representing the hydrodealkylation of toluene yielding benzene and methane demonstrates (Björnbom, 1977; Cavallotti *et al.*, 1980),

$$C_6H_5CH_3 + H_2 \rightarrow C_6H_6 + CH_4. \tag{3.40}$$

This system involves only two atomic species, but $n - \text{rank}(\mathbf{N}) = 3$. The system has in fact three maximal conserved moieties, notably the phenyl group, the methyl group, and the hydrogen atoms initially contained in the H_2 molecule.

Interestingly, even the number of conserved moieties may be less than $n - \text{rank}(\mathbf{N})$, both in open and in closed reaction systems. Among other examples, Alberty (1994) analyzed the ATP citrate (*pro*-S)-lyase reaction (EC 4.1.3.8),

$$\text{ATP} + \text{citrate} + \text{CoA} \rightarrow \text{ADP} + P_i + \text{acetyl} - \text{CoA} + \text{oxaloacetate}. \tag{3.41}$$

This system has six linearly independent conservation relations. On the other hand, it involves only five maximal conserved moieties, namely the ADP, P_i, CoA, acetyl, and oxaloacetate groups. This discrepancy is due to the fact that for the given set of conserved moieties, not all of the possible reactions are realized. A second reaction among the given compounds might be, for example,

$$\mathrm{ATP} \rightarrow \mathrm{ADP} + \mathrm{P_i}, \tag{3.42}$$

which is an example of a slippage reaction [for the notion of enzyme slip, cf. Pietrobon and Caplan (1985)]. If this reaction is included into the system (3.41), the system has only five linearly independent conservation relations, which we can choose so as to correspond to maximal conserved moieties.

The above reasoning on the number of conserved moieties becomes clearer by an alternative approach, in which the possible reactions among a given set of compounds are to be identified. From the set of substances, the conserved-moiety matrix **Z** can be derived. Now we are looking for a stoichiometry matrix **N** that fulfills the conservation equation $\mathbf{ZN} = \mathbf{0}$. We can conclude that **N** can have at most $n - \mathrm{rank}(\mathbf{Z})$ linearly independent columns. As for example (3.41), $n - \mathrm{rank}(\mathbf{Z})$ equals 2. Therefore, two independent reactions can proceed with the set of substances given above (ATP, citrate, etc.); for example, reactions (3.41) and (3.42).

For many chemically admissible reactions between macromolecules and/or metabolites, no enzyme is present to catalyze them, so they do not proceed in living cells at a measurable rate. Therefore, systems in which not all of the admissible reactions take place [incomplete systems in the terminology of Érdi and Tóth (1989)] and, hence, additional conservation relations occur, are of special importance in biochemistry.

It can easily be seen that for any reaction system, all reduced generating vectors of cone $\mathcal{K}$ fulfill conditions (P1)–(P4) and are, hence, elementary conserved-moiety vectors. For example, computation of the generating vectors for the pyruvate decarboxylase reaction (3.14) by the algorithm outlined in Section 3.1.2 yields the three rows of the matrix **G** given in Eq. (3.16) and, in addition, $\boldsymbol{e}_4 = (0\ 1\ 0\ 1)^{\mathrm{T}}$, all of them fulfilling conditions (P1)–(P4). The latter vector does not, however, correspond to a conserved chemical unit, whereas the three former vectors do.

Conversely, there are systems containing maximal conserved moieties that do not directly correspond to generating vectors of cone $\mathcal{K}$. For the example of the superoxide dismutase reaction (3.18), the generating vectors are $(1\ 0\ 2\ 0)^{\mathrm{T}}$, $(1\ 0\ 0\ 2)^{\mathrm{T}}$, $(0\ 1\ 2\ 0)^{\mathrm{T}}$, and $(0\ 1\ 0\ 2)^{\mathrm{T}}$. The first and third of them correspond to the extra electron initially belonging to the superoxide radical and to the proton, respectively, whereas the second and fourth of them have no immediate physical meaning. The system has, in addition, a third maximal conserved moiety, namely the O_2 moiety, which corresponds to the vector $(1\ 0\ 1\ 1)^{\mathrm{T}}$. This vector also satisfies conditions (P1)–(P4) although it is not a generating vector. It can be calculated by taking half of the sum of the first and second generating vectors given above. In contrast, for many other systems, such as reaction (3.14), no conserved-moiety vectors besides those given by the reduced generating vectors of cone $\mathcal{K}$ exist.

Recently, an algorithm for computing the complete conserved-moiety matrix based on knowledge of the stoichiometry matrix has been proposed and implemented as a Turbo-Pascal program for the PC (S. Schuster and Hilgetag, 1995). It is based on the algorithm for determining the extreme vectors to cone $\mathcal{K}$. To find possible additional conserved-moiety vectors, additional tableaux are calculated by linearly combining pairs of extreme vectors and testing whether the combined vectors can be divided by integers greater than 1 to give additional moiety vectors, which, upon calculation of the next tableau, are tried to combine to give still further moiety vectors.

As there may be more elementary conserved-moiety vectors than conserved chemical units really occurring, matrix **Z** contains all vectors that represent maximal conserved moieties and possibly, in addition, other vectors satisfying conditions (P1)–(P5). This is in line with results of Park Jr. (1988) saying that there may exist more than one moiety structure that conforms with a particular outward stoichiometry. This situation is somehow similar to the multiple solutions of polynomial equations, where usually only some of them are physically meaningful. The four properties (P1)–(P4) are well suited for distinguishing, for any given stoichiometry matrix, between those conservation relations that can, in principle, correspond to maximal conserved moieties and those that cannot.

3.2. ADMISSIBLE STEADY-STATE VECTORS AND THE NULL-SPACE

3.2.1. Linear Dependencies Between the Columns of the Stoichiometry Matrix

As stated in Section 2.3, analysis of stationary states plays an outstanding role in biological modeling. For metabolic systems, the central equation to describe steady states is Eq. (2.9). For fixed values of the parameters, p_k, this is an equation in the unknowns S_i. In addition, Eq. (2.9) can also be regarded as an equation in the unknowns v_j, which is of importance when the parameters are incompletely known or when the assumption that they are constant is no longer fulfilled [e.g., in the case of activation or inhibition of enzymes or for processes on evolutionary time scales (see Chapter 6)]. For all admissible parameter values, the reaction rates obey the equation

$$\mathbf{N}\boldsymbol{v} = \boldsymbol{0}. \tag{3.43}$$

This can now be regarded as an equation system restricting the admissible values of v_i. It is particularly easy to analyze because of its linearity.

Nontrivial solutions of Eq. (3.43) for the vector $\boldsymbol{v}$ only exist if there are linear dependencies between the columns of **N**, that is, if the rank of **N** is smaller than the number of reactions, r. These dependencies can be expressed by a matrix, **K**,

$$\mathbf{NK} = \mathbf{0}. \tag{3.44}$$

The $r - \mathrm{rank}(\mathbf{N})$ columns, $\boldsymbol{k}_i$, of **K** are particular, linearly independent solutions of Eq. (3.43). They span the null-space (also called kernel) of matrix **N** (cf. Groetsch and King, 1988), that is, the subspace of all vectors satisfying Eq. (3.43) within the space of reaction rates. These are the steady-state flux vectors mathematically compatible with the stoichiometric structure of the system. Accordingly, any steady-state flux vector, $\boldsymbol{J}$, can be written as a linear combination of the vectors $\boldsymbol{k}_i$,

$$\boldsymbol{J} = \sum_i a_i \boldsymbol{k}_i, \tag{3.45}$$

where the sum runs from 1 to $r - \mathrm{rank}(\mathbf{N})$.

The null-space matrix **K** is not uniquely determined. It can be postmultiplied by a nonsingular matrix, **Q**, of dimension $[r - \mathrm{rank}(\mathbf{N})] \times [r - \mathrm{rank}(\mathbf{N})]$ to give another admissible null-space matrix, $\hat{\mathbf{K}}$,

$$\hat{\mathbf{K}} = \mathbf{KQ}. \tag{3.46}$$

This follows immediately from Eq. (3.44).

In many applications, for example, in metabolic control analysis (cf. Section 5.3), one is interested in finding an appropriate, preferably simple representation of the null-space matrix. Of special interest is the representation containing an identity matrix

$$\mathbf{K} = \begin{pmatrix} \mathbf{K}' \\ \mathbf{I} \end{pmatrix}, \tag{3.47}$$

because it contains a large number of zeros. $\mathbf{K}'$ has dimension $\mathrm{rank}(\mathbf{N}) \times [r - \mathrm{rank}(\mathbf{N})]$. This representation may be obtained by the Gaussian elimination method.

3.2.2. Block-Diagonalization of the Null-Space Matrix

As stated above, the choice of the null-space matrix is not unique. In several instances, it is useful to seek a representation of **K** that has a block-diagonal structure,

$$\mathbf{K} = \begin{pmatrix} \mathbf{K}_1 & \mathbf{0} & \cdots & \mathbf{0} \\ \mathbf{0} & \mathbf{K}_2 & \cdots & 0 \\ \vdots & \vdots & \ddots & \vdots \\ \mathbf{0} & \mathbf{0} & \cdots & \mathbf{K}_p \\ \mathbf{0} & \mathbf{0} & \cdots & \mathbf{0} \end{pmatrix}, \tag{3.48}$$

where ρ denotes the maximum number of diagonal blocks in **K** for the reaction system under consideration. If ρ equals unity, **K** is not block-diagonalizable. If the stoichiometry matrix contains columns that are linearly independent of all other columns of **N**, whereas these other columns are linearly dependent on each other, the corresponding components in the columns of any **K** determined by Eq. (3.44) are zero. These null rows (if any) have been transferred to the bottom of **K** in Eq. (3.48). The steady-state flux through the reactions corresponding to such null rows is always zero, because of Eq. (3.45) (cf. Section 3.3.1).

The blocks of **K** correspond to subsystems of the reaction network, the fluxes of which are completely independent; that is, the fluxes within one subsystem can be changed by appropriate parameter changes without alteration of the fluxes in other subsystems.

Consider, for example, the reaction system shown in Figure 3.4, which includes the main reactions of glycolysis (with some of them lumped) and some adjacent reactions occurring, for example, in liver cells. ATP and ADP are here to be

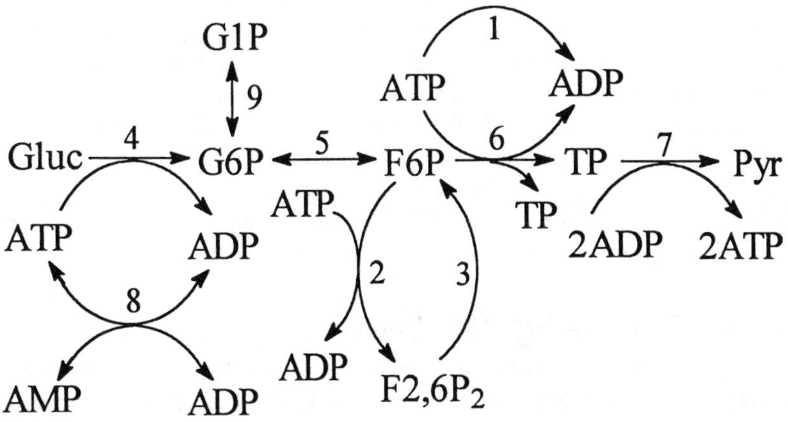

Figure 3.4 Scheme of the main reactions of glycolysis and some adjacent reactions. Reactions 1, 4, 5, and 8 are also involved in the scheme depicted in Figure 3.1 (with a different numbering). Numbers 6 and 7 stand for lumped reactions. Additional reactions: 2) 6-phosphofructo-2-kinase (EC 2.7.1.105); 3) fructose-2,6-bisphosphatase (EC 3.1.3.46); 9) phosphoglucomutase (EC 5.4.2.2). G1P, glucose-1-phosphate; $F2,6P_2$, fructose-2,6-bisphosphate; TP, pool of triose phosphates.

considered to be external metabolites with fixed concentrations. For this example, matrix **K** can be chosen to be

$$\mathbf{K}=\left(\begin{array}{c:c:c} 1 & 0 & 0 \\ \hdashline 0 & 1 & 0 \\ 0 & 1 & 0 \\ \hdashline 0 & 0 & 1 \\ 0 & 0 & 1 \\ 0 & 0 & 1 \\ 0 & 0 & 2 \\ \hdashline 0 & 0 & 0 \\ 0 & 0 & 0 \end{array}\right). \tag{3.49}$$

Because the two bottom rows are null vectors, reactions 8 and 9 catalyzed by adenylate kinase and phosphoglucomutase, respectively, subsist in equilibrium whenever the system is at steady state. The remaining submatrix consists of three diagonal blocks. They correspond to the ATPase reaction, the $F2,6P_2$ cycle, and the main glycolytic pathway. The fluxes in any one of these subsystems are independent of the fluxes in the other subsystems, provided that ATP and ADP are considered as external metabolites. Note that the adenylate kinase reaction has zero net flux even if ATP and ADP were treated as internal.

The representation of **K** as given by Eq. (3.48) is of importance for detecting strictly detailed balanced subsystems; that is, subsystems composed of reactions the fluxes of which are always zero when the whole system is at steady state (see Section 3.3.2). The representation of **K** in block-diagonal form is also of importance for metabolic control analysis (see Sections 5.11 and 5.13).

For computation of the block-diagonal form of the null-space matrix by computer, it is of importance that the representation of matrix **K** as given by Eq. (3.48) obtains, by rearranging rows and columns, from the form given by Eq. (3.47). The proof to this assertion and a source code of a program performing this computation were given by S. Schuster and R. Schuster (1991).

3.2.3. Non-negative Flux Vectors

In many situations, all the reaction rates are known to have fixed signs. Without loss of generality, we can, in this case, prescribe the orientation of reactions so that their fluxes are non-negative,

$$\boldsymbol{v} \geq \boldsymbol{0}. \tag{3.50}$$

This constraint is of importance in particular when the reaction rates are defined as opposite unidirectional rates rather than as net rates (Clarke, 1981, 1988; Fell and Small, 1986; Érdi and Tóth, 1989). For example, in the description of the dynamics of radioactive tracers or nuclear magnetic resonance (NMR) labels, the forward and backward reaction rates enter the equations separately (Holzhütter, 1985; R. Schuster *et al.,* 1992). Furthermore, in many biochemical models, even the net rates of some reactions are practically restricted to be non-negative, notably when they are quasi-irreversible, that is, when the forward reaction is much faster than the backward reaction (Heinrich *et al.,* 1977; Leiser and Blum, 1987; Joshi and Palsson, 1989a) or when this is implied by the biological function of the pathway (e.g., ATP production by glycolysis).

On calculating the null-space matrix **K** by standard methods of linear algebra [e.g., according to formula (3.47)], it may occur that some of its elements are negative although the corresponding fluxes should be non-negative for some of the reasons mentioned above. It is therefore of interest to find a non-negative representation of **K**.

Equation (3.43) gives, upon transposition, $\boldsymbol{v}^{\mathrm{T}}\mathbf{N}^{\mathrm{T}} = \boldsymbol{0}^{\mathrm{T}}$. Clearly, this equation together with the inequality system (3.50) is isomorphic to the equation/inequality system (3.4) and (3.17a) after replacing **N** by its transpose. Therefore, its solution set for $\boldsymbol{v}$ can be found by the algorithm given in Section 3.1.2. As in the case of the cone $\mathcal{K}$ given in Eq. (3.21), we can write the cone of all non-negative steady-state fluxes, $\mathcal{F}$, as non-negative linear combination of generating vectors,

$$\mathcal{F} = \{\boldsymbol{v} \in \mathrm{R}^r | \boldsymbol{v} = \sum_k \eta_k \boldsymbol{f}_k, \eta_k \geq 0\}. \tag{3.51}$$

The cone $\mathcal{F}$ is the intersection of the null-space of **N** and the non-negative orthant. It can therefore have any dimension from zero to $r - \mathrm{rank}\,(\mathbf{N})$. As in the case of the cone $\mathcal{K}$, the number of generating vectors of $\mathcal{F}$ may be greater than the dimension of the cone (see Fig. 3.2).

If, in addition to the non-negativity condition (3.50), the constraint that some rates have fixed values is imposed, the admissible region for all fluxes is a convex polyhedron. This more general situation is treated in R. Schuster and S. Schuster (1993). Another generalization is by restricting the signs of only some of the rates, whereas the others are allowed to have any sign. This situation will be dealt with in the following section.

3.2.4. Elementary Flux Modes

Some reactions may proceed in either direction under physiological conditions, such as the reactions shared by glycolysis and gluconeogenesis and the reversible reactions of the pentose phosphate pathway. In the case that some reactions are reversible and some are irreversible, we decompose the flux vector into the sub-vectors $\mathbf{v}^{\mathrm{irr}}$ and $\mathbf{v}^{\mathrm{rev}}$. The irreversibility constraint can be written as

$$\boldsymbol{v}^{\text{irr}} \geq \boldsymbol{0}. \tag{3.52}$$

Now, the situation may occur that depending on the kinetic parameters and the concentrations of external metabolites, $\boldsymbol{v}^{\text{rev}}$ as well as its opposite $-\boldsymbol{v}^{\text{rev}}$ are realizable. Accordingly, two classes of generating vectors are, in general, needed to span the cone (which is now to be denoted by C), basis vectors, $\boldsymbol{b}$, which have the property that their opposites, $-\boldsymbol{b}$, are also situated in the cone, and fundamental vectors, $\boldsymbol{f}$, which do not have this property (see Rockafellar, 1970; Nožička *et al.*, 1974). We then have

$$C = \{\boldsymbol{v} \in \mathrm{R}^r \mid \boldsymbol{v} = \sum_k \eta_k \boldsymbol{f}_k + \sum_m \lambda_m \boldsymbol{b}_m,\ \eta_k \geq 0\}. \tag{3.53}$$

Consider, for example, the branched reaction system shown in Scheme 7 and assume here that reaction 1 is irreversible in the direction from P_1 to S_1, whereas reactions 2 and 3 can proceed in any direction.

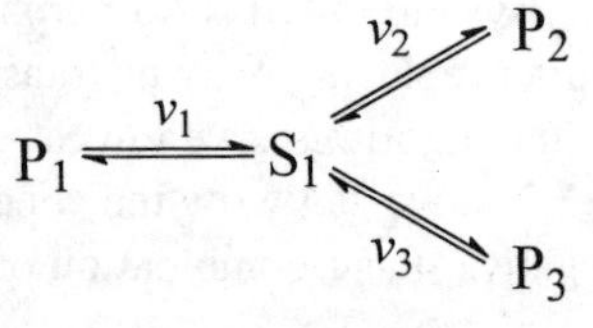

Scheme 7

The cone of admissible steady-state fluxes for this system is a half-plane, which is shown in Figure 3.5. This cone has the basis vector $(0\ -1\ 1)^{\mathrm{T}}$. Note that multiplication of this vector by any real number gives another admissible basis vector [e.g., $(0\ 1\ -1)^{\mathrm{T}}$]. As soon as the cones defined by Eq. (3.53) contain basis vectors, they are not pointed. For nonpointed cones, some of the fundamental vectors lie in their interior, so that the favorable uniqueness property characteristic for pointed cones as defined by Eq. (3.51) is lost. The set of basis vectors is not, in general, unique either, because any linear combination of these may also serve as a basis. For the system in Scheme 7, any one vector situated within the half-plane C could serve as a fundamental vector, for example, the vectors $\boldsymbol{f}_1 = (1\ 1\ 0)^{\mathrm{T}}$ or $\boldsymbol{f}_2 = (1\ 0\ 1)^{\mathrm{T}}$ shown in Figure 3.5. The two mentioned vectors correspond to the situations that the entire flux goes from P_1 to P_2 or from P_1 to P_3, respectively. Another possible choice would be to select the vector $\boldsymbol{f}_3 = (2\ 1\ 1)^{\mathrm{T}}$, which is the sum of $\boldsymbol{f}_1$ and $\boldsymbol{f}_2$ and is orthogonal to $\boldsymbol{b}_1$. It corresponds to a situation where the flux coming from P_1 is equally distributed between branches 2 and 3.

Owing to the nonuniqueness of fundamental and basis vectors, it is of interest to find those vectors that can be interpreted in biochemical terms. Guided by the principle of Ockham's razor, one may seek the simplest biochemically meaningful flux vectors possible. They should be chosen so that all other admissible flux patterns are superpositions of these elementary modes.

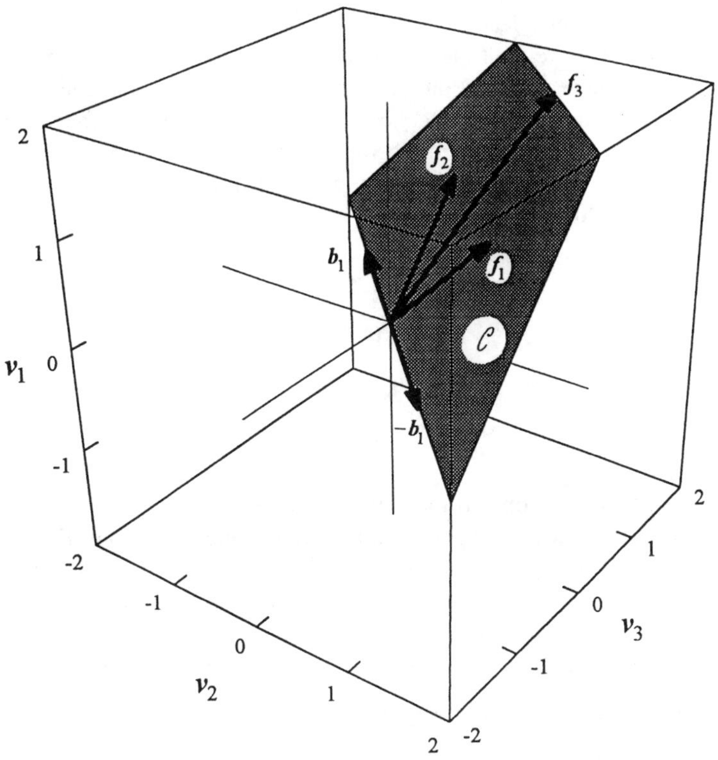

Figure 3.5 Cone C of admissible steady-state fluxes for the system in Scheme 7 with reactions 2 and 3 being reversible and reaction 1 being irreversible. C is here a half-plane. $\boldsymbol{b}_1 = (0\ -1\ 1)^{\mathrm{T}}$, a basis vector; $\boldsymbol{f}_1 = (1\ 1\ 0)^{\mathrm{T}}$ and $\boldsymbol{f}_2 = (1\ 0\ 1)^{\mathrm{T}}$, the fundamental vectors representing elementary modes. $\boldsymbol{f}_3 = (2\ 1\ 1)^{\mathrm{T}}$ is the sum of $\boldsymbol{f}_1$ and $\boldsymbol{f}_2$.

Leiser and Blum (1987) proposed to identify cyclic and noncyclic *fundamental modes* of systems containing substrate cycles, by invoking that any steady-state flux pattern could be decomposed as a linear superposition of these modes and that these modes are all thermodynamically realizable, that is, that they comply with possibly imposed sign constraints for fluxes [relation (3.52)]. Fell (1990, 1993) proposed to define fundamental modes as the simplest relevant ways of connecting the inputs to the outputs of the system and to represent them by a proper choice of basis vectors of the null-space. He observed that this method meets with the difficulties that irreversibility constraints may be violated and that there may be a greater number of fundamental modes than basis vectors of the null-space. Some authors (Seressiotis and Bailey, 1988; Mavrovouniotis *et al.*,

1990; Mavrovouniotis, 1992) developed methods for constructing, by computer, simple metabolic routes leading from a given substrate to a given product.

In what follows, we treat the problem of finding simple flux modes by using the theory of convex cones as outlined above. This leads to the concept of *elementary modes.* To illustrate this concept we again consider the branched system shown in Scheme 7 with reaction 1 being irreversible. The simplest flux patterns possible in steady state and qualitatively different from each other can be represented by the vectors

$$\boldsymbol{v}_1 = (1\ 1\ 0)^{\mathrm{T}}, \tag{3.54a}$$

$$\boldsymbol{v}_2 = (1\ 0\ 1)^{\mathrm{T}}, \tag{3.54b}$$

$$\boldsymbol{v}_3 = (0\ -1\ 1)^{\mathrm{T}}, \tag{3.54c}$$

$$\boldsymbol{v}_4 = (0\ 1\ -1)^{\mathrm{T}}. \tag{3.54d}$$

Although $\boldsymbol{v}_3$ and $\boldsymbol{v}_4$ belong to the same basis vector, we take them separately because opposite fluxes correspond to different biological functions. For example, the two directions of operation of the H^+-ATPase are related to ATP production and proton transport. On the other hand, flux vectors differing by a positive factor are considered to belong to the same flux mode.

We now formalize the above reasoning by the following definitions.

1. A *flux mode,* $\boldsymbol{M}$, is defined as the set

$$\boldsymbol{M} = \{\boldsymbol{v} \in \mathrm{R}^r \mid \boldsymbol{v} = \lambda \boldsymbol{v}^*, \lambda > 0\}, \tag{3.55}$$

where $\boldsymbol{v}^*$ is an $\boldsymbol{r}$-dimensional vector (unequal to the null vector) fulfilling the following two conditions:

(C1) Steady-state condition. $\boldsymbol{v}^*$ satisfies Eq. (3.43).

(C2) Sign restriction. If the system involves irreversible reactions, then the corresponding subvector, $\mathbf{v}^{\mathrm{irr}}$, of $\boldsymbol{v}^*$ fulfills inequality (3.52).

According to this definition, a flux mode is sufficiently characterized by one representative of $\boldsymbol{M}$.

2. A flux mode M with a representative $\boldsymbol{v}^*$ is called an *elementary flux mode* if, and only if, $\boldsymbol{v}^*$ fulfills the condition:

(C3) Simplicity (nondecomposability). For any couple of vectors $\boldsymbol{v}'$ and $\boldsymbol{v}''$ (unequal to the null vector) with the following properties:

(i) $\boldsymbol{v}'$ and $\boldsymbol{v}''$ obey restrictions (C1) and (C2),

(ii) both $\boldsymbol{v}'$ and $\boldsymbol{v}''$ contain zero elements wherever $\boldsymbol{v}^*$ does so, and they include at least one additional zero component each,

$\boldsymbol{v}^*$ is not a non-negative linear combination of $\boldsymbol{v}'$ and $\boldsymbol{v}''$,

$$v^* \neq \lambda_1 v' + \lambda_2 v'', \quad \lambda_1, \lambda_2 > 0. \tag{3.56}$$

Condition (C3, ii) is a formalization of the concept of *genetic independence* introduced by Seressiotis and Bailey (1988). This condition says that a decomposition into two other modes should not involve additional enzymes.

3. A flux mode $\boldsymbol{M}$ is called a *reversible flux mode* if, and only if, $\boldsymbol{M}' = \{-\boldsymbol{v} | \boldsymbol{v} \in \boldsymbol{M}\}$ is a flux mode as well. Otherwise, $\boldsymbol{M}$ is called an *irreversible flux mode.* The same distinction can then be made for elementary flux modes.

For the example depicted in Scheme 7 with reaction 1 being irreversible, the elementary modes represented by the flux vectors given in Eqs. (3.54a) and (3.54b) are irreversible elementary modes, whereas the vectors given in Eqs. (3.54c), and (3.54d) represent reversible elementary modes.

The reaction system shown in Figure 3.6 containing a cycle of irreversible reactions, can serve for illustration of the concept of elementary modes. This scheme was also studied by Leiser and Blum (1987).

It can easily be seen that the following seven flux vectors represent elementary flux modes:

$$v_1 = (-1 \; 1 \; 0 \; 1 \; 0 \; 0)^{\mathrm{T}}, \tag{3.57a}$$

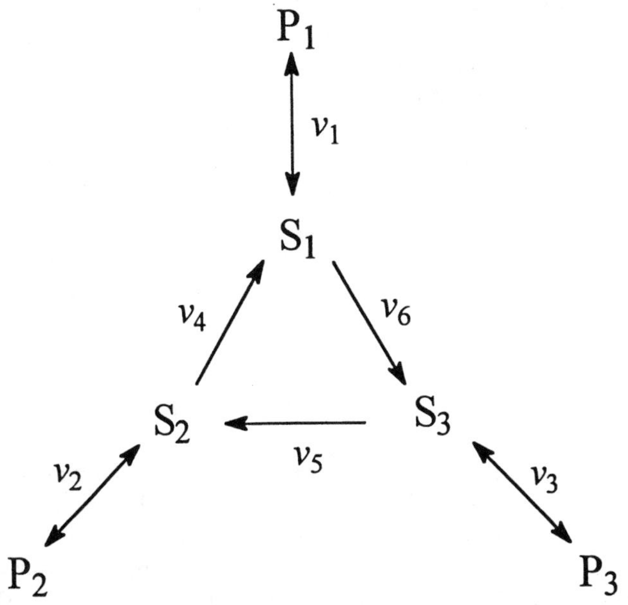

Figure 3.6 Cyclic reaction system. This scheme can stand, for example, for the pyruvate/oxaloacetate/phosphoenolpyruvate cycle and was also studied by Leiser and Blum (1987).

$$\boldsymbol{v}_2 = (0 \ \ -1 \ 1 \ 0 \ 1 \ 0)^{\mathrm{T}}, \tag{3.57b}$$

$$\boldsymbol{v}_3 = (1 \ 0 \ \ -1 \ 0 \ 0 \ 1)^{\mathrm{T}}, \tag{3.57c}$$

$$\boldsymbol{v}_4 = (0 \ 1 \ \ -1 \ 1 \ 0 \ 1)^{\mathrm{T}}, \tag{3.57d}$$

$$\boldsymbol{v}_5 = (-1 \ 0 \ 1 \ 1 \ 1 \ 0)^{\mathrm{T}}, \tag{3.57e}$$

$$\boldsymbol{v}_6 = (1 \ \ -1 \ 0 \ 0 \ 1 \ 1)^{\mathrm{T}} \tag{3.57f}$$

$$\boldsymbol{v}_7 = (0 \ 0 \ 0 \ 1 \ 1 \ 1)^{\mathrm{T}}. \tag{3.57g}$$

They are represented in Figure 3.7. Four modes involve three reactions and three modes comprise four reactions. The mode given in Eq. (3.57g) represents a cycle. For thermodynamic reasons (cf. Section 3.3), a nonzero cyclic flux is only possible if external metabolites participate in the reactions in the cycle. In fact, in the pyruvate/oxaloacetate/phosphoenolpyruvate cycle, ATP is hydrolyzed to ADP (not shown in Fig. 3.6).

To illustrate the decomposability condition (C3), consider the flux vector $\boldsymbol{v} = (0 \ 1 \ -1 \ 2 \ 1 \ 2)^{\mathrm{T}}$, which fulfills the steady-state condition. It is the sum of $\boldsymbol{v}_4$ and $\boldsymbol{v}_7$ given in Eqs. (3.57d) and (3.57g), both of which have more zeros than $\boldsymbol{v}$ and do not involve additional reactions. $\boldsymbol{v}$ is therefore not an elementary mode in the sense of definition 2.

When all reactions are irreversible, all elementary modes correspond to generating vectors of the flux cone $\mathcal{F}$ determined by Eq. (3.51) and *vice versa.* This can be rationalized by the reasoning that all generating vectors satisfy, by definition, conditions (C1) to (C3).

It is worth noting that there are systems that have irreversible elementary modes only, although some reactions of the system are reversible. For the system in Scheme 7 with only one reaction treated reversible, no reversible elementary mode occurs.

An algorithm for detecting the elementary modes for systems of any complexity was given by S. Schuster and Hilgetag (1994). This algorithm starts from a tableau containing the transposed stoichiometry matrix and the identity matrix,

$$\mathbf{T}^{(0)} = \begin{pmatrix} \mathbf{T}^{(0)}_{\mathrm{rev}} \\ \mathbf{T}^{(0)}_{\mathrm{irr}} \end{pmatrix} = \left(\begin{array}{c:cc} \mathbf{N}^{\mathrm{T}}_{\mathrm{rev}} & \mathbf{I} & \mathbf{0} \\ \mathbf{N}^{\mathrm{T}}_{\mathrm{irr}} & \mathbf{0} & \mathbf{I} \end{array}\right), \tag{3.58}$$

where the decomposition of $\mathbf{N}$ into $\mathbf{N}_{\mathrm{rev}}$ and $\mathbf{N}_{\mathrm{irr}}$ is done according to the decomposition of $\boldsymbol{v}$ into $\boldsymbol{v}^{\mathrm{rev}}$ and $\boldsymbol{v}^{\mathrm{irr}}$. At the beginning, the hypothetical situation is considered that all metabolites are external. In this case, every reaction represents an elementary mode on its own. This is reflected by the submatrices $(\mathbf{I}\ \mathbf{0})$ (reversible modes) and $(\mathbf{0}\ \mathbf{I})$ (irreversible modes) in the initial tableau, $\mathbf{T}^{(0)}$. In each step of the algorithm, preliminary elementary modes have to be linearly combined to give new preliminary elementary modes in the next tableau. The coefficients

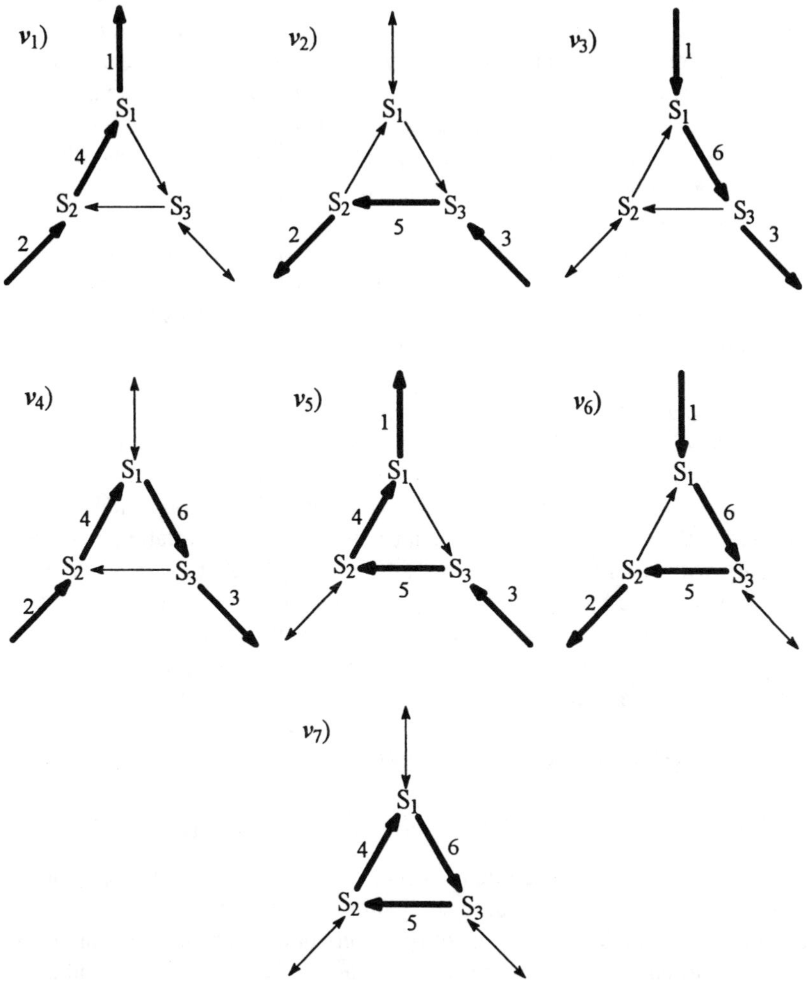

Figure 3.7 Elementary modes of the reaction system depicted in Figure 3.6. The thick arrows indicate the reactions involved in the elementary modes $\mathbf{v}_i$ given in Eqs. (3.57a)–(3.57g).

of these combinations are chosen so that the columns of the transposed stoichiometry matrix are consecutively transformed into null columns. The final tableau, $\mathbf{T}^{(n)}$, contains a submatrix (the columns on the right-hand side) whose rows represent the elementary modes. Finally, one has to take into account that for all reversible elementary modes which are obtained by the algorithm, its negative is such a mode also. A similar algorithm was given by Mavrovouniotis (1992). It does not, however, immediately yield all elementary modes, but only the set of all routes leading from a specified substrate to a specified metabolic product.

The number of elementary modes may be an important index characterizing biochemical systems. It indicates the richness of the system considered, by showing the variety of its physically realizable functions. Which of these functions are operative or in what proportions they operate simultaneously is determined by the extent of inhibition and activation of enzymes (i.e., by the actual values of kinetic parameters).

The present analysis serves to detect essential structural features of any given biochemical network not just by inspecting the reaction scheme but by algebraically analyzing the stoichiometry matrix. This method widens the approach of calculating null-space vectors to that matrix.

Although there are many biochemical reactions that can proceed in both directions, it seems that in living cells, reversible flux modes rarely occur. Nevertheless, many biochemical transformations can proceed in opposite direction, but not on exactly the inverse routes. Atkinson (1986) stressed that metabolism is organized so that nearly every pathway is paired with an oppositely directed conversion that involves different reactions and a different overall stoichiometry, especially with regard to the coupling agents, ATP/ADP and NAD/NADH. An example is provided by glycolysis and gluconeogenesis, which use phosphofructokinase and fructose-1,6-bisphosphatase, respectively. It seems that, in biochemical systems, irreversible reactions are located in sufficient number and at appropriate positions to exclude the occurrence of reversible flux modes.

3.3. THERMODYNAMIC ASPECTS

3.3.1. A Generalized Wegscheider Condition

Although biochemical reaction networks are usually open with respect to flow of energy and matter, it is useful to study properties of closed systems, as a limit situation. For example, subsystems of open networks can be approximately considered as closed in a fast time scale if they are fast compared to the other subsystems (cf. Chapter 4). As closed reaction systems have no inputs and outputs, the only steady state possible is the thermodynamic equilibrium state. Moreover, certain subnetworks of open systems may subsist in thermodynamic equilibrium irrespective of the separation of time scales, such as the phosphoglucomutase reaction in Fig. 3.4.

A network of biochemical reactions is called *detailed balanced* if in every steady state all net reaction rates are zero (cf. Horn and Jackson, 1972). Closed reaction systems are always detailed balanced owing to the principle of microreversibility (Lewis, 1925, cf. Wei, 1962), that is, in each steady state,

$$\boldsymbol{v}(\boldsymbol{S},\boldsymbol{p}) = \boldsymbol{0}. \tag{3.59}$$

For the special situation of cycles of monomolecular reactions, Wegscheider (1902) showed that detailed balance implies that the product of equilibrium constants around any cycle must be equal to unity (cf. also Hearon, 1953),

$$\prod_j q_j = 1. \tag{3.60}$$

Consider, for example, the cyclic reaction system shown in Scheme 8. When this system does not involve any external metabolites, it is closed and, hence, detailed balanced. For this system, Wegscheider's condition can be written as $q_1 q_2 q_3 = 1$.

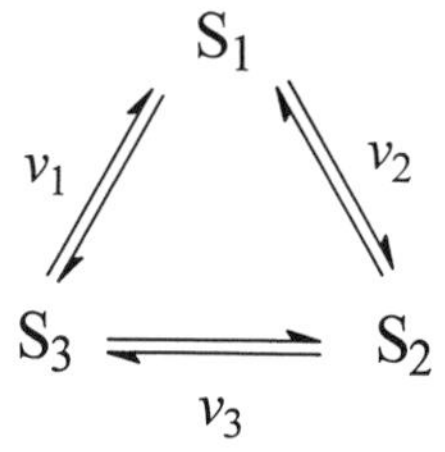

Scheme 8

Using the thermodynamic definition of the equilibrium constant in terms of the change in standard Gibbs free energy,

$$q_j = \exp\left(-\frac{\Delta G_j^0}{RT}\right), \tag{3.61}$$

Wegscheider's condition (3.60) can be derived from the fact that the change in free energy that accompanies the turnover of a complete cycle is zero.

Wegscheider's condition can be generalized for closed systems of any complexity with reversible reactions endowed with the generalized mass-action kinetics (2.15), which is equivalent to Eq. (2.17). As, in open systems also, thermodynamic equilibria may occur for special values of the external metabolites, the generalization of Eq. (3.60) may even comprise open systems (cf. Vol'pert and Khudyaev, 1975; Feinberg, 1989; S. Schuster and R. Schuster, 1989). We distinguish two cases according to whether the rank of the stoichiometry matrix is smaller than, or equal to, the number of reactions, r. In the latter case, the null-space of $\mathbf{N}$ is void, so that $\boldsymbol{v} = \boldsymbol{0}$ is the only steady-state solution.

In the case rank($\mathbf{N}$) $< r$, we can construct a null-space matrix $\mathbf{K}$, which fulfills Eq. (3.44). We now take into account that the functions $G_j(\boldsymbol{S})$ in Eq. (2.17) are positive throughout. Strictly speaking, these functions may be equal to zero if some concentration is zero. This can lead to "false equilibria," in which rates but not all affinities are zero (cf. Othmer, 1981). We will exclude occurrence of false equilibria here. At any proper equilibrium state, the generalized mass-action rate law (2.17) implies that all affinities are zero. Equation (2.16) then gives

$$\ln \tilde{q}_j = \sum_i n_{ij} \ln S_i. \tag{3.62}$$

This can be written in matrix notation as

$$\ln \tilde{\boldsymbol{q}} = \mathbf{N}^{\mathrm{T}} \ln \boldsymbol{S}. \tag{3.63}$$

Premultiplication of this equation by $\mathbf{K}^{\mathrm{T}}$ from the left yields, owing to Eq. (3.44),

$$\mathbf{K}^{\mathrm{T}} \ln \tilde{\boldsymbol{q}} = \boldsymbol{0}, \tag{3.64a}$$

which can also be written as

$$\prod_{j=1}^{r} \tilde{q}_j^{k_{ji}} = 1, \quad i = 1, \ldots, r - \operatorname{rank}(\mathbf{N}), \tag{3.64b}$$

where k_{ji} are the elements of matrix $\mathbf{K}$. Because Eq. (3.64) results from the assumption $\boldsymbol{v} = \boldsymbol{0}$, it is a necessary condition for an equilibrium state to exist. We now prove that the condition is also sufficient. As the steady-state flux vector, $\boldsymbol{J}$, is situated in the null-space of $\mathbf{N}$, it is a linear combination of the columns of $\mathbf{K}$ [cf. Eq. (3.45)]. Equation (3.64) therefore implies

$$\boldsymbol{J}^{\mathrm{T}} \ln \tilde{\boldsymbol{q}} = 0. \tag{3.65}$$

In steady state, the generalized mass action kinetics (2.17) can be written as

$$J_j = G_j\left(\tilde{q}_j \prod_i S_i^{-n_{ij}} - 1\right). \tag{3.66}$$

Multiplying this equation by the term $J_j G_j^{-1} Q_j$ with

$$Q_j = \frac{\ln \tilde{q}_j - \sum_i n_{ij} \ln S_i}{\tilde{q}_j \prod_i S_i^{-n_{ij}} - 1} \tag{3.67}$$

and summing up over all j gives

$$\sum_j J_j^2 G_j^{-1} Q_j = \sum_j J_j\left(\ln \tilde{q}_j - \sum_i n_{ij} \ln S_i\right). \tag{3.68}$$

It can be seen easily that the numerator and denominator in Q_j always have the same sign. In the case they are both zero, Q_j has the limit $\tilde{q}_j \prod_i S_i^{n_{ij}}$, as can be derived by l'Hôpital's rule. In any case, Q_j is positive. Because the terms G_j^{-1} are positive also, the left-hand side of Eq. (3.68) is a positive-definite function of the fluxes. The right-hand side of this equation vanishes because of Eq. (3.65) and the steady-state equation $\mathbf{N}\boldsymbol{J} = \boldsymbol{0}$. Therefore, condition (3.64) entails that all fluxes are zero, which completes the proof of the assertion that Eq. (3.64) is necessary and sufficient for the equivalence of steady state and thermodynamic equilibrium.

Equation (3.64) consists of $r - \text{rank}(\mathbf{N})$ particular equations. They can be considered as a generalized Wegscheider condition. Because in closed systems no external metabolites occur, the apparent equilibrium constants coincide with the "real" equilibrium constants, and condition (3.64) reads

$$\mathbf{K}^{\mathrm{T}} \ln \boldsymbol{q} = \boldsymbol{0}. \tag{3.69}$$

As the only steady state in closed systems is thermodynamic equilibrium, the generalized Wegscheider condition (3.69) is always fulfilled in such systems. In contrast, for open systems, Eq. (3.64) is only fulfilled for special values of the external concentrations.

For the reaction cycle shown in Scheme 8, the null-space matrix reads $\mathbf{K}^{\mathrm{T}} = (1\ 1\ 1)^{\mathrm{T}}$. Thus, Eq. (3.64) reads $\ln q_1 + \ln q_2 + \ln q_3 = 0$, which is equivalent to Eq. (3.60) for this system.

A more complex example is the reaction system depicted in Figure 2.12. in the more general situation that all reactions are reversible (cf. Section 2.4). The null-space matrix can be chosen to contain the vectors $\boldsymbol{k}_1 = (1\ 1\ 0\ 0\ 0)^{\mathrm{T}}$ and $\boldsymbol{k}_2 = (1\ 0\ 1\ 1\ 1)^{\mathrm{T}}$. The generalized Wegscheider condition (3.64) then consists of the two equations

$$P_1 q_1 \frac{q_2}{P_2} = 1, \tag{3.70a}$$

$$P_1 q_1 \frac{q_3}{P_3} q_4 q_5 = 1. \tag{3.70b}$$

If and only if both of these equations are fulfilled, the considered reaction system is detailed balanced. For given equilibrium constants, Eqs. (3.70a) and (3.70b) represent two equations for the three external concentrations. They determine a one-dimensional manifold in the state space of external concentrations.

In some enzyme-kinetic reaction schemes, the interesting situation occurs that the concentrations of external metabolites drop out upon multiplication of apparent equilibrium constants around a cycle. An example is provided in Figure 3.8. P_1, P_2, and P_3 are considered to be external species, because upon derivation

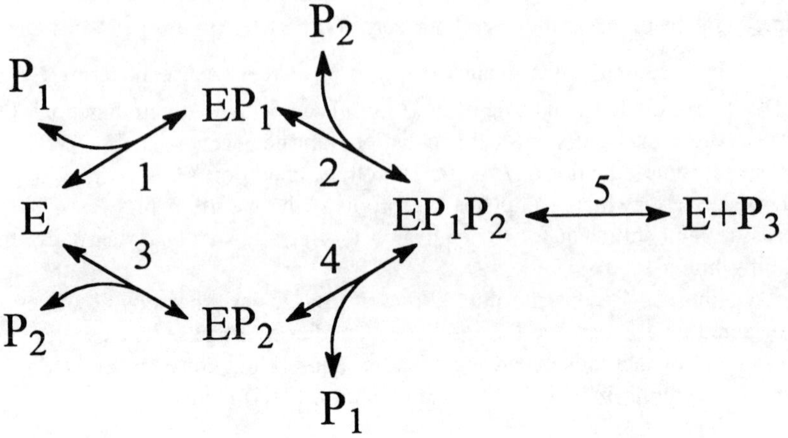

Figure 3.8 Catalytic scheme of a bi-uni enzyme with random mechanism.

of enzyme-kinetic rate laws, the substrates and products are usually treated in this way. An admissible null-space vector reads $\boldsymbol{k}_1 = (1\ 1\ -1\ -1\ 0)^{\mathrm{T}}$. The corresponding generalized Wegscheider condition reads

$$\frac{\tilde{q}_1\tilde{q}_2}{\tilde{q}_3\tilde{q}_4} = \frac{P_1 q_1 P_2 q_2}{P_2 q_3 P_1 q_4}, \tag{3.71}$$

Here, the concentrations P_1 and P_2 can be canceled. So we are left with a ratio of equilibrium constants that refers to the cycle containing reactions 1–4. Therefore, this equation is always fulfilled, irrespective of the values of P_1 and P_2.

A second null-space vector reads $\boldsymbol{k}_2 = (1\ 1\ 0\ 0\ 1)^{\mathrm{T}}$. Equation (3.64) implies

$$\tilde{q}_1\tilde{q}_2\tilde{q}_5 = q_1 q_2 q_5 \frac{P_3}{P_1 P_2} = 1. \tag{3.72}$$

This equation is only satisfied for special values of substrate and product concentrations, in contrast to Eq. (3.71). Therefore, the system shown in Figure 3.8 can, in general, reach a steady state with nonzero fluxes.

For further applications of Wegscheider's condition in enzyme kinetics, see the works of Ricard (1978), Walz and Caplan (1988), and Kuby (1991). This principle has also to been taken into account in analyzing the effects of metabolic channeling (Mendes *et al.*, 1992; Cornish-Bowden and Cárdenas, 1993).

It can be shown that when the generalized Wegscheider condition is fulfilled (in particular in closed systems), there is exactly one equilibrium state in the interior of each reaction simplex of full dimension. Furthermore, this state is asymptotically stable with the interior of this simplex being the basin of attraction. For the usual mass-action kinetics, the proof to this statement was given by Horn and Jackson (1972). It makes use of the Lyapunov function

$$V_L(\boldsymbol{S}) = \sum_{i=1}^{n} (S_i \ln S_i + a_i S_i) + b, \tag{3.73}$$

where a_i and b are constants chosen appropriately. This proof can easily be modified for the case of the generalized kinetics (2.17) (S. Schuster and R. Schuster, 1989). For reactions with more complex kinetics [e.g., in regular (nondilute) solutions], equilibrium states are not, however, always globally stable (cf. Othmer, 1981).

Checking the stability of equilibrium is important for fast subsystems. Provided that the generalized Wegscheider condition (3.64) is fulfilled by a fast subsystem under consideration, it has a unique, globally asymptotically stable equilibrium state. This is of importance for the applicability of the rapid-equilibrium approximation (cf. Section 4.3.).

The issue of detailed balancing can also be approached from the viewpoint of irreversible thermodynamics. Entropy production is defined as

$$\sigma = \sum_{j=1}^{r} \frac{v_j A_j}{T} \tag{3.74}$$

with T denoting temperature. Inserting the definitions of apparent equilibrium constants and affinities [Eqs. (2.14) and (2.16), respectively] gives

$$\sigma = -R \sum_j v_j \sum_k n_{kj} \ln S_k + R \sum_j v_j \ln \tilde{q}_j. \tag{3.75}$$

When the system subsists in steady state, the first term on the right-hand side of Eq. (3.75) equals zero. Inserting the vector $\boldsymbol{J}$ of steady-state fluxes, we obtain

$$\sigma = R \sum_j J_j \ln \tilde{q}_j = R\boldsymbol{J}^{\mathrm{T}} \ln \tilde{\boldsymbol{q}}. \tag{3.76}$$

Because the vector $\boldsymbol{J}$ is situated in the null-space of the stoichiometry matrix (see Section 3.2.1), we can express it as a linear combination of the columns of the matrix $\mathbf{K}$, as given in Eq. (3.45). This equation can be written more concisely as

$$\boldsymbol{J} = \mathbf{K}\boldsymbol{\alpha} \tag{3.77}$$

with $\boldsymbol{\alpha}$ being a vector of dimension r − rank (**N**). Equations (3.76) and (3.77) yield

$$\frac{\sigma}{R} = \boldsymbol{\alpha}^{\mathrm{T}}\mathbf{K}^{\mathrm{T}} \ln \tilde{\boldsymbol{q}}. \tag{3.78}$$

In this equation, the terms are amenable to an interesting new interpretation. The vector

$$\boldsymbol{A}' = RT\mathbf{K}^{\mathrm{T}} \ln \tilde{\boldsymbol{q}} \tag{3.79}$$

can be regarded as containing the *overall affinities* of the reaction system (i.e., generalized forces). The vector $\boldsymbol{\alpha}$ encompasses the corresponding overall flows (independent fluxes).

The definition of the affinity [Eq. (2.16)] can be written in matrix notation as

$$\boldsymbol{A} = RT(\ln \tilde{\boldsymbol{q}} - \mathbf{N}^{\mathrm{T}} \ln \boldsymbol{S}). \tag{3.80}$$

Due to Eq. (3.44), Eq. (3.79) can be written as

$$\boldsymbol{A}' = \boldsymbol{K}^{\mathrm{T}}\boldsymbol{A}. \tag{3.81}$$

This equation can be illustrated by the example of a chain of consecutive monomolecular reactions. The overall affinity is here simply the sum of the particular reaction affinities, and we indeed have $\mathbf{K}^{\mathrm{T}} = (1\ 1\ \ldots\ 1)$. In steady state, entropy production can therefore be written in terms of a smaller number of reaction rates than for nonstationary dynamics. The generalized Wegscheider condition (3.64) can then be interpreted in that all overall affinities in the system (assumed to be in steady state) are zero. We have shown above that when this is the case, all rates are zero. This can be shown using the generalized flows in the following way.

The generalized mass action kinetics (2.17) can be written as

$$\boldsymbol{v} = (\mathrm{dg}\ \boldsymbol{G}')(\ln \tilde{\boldsymbol{q}} - \mathbf{N}^{\mathrm{T}} \ln \boldsymbol{S}) \tag{3.82}$$

with $\boldsymbol{G}'$ being a vector with the components $G_j Q_j^{-1}$. Under steady-state conditions, Eqs. (3.77) and (3.82) lead to

$$(\mathrm{dg}\ \boldsymbol{G}')(\ln \tilde{\boldsymbol{q}} - \mathbf{N}^{\mathrm{T}} \ln \boldsymbol{S}) = \mathbf{K}\boldsymbol{\alpha}. \tag{3.83}$$

Premultiplying this equation by $\boldsymbol{\alpha}^{\mathrm{T}}\boldsymbol{K}^{\mathrm{T}}(\mathrm{dg}\boldsymbol{G}')^{-1}$, we obtain by consideration of $\mathbf{KN} = \mathbf{0}$

$$\boldsymbol{a}^{\mathrm{T}}\mathbf{K}^{\mathrm{T}} \ln \tilde{\boldsymbol{q}} = \boldsymbol{a}^{\mathrm{T}}\mathbf{K}^{\mathrm{T}}(\mathrm{dg}\, \boldsymbol{G}')^{-1}\mathbf{K}\boldsymbol{a}. \tag{3.84}$$

When condition (3.64) is fulfilled, the left-hand side of Eq. (3.84) equals zero. Because all G_j' are positive, the right-hand side of Eq. (3.84) represents a positive-definite quadratic form in the generalized fluxes, a_j. Therefore, all generalized flows equal zero in steady state when the generalized Wegscheider condition (3.64) is fulfilled. Consequently, this condition implies, for any steady state,

$$\boldsymbol{J} = \mathbf{K}\boldsymbol{a} = \boldsymbol{0}. \tag{3.85}$$

3.3.2. Strictly Detailed Balanced Subnetworks

Zero fluxes can be relevant in certain subsystems of open reaction networks. Consider, for example, the scheme of glycolysis and of some adjacent reactions shown in Figure 3.4. It has been shown in Section 3.2.2 that the phosphoglucomutase reaction is a dead-end branch so that its net reaction rate always equals zero when the considered network has attained a stationary state. This feature is independent of the kinetic parameters of all reactions involved and will therefore be called strict detailed balancing. Other examples of dead-end branches are the reactions leading to the complexes EI and ESI in Scheme 3 (Section 2.2.2).

The situation changes when G1P is treated as an external species. The flux through the phosphoglucomutase reaction can then be zero in an exceptional situation only, namely for very special values of the kinetic parameters of those reactions affecting the concentration of G6P. This reaction is then detailed balanced, but not strictly detailed balanced.

Necessary and sufficient conditions for strict detailed balancing are given by the following theorem (the proof was given by S. Schuster and R. Schuster, 1991):

Theorem 3A. *A subnetwork, $\boldsymbol{\Gamma}_i$, of a given reaction network is strictly detailed balanced if and only if the following two conditions are fulfilled:*

- **(i)** *The null-space matrix* $\mathbf{K}$ *can be chosen to be block-diagonal, as given in Eq.* (3.48).
- **(ii)** *Either the reactions of $\boldsymbol{\Gamma}_i$ correspond to some or all rows of the null submatrix on the bottom of* $\mathbf{K}$, *or they correspond to a submatrix* $\mathbf{K}^{(i)}$ *and the equation*

$$\mathbf{K}^{(i)\mathrm{T}} \ln \tilde{\boldsymbol{q}}^{(i)} = \boldsymbol{0} \tag{3.86}$$

is fulfilled, where $\tilde{\boldsymbol{q}}^{(i)}$ is the vector of apparent equilibrium constants of $\boldsymbol{\Gamma}_i$.

Equation (3.86) can be considered as a generalized Wegscheider condition for subnetwork $\boldsymbol{\Gamma}_i$.

Detection of strictly detailed balanced subnetworks is helpful when steady states of the whole network are analyzed. This analysis is simplified if all reactions that have, at any steady state, zero net reaction rates are detected at the very beginning. Very frequently, simulation of reaction systems is hampered by incompleteness or unsatisfactory accuracy of the known data. However, the effect of strictly detailed balanced reactions on the concentrations of internal metabolites is fully determined by their equilibrium constants. Usually, thermodynamic parameters can be measured more accurately than kinetic ones (e.g., rate constants), especially for very fast reactions.

Note that for the system shown in Figure 3.8, the null-space matrix is not diagonalizable. Therefore, no strictly detailed balanced reactions occur, although one of the two generalized Wegscheider conditions involved in Eq. (3.64) is fulfilled.

3.3.3. Onsager's Reciprocity Relations for Coupled Enzyme Reactions

The concept of null-space can be used to prove the Onsager reciprocity relation (2.59) for the case of chemical reaction systems at steady state. This is of importance, in particular, for enzymes coupling endergonic to exergonic processes (see Section 2.2.3). At steady state, the flux vector can be written as a linear combination of the columns of the null-space matrix [cf. Eq. (3.77)]. Because the components of the vector $\boldsymbol{a}$ can be considered as containing the overall reaction rates (cf. Section 3.3.1), we rename it $\boldsymbol{J}'$. For the enzyme scheme shown in Figure 2.2, for example, admissible null-space vectors are $\boldsymbol{k}_1 = (1\ 1\ 1\ 1\ 1\ 1\ 0)^{\mathrm{T}}$ and $\boldsymbol{k}_2 = (1\ 0\ 0\ 0\ 1\ 1\ 1)^{\mathrm{T}}$. Accordingly, the velocities of ATP production and reaction slip can be taken as overall fluxes, J_1' and J_2'.

In correspondence to these new rate variables, overall affinities can be defined, which are to be gathered in the vector $\boldsymbol{A}'$ defined in Eq. (3.79). Equation (3.78) means that entropy production can also be written in terms of the overall affinities and fluxes,

$$\sigma = \frac{(\boldsymbol{J}')^{\mathrm{T}}\boldsymbol{A}'}{T}. \tag{3.87}$$

Equations (2.57) and (3.77) give

$$\boldsymbol{A} = (\mathrm{dg}\,\boldsymbol{L})^{-1}\mathbf{K}\boldsymbol{J}'. \tag{3.88}$$

Due to Eq. (3.81), this leads to

$$\boldsymbol{A}' = \mathbf{K}^{\mathrm{T}}(\mathrm{dg}\,\boldsymbol{L})^{-1}\mathbf{K}\boldsymbol{J}'. \tag{3.89}$$

With the help of Eq. (2.58), we obtain

$$\boldsymbol{J}' = \mathbf{L}'\mathbf{K}^{\mathrm{T}}(\mathrm{dg}\, \boldsymbol{L})^{-1}\mathbf{K}\boldsymbol{J}'. \tag{3.90}$$

Because $\boldsymbol{J}'$ can be changed to assume a manifold of different values, by alteration of the external metabolite concentrations, Eq. (3.90) can only hold if

$$\mathbf{L}' = [\mathbf{K}^{\mathrm{T}}(\mathrm{dg}\, \boldsymbol{L})^{-1}\mathbf{K}]^{-1}. \tag{3.91}$$

This equation gives the sought matrix of transformed Onsager coefficients, which link the vectors $\boldsymbol{J}'$ and $\boldsymbol{A}'$. $\mathbf{L}'$ is not normally diagonal. Indeed, there are in general strong cross-effects between the forces A_j', because the exergonic and endergonic processes are coupled to each other by the enzyme.

Using the rule for transposition of matrix products, one obtains that $[\mathbf{K}^{\mathrm{T}}\, (\mathrm{dg}\, \boldsymbol{L})^{-1}\, \mathbf{K}]^{\mathrm{T}} = \mathbf{K}^{\mathrm{T}}\, (\mathrm{dg}\, \boldsymbol{L})^{-1}\, \mathbf{K}$. The latter matrix is therefore symmetric. As the inverse of a symmetric matrix is also symmetric, the matrix on the right-hand side of Eq. (3.91) and, hence, matrix $\mathbf{L}'$ have this property as well. This completes the proof of Eq. (2.59), which is a particular case of Onsager's reciprocity relations (Onsager, 1931; cf. Guggenheim, 1967). They are usually proved in a more complicated way on the basis of the symmetry of time.

4

Time Hierarchy in Metabolism

4.1. TIME CONSTANTS OF METABOLIC PROCESSES

A time constant is a measure of the time span over which significant changes occur in a given system, generally during the relaxation after perturbation of a stable steady state. Basically, one can distinguish between time constants of reactions, of substances, and of a whole system. For none of them, however, a unique definition in mathematical terms has been agreed upon. For the time constant of (individual) reactions with linear rate laws, a widely used definition is

$$\tau = \frac{1}{k_+ + k_-}, \tag{4.1}$$

where k_+ and k_- denote the forward and backward rate constants, respectively. This definition results from the solution to the differential equation governing the relaxation of such reactions, which is proportional to $\exp(-(k_+ + k_-)t)$.

Equation (4.1) also applies to reactions with Michaelis–Menten kinetics [Eq. (2.20)] when substrate and product concentrations are low. Under this condition, that equation can be approximated by the linear rate equation

$$v(S_1,S_2) = \frac{V_m^+}{K_{m1}} S_1 - \frac{V_m^-}{K_{m2}} S_2. \tag{4.2}$$

Equation (4.1) then gives

$$\tau = \frac{1}{V_m^+/K_{m1} + V_m^-/K_{m2}}. \tag{4.3}$$

Following a suggestion by Higgins (1965) one can estimate the response time for reactions with nonlinear rate laws as

$$\tau_j = \left(\sum_i (-n_{ij}) \frac{\partial v_j}{\partial S_i} \right)^{-1}, \tag{4.4}$$

at an operating point under consideration. This can be derived by linearization of the system equation around this operating point (see Section 2.3.2).

For the reaction A + B ↔ C + D with standard mass-action kinetics, for example, the response time reads

$$\tau = [k_+(A + B) + k_-(C + D)]^{-1}. \tag{4.5}$$

Equation (4.3) can be derived from Eq. (4.4) by applying it to the reversible Michaelis–Menten kinetics and taking the derivatives at $S_1 = 0$ and $S_2 = 0$. As the first derivatives of this kinetics with respect to S_1 and $-S_2$ are monotonic decreasing functions of S_1 and S_2, respectively, Eq. (4.3) underestimates the time constant as computed by Eq. (4.4). For example, applying Eq. (4.3) to three glycolytic enzymes in human erythrocytes (using data of several authors cited by Liao and Lightfoot Jr., 1987) gives Na/K–ATPase (EC 3.6.1.37), $\tau = 50$ min; 2,3-bisphosphoglycerate phosphatase (EC 3.1.3.13), 23 min; bisphosphoglycerate mutase (EC 5.4.2.4), 0.013 s. By this approximation, the relaxation time of 2,3-bisphosphoglycerate phosphatase is, however, extremely underestimated, because this enzyme is nearly saturated with its substrate, so that the derivative entering Eq. (4.4) is nearly zero. By consideration of the degradation kinetics of 2,3P_2G, an estimate of $\tau = 10$ h was evaluated for this enzyme (Rapoport *et al.,* 1976). For bisphosphoglycerate mutase, Heinrich *et al.* (1977) calculated the relaxation time of 3.9 s. The difference to the value given above results from consideration of the inhibition by 2,3P_2G. For pyruvate kinase (EC 2.7.1.40), hexokinase (EC 2.7.1.1), and phosphofructokinase (EC 2.7.1.11), relaxation times of 28 s, 36 min, and 74 s, respectively, were calculated. Accordingly, the relaxation time constants of the glycolytic enzymes in human erythrocytes cover at least a range of four orders of magnitude, let alone the enzymes so fast as to be near equilibrium, for which the time constants are very low and difficult to measure or calculate. This separation of time constants is accompanied by the fact that the fast enzymes are so efficient that they can catalyze rates much higher than maximum pathway flux (e.g., 100-fold in glycolysis in muscle, see Betts and Srivastava, 1991). A biochemical reaction is usually said to be fast if τ is less than 1 s.

From among the 20 enzymes considered in a model of the tricarboxylic acid cycle in *Dictyostelium discoideum* (Wright *et al.,* 1992), two enzyme rates are described by a reversible uni-uni Michaelis–Menten rate law so that Eq. (4.3) can be applied. This gives the relaxation times of 1.73 s for succinate dehydrogenase

(EC1.3.99.1) and 0.21 s for fumarase (EC4.2.1.2) (with the above-mentioned uncertainty due to saturation). Another nine enzymes were modeled by irreversible, unimolecular mass-action kinetics so that the relaxation times are equal to the reciprocals of the rate constants, which gives values in a range from 0.075 s to 10 min. Accordingly, the modeled system also exhibits a distinct separation of time scales, all the more as it encompasses other enzymes with even shorter relaxation times, which were not explicitly included in the model.

Separation of time constants is also observed for membrane transport processes. For the membrane fluxes through the erythrocyte membrane, time constants lie between 10^{-2} s (water exchange) and 10^5 s (passive Na^+ and K^+ exchange) (Glaser *et al.,* 1983; Brumen and Heinrich, 1984).

The wide separation of time constants is often called *time hierarchy* (Park, 1974; Reich and Selkov, 1975; Heinrich *et al.,* 1977). Hierarchic organization is a striking feature of living matter in general. Living organisms are built up of nested spatial structures (organelles, cells, tissues, organs, etc.). Control and regulation act at different levels (metabolic regulation, epigenetic regulation and hormone system, etc.). Evolutionary processes clearly run much slower than processes in metabolism. By time hierarchy *sensu stricto,* however, we mean the operation on distinct time scales at one and the same spatial level of organization, for example, within the living cell.

The occurrence of time hierarchy has important consequences for the mathematical modeling, because the resulting differential equation systems are then stiff. The usual integration routines, such as the Runge–Kutta procedure, are then only stable when operating with very small step sizes of the order of magnitude of the time constants of the fast processes.

To reflect the *systemic* properties of metabolic pathways, time constants which take into account the interactions within the system rather than relaxation times of isolated reactions should be used. This can be seen, for example, when calculating the relaxation time of the hexokinase-phosphofructokinase system in glycolysis (about 1.5 h in human erythrocytes), which exceeds the values for the particular enzymes given above, due to the glucose-6-phosphate inhibition of hexokinase (Heinrich *et al.,* 1977).

When small perturbations of a steady state are considered ($S_i(t) = \bar{S}_i + \delta S_i(t)$) the solutions for $\delta S_i(t)$ may be expressed by the eigenvectors and the eigenvalues, λ_i, of the Jacobian (see Section 2.3.2). If $\mathrm{Re}(\lambda_i) < 0$, an appropriate measure for characterizing relaxation processes is provided by the characteristic times

$$\tau_i = \frac{1}{|\mathrm{Re}(\lambda_i)|}, \quad i = 1, \ldots, n. \tag{4.6}$$

For red blood cell glycolysis, for example, characteristic times were calculated by Eq. (4.6) to cover a range from 0.9 ms to 12 h (Liao and Lightfoot Jr., 1987).

Note that the above examples of time-scale separation concern systems at the metabolic level. It is clear that an even more pronounced separation applies to cellular physiology when genetic processes are included also, which normally have time constants of hours up to years. When interested in characterizing the long-term behavior of a metabolic system after perturbations, one may use the largest time constant defined by Eq. (4.6) (cf. Schuster and Heinrich, 1987).

A time constant for concentrations is the turnover time introduced by Reich and Selkov (1981, Chap. III). To define this quantity, one should split up the reaction rates into the rates of forward and reverse reactions,

$$v_j = v_j^+ - v_j^- . \tag{4.7}$$

The system equation (2.8) can now be written as

$$\frac{dS_i}{dt} = \sum_{j=1}^{r} (n_{ij}^+ v_j^+ + n_{ij}^- v_j^- - n_{ij}^- v_j^+ - n_{ij}^+ v_j^-), \tag{4.8}$$

where n_{ij}^+ and n_{ij}^- are defined by Eqs. (2.11a) and (2.11b). The turnover time is to characterize the time needed to convert the pool of a given substance S_i once. A possible definition is

$$\tau_i^{\text{turn}} = \frac{S_i}{\sum_{j=1}^{r} (n_{ij}^- v_j^+ + n_{ij}^+ v_j^-)}, \tag{4.9}$$

which includes the effects of all unidirectional processes utilizing S_i. The original definition of Reich and Selkov (1981) was more restrictive in that only the largest of the unidirectional rates entered the denominator in Eq. (4.9).

Clearly, the turnover times reflect the level of aggregation used in the model. When, for example, two metabolites isomerizing into each other by a fast reaction are combined into one pool (e.g., the pool comprising glucose-6-phosphate and fructose-6-phosphate), the latter has a much larger turnover time than either of the particular metabolites. Aggregate pools can often be defined in a way that they are produced or consumed only by irreversible reactions, so that the unidirectional rates are equal to the net rates. This simplifies to calculate turnover times because for metabolic systems, unidirectional rates of reversible reactions are generally known to much less an extent then net rates.

To illustrate definition (4.9), we again consider the data on the tricarboxylic acid cycle in *Dictyostelium discoideum* given by Wright *et al.* (1992). For the reversible isomerization between the tricarboxylic acids citrate, *cis*-aconitate and isocitrate, only the net rate was calculated. Therefore, the turnover times of the particular acids cannot be determined from this model, but the turnover time of the tricarboxylic acid pool can, because the isocitrate dehydrogenase reaction (EC1.1.1.42) is virtually irreversible. Its rate (given to be 2 mM/min, which is at steady state equal to the net rate of isomerization) can be used in the denominator of definition (4.9). The concentrations of isocitrate and citrate are given to be 0.0099 mM and 0.025 mM, respectively. *cis*-Aconitate can be neglected in comparison with these values. So we obtain $\tau^{\text{turn}} = (0.0099 + 0.025)/2$ min $= 1.047$s for the tricarboxylic acid pool.

The turnover time τ_i^{turn} is related to the transition time $\tau_i = S_i/J$ introduced by Easterby (1973, 1981) [cf. Section 5.8.4, in particular Eq. (5.278)], with the difference that in the latter definition, the net rates instead of unidirectional rates are considered. For unbranched reaction chains with the first step being irreversible, τ_i characterizes the contribution of S_i to the time needed to establish all the steady-state concentrations after "switching on" the reaction chain. A further definition of a time constant (slow substrate time scale) was given by Segel (1988).

As a quantitative measure of time hierarchy, Heinrich and Sonntag (1982) proposed the following definition, which takes into account the mean difference, $\langle \Delta\tau_i \rangle$, between the largest time constant and all the others:

$$H_\tau = \frac{\langle \Delta\tau t_i \rangle}{\langle \tau_i \rangle} = \frac{r}{r-1} \frac{\sum_i (\tau_s - \tau_i)}{\sum_i \tau_i}, \tag{4.10}$$

where

$$\tau_s = \max_i \tau_i. \tag{4.11}$$

The mean value of the time constants, $\langle \tau_i \rangle$, in Eq. (4.10) serves for normalization. It is readily verified that H_τ is bounded by

$$H_\tau \leq r \tag{4.12}$$

and attains this maximum value if one reaction is slow and all remaining reactions are as fast as possible. This is in line with the common idea of a hierarchy that there are more dominated items (here: reactions) than dominating ones.

4.2. THE QUASI-STEADY-STATE APPROXIMATION

When analyzing any time-dependent system, one can usually discern three different classes of processes according to their time constants. The "central" class comprises the processes moving in the time scale of interest. A second class comprises the processes so slow that they can, in the experiment or theoretical study, be neglected or the concentrations of the substances involved can be treated as parameters (external metabolites, see Section 2.1). The third class is made up by the processes which are so fast that they can be considered to have run off in the time scale of interest. As we will show later in this section, this relaxation only occurs under some stability conditions. If these conditions are not fulfilled, the fast processes may oscillate or exhibit an even more complex behavior. They may then also be eliminated from the analysis by appropriate averaging techniques.

$$P_1 \xrightarrow{v_1} S_1 \overset{v_2}{\rightleftharpoons} S_2 \xrightarrow{v_3} P_2 \qquad \text{Scheme 9}$$

The reaction chain depicted in Scheme 9 may serve as an example. Here, we assume the three reactions to be irreversible and endowed with linear kinetics,

$$v_1 = k_1 P_1, \tag{4.13a}$$

$$v_2 = k_2 S_1, \tag{4.13b}$$

$$v_3 = k_3 S_2. \tag{4.13c}$$

Assume that

$$k_3 >> k_2. \tag{4.14}$$

After a brief initial relaxation period, the concentration S_2 will approximately have the value

$$S_2 = \frac{k_2 S_1}{k_3}. \tag{4.15}$$

because any deviation from this value will rapidly vanish; that is, S_2 attains a state approximately given by

$$\frac{dS_2}{dt} = k_2 S_1 - k_3 S_2 = 0. \tag{4.16}$$

Note that as long as S_1 does not reach a steady state, S_2 does not really reach one either. However, as Segel (1988) put it, S_2 can "keep up" with the changing concentration of S_1, because Eq. (4.15) couples S_2 to S_1. For such behavior, the term quasi-steady-state is used. Figure 4.1 shows a typical phase curve S_1 versus

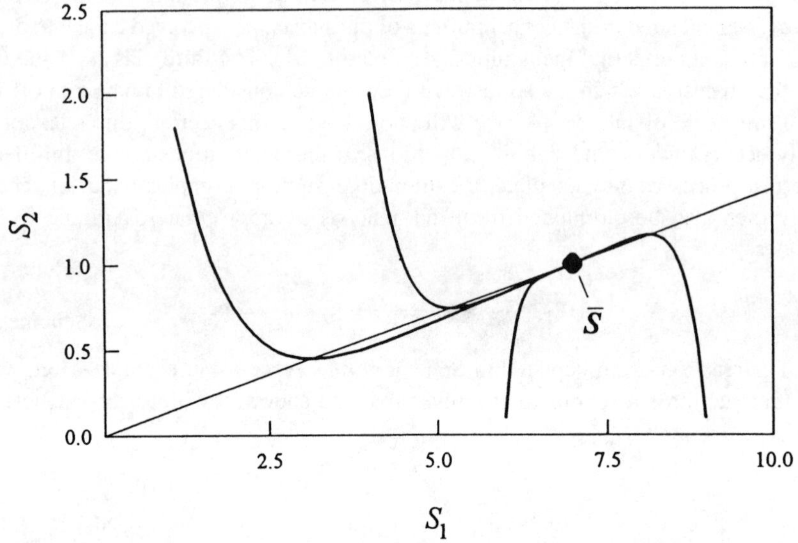

Figure 4.1 Phase plot S_1 versus S_2 illustrating the quasi-steady-state approximation for the reaction chain shown in Scheme 9 for the case that reaction 3 is faster than reaction 2. Parameter values: $P_1 = 1, k_1 = 7, k_2 = 1, k_3 = 7$. The thick lines represent trajectories for different initial conditions. The thin straight line is the nullcline for S_2. $\bar{S}$ denotes the final steady state.

S_2 computed numerically. It can be seen that any trajectory goes, in a first period, toward the line given by Eq. (4.15). As this initial relaxation is very rapid, the long-term behavior can be described by the differential equation

$$\frac{dS_1}{dt} = k_1P_1 - k_2S_1 \tag{4.17}$$

together with the algebraic equation (4.15). The number of differential equations is therefore reduced from two to one.

Such a reduction of the system equations is the basic idea of the *Bodenstein method* or *quasi-steady-state approximation* (QSSA) (Bodenstein, 1913; cf. Heineken *et al.,* 1967; Kondratiev, 1969). This method is widely used in chemical kinetics for systems that exhibit large differences in concentrations among the different substances involved, due to widely separated rate constants. For the example considered above, S_2 as given by Eq. (4.15) is, in fact, much smaller than S_1, owing to inequality (4.14).

In a general presentation, the Bodenstein method can be outlined as follows. Assume that the concentration vector $\boldsymbol{S}$ can be decomposed as

$$\mathbf{S} = \begin{pmatrix} \mathbf{S}^{(1)} \\ \mathbf{S}^{(2)} \end{pmatrix} \tag{4.18}$$

with

$$S_i^{(1)} >> S_j^{(2)} \quad \text{for all } i \text{ and } j. \tag{4.19}$$

Note that concentrations may be small for two distinct reasons. First, some conservation quantities such as total enzyme concentrations may be small and, second, some rate constants may be high (such as k_3 in the system shown in Scheme 9), so that the concentrations of the substances utilized by the corresponding fast reactions become very small after an initial time span. Note that inequality (4.19) generally holds only after this initial transient.

Under condition (4.19), it is reasonable to normalize the concentrations by some typical value of each subvector:

$$s_i^{(k)} = \frac{S_i^{(k)}}{S^{(k)}}, \quad k = 1, 2, \tag{4.20}$$

with $S^{(1)} >> S^{(2)}$. For the large concentrations, $S_i^{(1)}$, the normalization factor, $S^{(1)}$, could be the largest initial value, $S_{\max}^{(1)}(0)$. For the small concentrations, $S^{(2)}$, a similar choice might be problematic because the $S_i^{(2)}$ may be rather large for $t = 0$ and relax to small values only after some initial period. Taking a value $S_i^{(2)}(t)$ for some small $t > 0$ is then more appropriate. Things are easier if the $S_i^{(2)}$ are involved in a conservation relation. Then the respective conservation quantities can be taken as normalization factors.

Partitioning the stoichiometry matrix **N** in accordance with Eq. (4.18), we can write the system Eq. (2.8) as

$$\frac{\mathrm{d}\mathbf{s}^{(1)}}{\mathrm{d}t} = \frac{1}{S^{(1)}} \mathbf{N}^{(1)}\mathbf{v}, \tag{4.21a}$$

$$\mu \frac{\mathrm{d}\mathbf{s}^{(2)}}{\mathrm{d}t} = \frac{1}{S^{(1)}} \mathbf{N}^{(2)}\mathbf{v} \tag{4.21b}$$

with the small parameter

$$\mu = \frac{S^{(2)}}{S^{(1)}}. \tag{4.22}$$

As μ is a small parameter, it is sensible to approximate the solutions of the equation system (4.21) by the solutions to the algebro-differential equation system composed of Eq. (4.21a) and

$$\mathbf{N}^{(2)}\boldsymbol{v}(\boldsymbol{S}) = \boldsymbol{0}. \tag{4.23}$$

The latter equation is the steady-state condition for the substances with small concentrations. Assume that this equation can be solved for $\boldsymbol{S}^{(2)}$ in terms of $\boldsymbol{S}^{(1)}$. Inserting the corresponding function into Eq. (4.21a) gives a differential equation system for $\boldsymbol{S}^{(1)}$ of dimension smaller than that of the original system [(4.21a) and (4.21b)]. This is, therefore, an example of a singularly perturbed differential equation system. A regularly perturbed differential equation system, in contrast, involves small parameters only on the right-hand sides, so that the dimension of the system (and, hence, the number of initial or boundary conditions necessary to solve the system) does not decrease as the small parameters tend to zero.

More rigorously, the outlined approximation only applies if some conditions phrased in a theorem given by Tikhonov (1948; cf. Wasow, 1965; Klonowski, 1983) are satisfied. We now give this theorem in a general form for any vector, $\boldsymbol{Y}$, of state variables.

Consider the ordinary differential equation system

$$\frac{\mathrm{d}\boldsymbol{Y}^{\mathrm{s}}}{\mathrm{d}t} = \boldsymbol{F}^{\mathrm{s}}(\boldsymbol{Y}^{\mathrm{s}},\boldsymbol{Y}^{\mathrm{f}}), \tag{4.24a}$$

$$\mu\frac{\mathrm{d}\boldsymbol{Y}^{\mathrm{f}}}{\mathrm{d}t} = \boldsymbol{F}^{\mathrm{f}}(\boldsymbol{Y}^{\mathrm{s}},\boldsymbol{Y}^{\mathrm{f}}), \tag{4.24b}$$

where μ is a small parameter and $\boldsymbol{Y}$ has been decomposed into two subvectors, $\boldsymbol{Y}^{\mathrm{s}}$ and $\boldsymbol{Y}^{\mathrm{f}}$, of slow and fast variables, respectively. Let

$$\boldsymbol{Y}^{\mathrm{f}} = \boldsymbol{\varphi}(\boldsymbol{Y}^{\mathrm{s}}) \tag{4.25}$$

denote a solution of the equation system

$$\boldsymbol{F}^{\mathrm{f}}(\boldsymbol{Y}^{\mathrm{s}},\boldsymbol{Y}^{\mathrm{f}}) = \boldsymbol{0}, \tag{4.26}$$

if such a solution exists. For every given vector $\boldsymbol{Y}^{\mathrm{s}}$, $\boldsymbol{\varphi}(\boldsymbol{Y}^{\mathrm{s}})$ is a fixpoint of the so-called adjoint system

$$\frac{\mathrm{d}\boldsymbol{Y}^{\mathrm{f}}}{\mathrm{d}t'} = \boldsymbol{F}^{\mathrm{f}}(\boldsymbol{Y}^{\mathrm{s}},\boldsymbol{Y}^{\mathrm{f}}), \tag{4.27}$$

where $\boldsymbol{Y}^{\mathrm{s}}$ is considered as a vector of parameters. t' can be interpreted as stretched time, t/μ.

Theorem 4A (Tikhonov's Theorem). *The solution $\boldsymbol{Y}(t)$ of the equation system*

(4.24) *tends to the solution* $(\boldsymbol{Y}^{s}(t)\ \varphi[\boldsymbol{Y}^{s}(t)])^{T}$ *of the "degenerate system"* (4.24a), (4.26) *as* μ *tends to zero, if:*

(T1) *These solutions exist and are unique, and the right-hand sides of the equation systems are unique*

(T2) *A solution* $\varphi(\boldsymbol{Y}^{s})$, *exists, which corresponds to an isolated, asymptotically stable fixpoint of the adjoint system* (4.27)

(T3) *The initial conditions* $\boldsymbol{Y}^{f}(0)$ *of the adjoint system* (4.27) *lie in the basin of attraction of the solution* $\varphi(\boldsymbol{Y}^{s}(0))$.

The proof of this theorem can be found in the works of Tikhonov (1948) and Wasow (1965).

Condition (T1) is normally fulfilled for biochemical systems because the right-hand sides of Eq. (4.21) involve kinetic rate laws, which are continuously differentiable at least once. The stability conditions (T2) and (T3) are not always met (cf. Section 2.3.2). Because usually only one or a few metabolites are treated as quasi-steady-state species, the stability analysis often bears no difficulties. For example, the fixpoint of Eq. (4.16) is globally asymptotically stable because the corresponding eigenvalue is negative ($-k_3$).

The uniqueness and stability of quasi-steady-states is always granted when the subsystem consists of reactions with linear kinetics and satisfies Wegscheider's condition (see Theorem 2G). Accordingly, applying the quasi-steady-state approximation to such subsystems that have these properties guarantees that the preconditions of Tikhonov's Theorem are fulfilled.

The quasi-steady-state approximation is of particular importance in the derivation of enzyme-kinetic rate laws. The total concentrations of enzymes are constant and normally much below substrate concentrations (Albe *et al.,* 1990) so that one can introduce the small parameter $\mu = E_T/S_1$. There are, however, a number of exceptions to this rule (cf. Betts and Srivastava, 1991), notably in the case of ribulose bisphosphate carboxylase (EC 4.1.1.39) (Farquhar, 1979). Consider a simple enzymic reaction with the catalytic mechanism shown in Scheme 1 (Section 2.2). As outlined above, it is helpful to use normalized concentrations [cf. Eq. (4.20)]. For the present example we choose $e = E/E_T$, $es = ES/E_T$, $s_1 = S_1/S_1(0)$ and $s_2 = S_2/S_1(0)$ with $S_1(0)$ denoting the initial concentration of S_1. The system equations can then be written as

$$\mu\frac{\mathrm{d}es}{\mathrm{d}t} = k_1S_1(0)s_1\cdot e - (k_{-1} + k_2)es + k_{-2}S_1(0)s_2\cdot e, \tag{4.28a}$$

$$\frac{\mathrm{d}s_1}{\mathrm{d}t} = -k_1S_1(0)s_1\cdot e + k_{-1}es, \tag{4.28b}$$

with the new time scale $tE_T/S_1(0) \rightarrow t$. It is not necessary to write down the equations governing e and S_2, because they result from the given system equations by the conservation relations $E + ES = E_T$ and $S_1 + S_2 + ES =$ const. of the scheme. For very small μ, Eq. (4.28a) can be approximated by replacing the left-hand side by zero, which gives, together with the conservation relation for the enzyme species, the Michaelis–Menten equation (2.20) with the phenomenological parameters given in Eqs. (2.21) and (2.22).

The new time scale has been introduced because in the limit $\mu = E_T/S_1(0) \to 0$ [which is to be thought of as diminution of E_T rather than increase of $S_1(0)$ (cf. Battelli and Lazzari, 1985)], the reaction would normally cease to proceed because of lack of enzyme. This can be compensated for by compressing the time scale or by rescaling the rate constants (see, e.g., Heineken *et al.*, 1967; Heinrich *et al.*, 1977; Battelli and Lazzari, 1985, 1986). In the latter case, any decrease in E_T is then accompanied by an increase in the rate constants.

Stability of the steady state of the adjoint system is always granted in enzyme kinetics, as long as there is no enzyme–enzyme interaction, because the slow variables (i.e., the non-enzymatic species) are considered constant in this system and simple mass-action kinetics applies satisfactorily well, so that the equation system is linear (cf. Theorem 2G).

It was shown by Segel (1988) and Frenzen and Maini (1988) that the condition $E_T/S_1(0) \ll 1$ for applicability of the quasi-steady-state hypothesis in deriving the Michaelis–Menten rate law is unnecessarily restrictive. It can be replaced by

$$\frac{E_T}{S_1(0) + K_{m1}} \ll 1, \tag{4.29}$$

provided that $k_1 \gg k_{-2}$. This can be understood by the reasoning that the change in substrate concentration during the initial transient period should be small compared to its initial value. As this change is approximately equal to the substrate sequestered in the ES complex, one can invoke that $E_T S_1(0)/[K_{m1} + S_1(0)] \ll S_1(0)$, which leads to condition (4.29). This condition can be fulfilled, for example, by an enzyme showing weak binding to the substrate (high K_{m1} value) even when the enzyme concentration is high.

In the case of competitive inhibition kinetics, the condition must be modified to

$$\frac{E_T}{S_1(0) + K_{m1} + I(0) + K_{Ia}} \ll 1 \tag{4.30}$$

(Segel, 1988), with K_{Ia} denoting the inhibition constant defined in Eq. (2.31). It is clear that the higher the inhibitor concentration, the smaller will be the effective enzyme concentration, so that the limit of validity of the quasi-steady-state hypothesis is shifted to higher total enzyme levels.

4.3. THE RAPID-EQUILIBRIUM APPROXIMATION

Consider again the reaction system shown in Scheme 9 (Section 4.2) and assume now that reaction 2 is reversible. This reaction scheme can then represent, for example, part of the glycolytic pathway with S_1 and S_2 standing for glucose-6-phosphate (G6P) and fructose-6-phosphate (F6P). For simplicity's sake, we assume the rate of reaction 2 to be given by

$$v_2 = k_2 S_1 - k_{-2} S_2. \tag{4.31}$$

The system equations read

$$\frac{dS_1}{dt} = k_1 P_1 - k_2 S_1 + k_{-2} S_2, \tag{4.32a}$$

$$\frac{dS_2}{dt} = k_2 S_1 - k_{-2} S_2 - k_3 S_2. \tag{4.32b}$$

Assume further that

$$k_2, k_{-2} >> k_3. \tag{4.33}$$

As can be seen in Figure 4.2, the concentrations S_1 and S_2 will reach, after a short initial time span, such values that the ratio S_2/S_1 approximately equals the equilibrium constant,

$$\frac{S_2}{S_1} \cong q_2 = \frac{k_2}{k_{-2}}. \tag{4.34}$$

This is because any excess of S_1 or S_2 deviating from this ratio will be removed very rapidly due to the high rate constants of reaction 2. Accordingly, this reaction can be considered to subsist nearly in equilibrium, although a nonzero flux may go through it. As can be seen from Eq. (4.31) and inequality (4.33), v_2 may differ considerably from zero even if the concentration ratio S_2/S_1 satisfies relation (4.34).

The quasi-equilibrium relation (4.34) entails that the system equation (4.32) can be simplified to a system of dimension one because the two variables are (approximately) proportional to each other. Summation of Eqs. (4.32a) and (4.32b) gives

$$\frac{d(S_1 + S_2)}{dt} = k_1 P_1 - k_3 S_2. \tag{4.35}$$

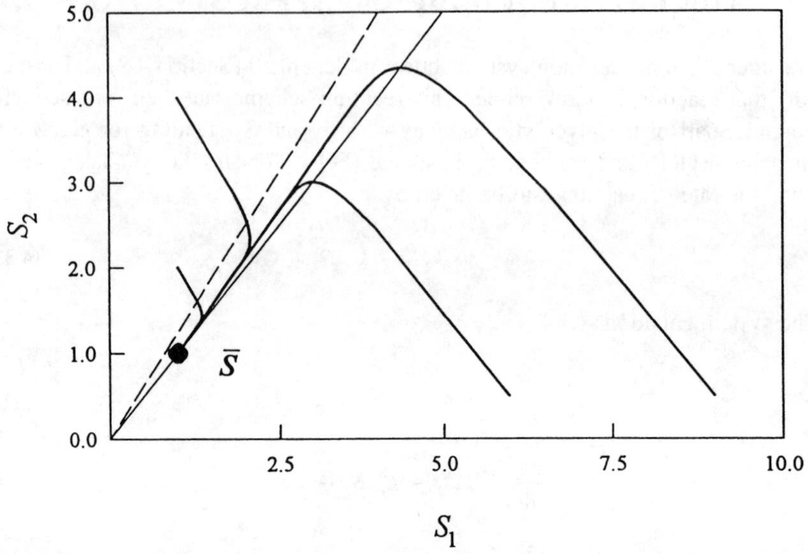

Figure 4.2 Phase plot S_1 versus S_2 illustrating the rapid-equilibrium approximation as applied to reaction 2 in the reaction chain shown in Scheme 9. Parameter values: $P_1 = 1$, $k_1 = 7$, $k_2 = 5$, $k_{-2} = 4$, $k_3 = 1$. The thick lines represent trajectories for different initial conditions. The thin straight line is the nullcline for S_2. The thin broken lines is given by the rapid-equilibrium condition (4.34).

Using the rapid-equilibrium relation (4.34), we can rewrite the left-hand side of Eq. (4.35) as $(1/q_2 + 1)\mathrm{d}S_2/\mathrm{d}t$. This gives the reduced system equation

$$\frac{\mathrm{d}S_2}{\mathrm{d}t} = \frac{q_2(k_1P_1 - k_3S_2)}{1 + q_2}. \tag{4.36}$$

The dynamics of the system is approximated by this equation after the initial relaxation period. It is worth noting that for the considered system, also the quasi-steady-state equation can be applied. $\mathrm{d}S_2/\mathrm{d}t \cong 0$ gives $S_2 = k_2S_1/(k_{-2} + k_3)$. As can be seen in Figure 4.2, this approximation is even better than the rapid-equilibrium approximation. However, construction of the quasi-steady-state line requires knowledge of three kinetic parameters, whereas the rapid-equilibrium line can be computed by knowledge of the equilibrium constant q_2 only. Parameter values in Figure 4.2 have been chosen so as to clearly show the difference between the two lines. Normally, one applies the rapid-equilibrium approximation only in case of a more distinct time hierarchy, so that the two lines lie more closely to each other.

The idea underlying the development of the rapid-equilibrium approximation dates back at least to the beginning of our century (Wegscheider, 1900). We will now outline, in a general way, the approximation method to treat systems involving several fast, reversible reactions (see also Schauer and Heinrich, 1983; Liao and Lightfoot, 1988b; and for the case of tracer kinetics, R. Schuster *et al.*, 1992).

Reaction velocities can be classified into slow rates, v_i^{s}, and fast rates, v_j^{f}, fulfilling the inequality

$$\left|v_i^{\mathrm{s}}\right| \ll \left|v_j^{\mathrm{f}}\right|. \tag{4.37}$$

Fast reactions are characterized by high rate constants (or analogous kinetic parameters in the case of rate laws more complicated than simple mass-action kinetics, such as maximal activities in the case of Michaelis–Menten rate laws) and low time constants. Clearly, relation (4.37) cannot be fulfilled in the whole concentration space because the rates also depend on concentrations. For example, there are submanifolds in the concentration space where one or more fast reactions are at equilibrium ($v_j^{\mathrm{f}} = 0$). In the largest part of this space, however, high rate constants do imply high reaction rates, so that inequality (4.37) holds. Accordingly, we can partition the rate vector $\boldsymbol{v}$ as

$$\boldsymbol{v} = \begin{pmatrix} \boldsymbol{v}^{\mathrm{s}} \\ \boldsymbol{v}^{\mathrm{f}} \end{pmatrix}, \tag{4.38}$$

where the components of $\boldsymbol{v}^{\mathrm{s}}$ and $\boldsymbol{v}^{\mathrm{f}}$ correspond to the slow and fast reactions, respectively. Likewise, we can decompose the stoichiometry matrix as

$$\mathbf{N} = (\mathbf{N}^{\mathrm{s}}\ \mathbf{N}^{\mathrm{f}}). \tag{4.39}$$

$\mathbf{N}^{\mathrm{f}}$ is then the stoichiometry matrix of the *fast subsystem.* Due to the decomposition (4.39), rank ($\mathbf{N}^{\mathrm{f}}$) is less than, or equal to, the rank of $\mathbf{N}$.

Now we rescale the fast rates by a small parameter, $\mu \ll 1$,

$$\hat{\boldsymbol{v}}^{\mathrm{f}} = \mu \boldsymbol{v}^{\mathrm{f}} \tag{4.40}$$

so that they get the same, or a smaller, order of magnitude as the components of $\boldsymbol{v}^{\mathrm{s}}$. Inserting Eqs. (4.38), (4.39), and (4.40) into the system equation (2.8) yields

$$\frac{\mathrm{d}\boldsymbol{S}}{\mathrm{d}t} = \mathbf{N}^{\mathrm{s}}\boldsymbol{v}^{\mathrm{s}}(\boldsymbol{S}) + \frac{1}{\mu}\mathbf{N}^{\mathrm{f}}\hat{\boldsymbol{v}}^{\mathrm{f}}(\boldsymbol{S}). \tag{4.41}$$

This is a singularly perturbed differential equation system because, in the limit $\mu \to 0$, the dimension of the system decreases, as will become clear below. The conditions under which such equation systems can be approximated by consideration of the limit $\mu \to 0$ are given in Tikhonov's Theorem (see Section 4.2). As far as Eq. (4.41) is concerned, application of this theorem is facilitated by a variable transformation which generates, out of this equation, some differential equations (say, a in number) that do not involve the factor $1/\mu$. An example of such a transformation is the summation of the Eqs. (4.32a) and (4.32b). As this independence is to hold irrespective of the special values of kinetic parameters, as long as they are consistent with condition (4.37), it must be a property resulting only from stoichiometry. Accordingly, an $n \times n$ transformation matrix $\mathbf{T}$ consisting of two submatrices is used,

$$\mathbf{T} = \begin{pmatrix} \mathbf{T}^{\mathrm{s}} \\ \mathbf{T}^{\mathrm{f}} \end{pmatrix}, \tag{4.42}$$

so that one submatrix transforms the matrix $\mathbf{N}^{\mathrm{f}}$ to a null matrix,

$$\mathbf{T}^{\mathrm{s}}\mathbf{N}^{\mathrm{f}} = \mathbf{0}. \tag{4.43}$$

We denote the new variable vector by $\boldsymbol{Y}$,

$$\boldsymbol{Y} = \begin{pmatrix} \boldsymbol{Y}^{\mathrm{s}} \\ \boldsymbol{Y}^{\mathrm{f}} \end{pmatrix} = \mathbf{T}\boldsymbol{S} = \begin{pmatrix} \mathbf{T}^{\mathrm{s}}\boldsymbol{S} \\ \mathbf{T}^{\mathrm{f}}\boldsymbol{S} \end{pmatrix}, \tag{4.44}$$

with dim $(\boldsymbol{Y}^{\mathrm{s}}) = a$ and dim $(\boldsymbol{Y}^{\mathrm{f}}) = n - a$. $\mathbf{T}$ must be nonsingular in order that the time course $\boldsymbol{S}(t)$ of the original variables can be determined from $\mathbf{Y}(t)$. Consequently, both $\mathbf{T}^{\mathrm{s}}$ and $\mathbf{T}^{\mathrm{f}}$ have to be full rank.

Equation (4.43) expresses linear dependencies among the rows of $\mathbf{N}^{\mathrm{f}}$. As there are n - rank $(\mathbf{N}^{\mathrm{f}})$ such dependencies and $\mathbf{T}^{\mathrm{s}}$ has full rank,

$$\alpha \leq n - \mathrm{rank}(\mathbf{N}^{\mathrm{f}}). \tag{4.45}$$

Under consideration of Eqs. (4.43) and (4.44), we transform Eq. (4.41) to

$$\frac{\mathrm{d}\boldsymbol{Y}^{\mathrm{s}}}{\mathrm{d}t} = \mathbf{T}^{\mathrm{s}}\frac{\mathrm{d}\boldsymbol{S}}{\mathrm{d}t} = \mathbf{T}^{\mathrm{s}}\mathbf{N}^{\mathrm{s}}\boldsymbol{v}^{\mathrm{s}}, \tag{4.46a}$$

$$\frac{\mathrm{d}\mathbf{Y}^{\mathrm{f}}}{\mathrm{d}t} = \mathbf{T}^{\mathrm{f}}\frac{\mathrm{d}\boldsymbol{S}}{\mathrm{d}t} = \mathbf{T}^{\mathrm{f}}\mathbf{N}^{\mathrm{s}}\boldsymbol{v}^{\mathrm{s}} + \frac{1}{\mu}\mathbf{T}^{\mathrm{f}}\mathbf{N}^{\mathrm{f}}\hat{\boldsymbol{v}}^{\mathrm{f}}. \tag{4.46b}$$

Multiplying Eq. (4.46b) by μ and taking the limit $\mu \to 0$, we obtain

$$\mathbf{T}^{\mathrm{f}}\mathbf{N}^{\mathrm{f}}\hat{\boldsymbol{v}}^{\mathrm{f}} = \boldsymbol{0}. \tag{4.47}$$

Equations (4.43) and (4.47) can be combined to

$$\begin{pmatrix} \mathbf{T}^{s} \\ \mathbf{T}^{f} \end{pmatrix} \mathbf{N}^{f} \hat{\boldsymbol{v}}^{f} = \boldsymbol{0}. \tag{4.48}$$

As **T** must be nonsingular, Eq. (4.48) implies

$$\mathbf{N}^{f} \hat{\boldsymbol{v}}^{f} = \boldsymbol{0}. \tag{4.49}$$

Based on Tikhonov's Theorem (see Section 4.2), the differential equation system (4.46a) and (4.46b) can, in the limit $\mu \to 0$, be replaced by the algebro-differential equation system (4.46a) and (4.49). Equation (4.49) may serve to eliminate the $n - a$ fast variables in Eq. (4.46a). At most rank ($\mathbf{N}^f$) variables can be eliminated, because this corresponds to the maximum number of independent equations in Eq. (4.49). This implies

$$n - a \leq \operatorname{rank}(\mathbf{N}^{f}), \tag{4.50}$$

which together with relation (4.45) gives

$$\alpha = n - \operatorname{rank}(\mathbf{N}^{f}). \tag{4.51}$$

The approximation under consideration can only be applied if condition (T2) of Theorem 4A is fulfilled. In the framework of linear stability analysis, this condition is equivalent to the condition that all eigenvalues of the Jacobian of the adjoint system

$$\frac{\mathrm{d}\boldsymbol{Y}^{f}}{\mathrm{d}t'} = \mathbf{T}^{f} \mathbf{N}^{f} \hat{\boldsymbol{v}}^{f} \tag{4.52}$$

have negative real parts at the considered fixpoint. Note that compared with Eq. (4.46b), the term $\mu \mathbf{T}^{f} \mathbf{N}^{s} \boldsymbol{v}^{s}$ has been omitted on the right-hand side of Eq. (4.52), because for sufficiently small μ, it does not have any effect on the signs of the eigenvalues (cf. Levin and Levinson, 1954).

Usually, the quasi-steady-state equation (4.49) is solved by concentration values which even fulfill the more restrictive equilibrium condition

$$\hat{\boldsymbol{v}}^{f}(\boldsymbol{S}) = \boldsymbol{0}. \tag{4.53}$$

However, such a solution only exists if the fast subsystem given by the differential equation $\mathrm{d}\boldsymbol{S}/\mathrm{d}t = \mathbf{N}^{f} \boldsymbol{v}^{f}$ is detailed balanced. Such a situation occurs when the fast subsystem is, upon canceling of the slow reactions, decoupled from the inputs and outputs and becomes closed, which implies detailed balancing. The steady state of the fast subsystem is then stable (see Section 3.3.1), so that condition (T2) is fulfilled.

According to Eqs. (4.43) and (4.51), we can choose $\mathbf{T}^s$ such that it expresses a complete set of independent conservation relations of the fast subsystem (see Section 3.1.1). The variables Y_i^s are therefore the conservation quantities of the fast subsystem. They are often called *slow moieties* or *pool variables*. Note that all metabolites not participating in any fast reactions represent pool variables on their own.

Park (1974) advocates to admit only transformation matrices $\mathbf{T}^s$ with non-negative, integer entries because the pool variables are to represent concentrations of real atom groups (moieties). Integer entries can always be guaranteed, as long as the stoichiometry matrix of the fast subsystem only contains integer elements. Non-negativity can be met if the fast subsystem is a closed system, otherwise it is not always fulfilled (see Section 3.1.2). Formally, the two conditions need not be satisfied in order that the rapid-equilibrium approximation be applied [cf. the model of erythrocyte metabolism presented by R. Schuster *et al.* (1988)], but the pool variables are easier to interpret if they are linear combinations of concentrations with non-negative, integer coefficients.

As was shown in Section 3.1.1, conservation matrices can always be chosen so as to contain the identity matrix,

$$\mathbf{T}^s = (\hat{\mathbf{T}}^s \ \mathbf{I}). \tag{4.54}$$

The matrix $\mathbf{T}^f$ necessary to calculate the fast variables, Y_i^f, can be chosen arbitrarily, only subject to the constraint that $\mathbf{T}$ be non-singular. If $\mathbf{T}^s$ is chosen as in Eq. (4.54), a feasible choice of $\mathbf{T}^f$ is

$$\mathbf{T}^f = (\mathbf{I} \ \mathbf{0}). \tag{4.55}$$

This means that some original variables S_i may be used as fast variables Y_i^f. It can easily be shown that the validity of conditions (T1), (T2), and (T3) is independent of the choice of matrices $\mathbf{T}^s$ and $\mathbf{T}^f$.

The elimination of variables by rapid-equilibrium approximation is feasible in two ways according to whether the algebro-differential equation system is expressed in terms of the original concentration variables, S_i, which directly enter the rate laws, or the transformed variables, Y_i. When this equation system is to be written in terms of the original variables, the set of concentration variables should be decomposed into subsets of independent and dependent variables (S^{ind} and S^{dep}, respectively). As long as all reactions are reversible ($0 < q_i < \infty$) and obey the law of mass action [Eq. (2.12)], we can write the equilibrium condition (4.53) in logarithmic form:

$$(\mathbf{N}^f)^{\mathrm{T}} \ln \boldsymbol{S} = \ln \tilde{\boldsymbol{q}}^f \tag{4.56}$$

[cf. Eq. (3.63)], where $\tilde{\boldsymbol{q}}^{\mathrm{f}}$ is the vector of apparent equilibrium constants of the fast reactions. We now rearrange the rows and columns of $\mathbf{N}^{\mathrm{f}}$ so that the submatrix of dimension rank($\mathbf{N}^{\mathrm{f}}$) $\times$ rank($\mathbf{N}^{\mathrm{f}}$) on the upper left of $\mathbf{N}^{\mathrm{f}}$ is nonsingular. Choosing the concentration variables corresponding to this submatrix as dependent variables, they can be calculated in terms of the independent concentrations, by solving Eq. (4.56) for $\boldsymbol{S}^{\mathrm{dep}}$. This gives

$$\boldsymbol{S}^{\mathrm{dep}} = \boldsymbol{S}^{\mathrm{dep}}(\boldsymbol{S}^{\mathrm{ind}}). \tag{4.57}$$

The fact that Eq. (4.56) is linear in the logarithmic concentrations is in favor of keeping to the original variables. In the alternative version, where the algebro-differential equation system is expressed in terms of the transformed variables, construction of the function $\boldsymbol{Y}^{\mathrm{f}}$ ($\boldsymbol{Y}^{\mathrm{s}}$) would have to be made on the basis of the conservation equation of the fast subsystem, $\boldsymbol{Y}^{\mathrm{s}} = \mathbf{T}^{\mathrm{s}}\,\boldsymbol{S}$, which is involved in Eq. (4.44), and the equilibrium condition (4.56). Because the former equation is linear in $\boldsymbol{S}$ and the latter equation is linear in ln$\boldsymbol{S}$, they cannot, in general, be combined to give an explicit expression $\boldsymbol{Y}^{\mathrm{f}}(\boldsymbol{Y}^{\mathrm{s}})$. Nevertheless, the values of the fast variables Y_i^{f} are uniquely determined by the slow variables on the basis of Eqs. (4.44) and (4.56). This is due to the fact that for systems of reactions endowed with the generalized mass-action kinetics (2.15), the equilibrium concentrations are uniquely determined in terms of the conservation quantities and equilibrium constants (Horn and Jackson, 1972; S. Schuster and R. Schuster, 1989). The computation of equilibrium concentrations is the subject of chemical reaction equilibrium analysis, which provides sophisticated computation algorithms (cf. Smith and Missen, 1992).

Returning to the original concentration variables, we now outline a method for treating the algebro-differential equation system resulting from the rapid-equilibrium assumption. Inserting Eq. (4.57) into Eq. (4.46a) gives

$$\mathbf{T}^{\mathrm{s}}\begin{pmatrix} \dfrac{\mathrm{d}\boldsymbol{S}^{\mathrm{ind}}}{\mathrm{d}t} \\[2ex] \dfrac{\partial \boldsymbol{S}^{\mathrm{dep}}}{\partial \boldsymbol{S}^{\mathrm{ind}}}\dfrac{\mathrm{d}\boldsymbol{S}^{\mathrm{ind}}}{\mathrm{d}t} \end{pmatrix} = \mathbf{T}^{\mathrm{s}}\mathbf{N}^{\mathrm{s}}\boldsymbol{v}^{\mathrm{s}}(\boldsymbol{S}^{\mathrm{ind}}, \boldsymbol{S}^{\mathrm{dep}}(\boldsymbol{S}^{\mathrm{ind}})). \tag{4.58}$$

Note that the partial derivative of the vector S^{dep} with respect to the vector S^{ind} is meant to denote the matrix $(\partial S_i^{\mathrm{dep}}/\partial S_j^{\mathrm{ind}})$. The left-hand side of the equation system (4.58) is linear in $\mathrm{d}S^{\mathrm{ind}}/\mathrm{d}t$ and can be solved for these variables by matrix inversion,

$$\frac{\mathrm{d}\boldsymbol{S}^{\mathrm{ind}}}{\mathrm{d}t} = \mathbf{H}^{-1}\mathbf{T}^{\mathrm{s}}\mathbf{N}^{\mathrm{s}}\boldsymbol{v}^{\mathrm{s}}(\boldsymbol{S}^{\mathrm{ind}}, \boldsymbol{S}^{\mathrm{dep}}(\boldsymbol{S}^{\mathrm{ind}})) \tag{4.59a}$$

with

$$\mathbf{H} = \mathbf{T}^{\mathrm{s}} \begin{pmatrix} \mathbf{I} \\ \dfrac{\partial S^{\mathrm{dep}}}{\partial S^{\mathrm{ind}}} \end{pmatrix}. \tag{4.59b}$$

Equation (4.59a) is a differential equation system that can be used to describe the system behavior under the rapid-equilibrium assumption. Numerical integration of this differential equation system has several advantages, compared to that of the original system. First, Eq. (4.59a) only has dimension n-rank($\mathbf{N}^{\mathrm{f}}$). Second, it is not as stiff as the original system because the high kinetic constants have been eliminated. The fast reactions only affect the system behavior via their equilibrium constants. Third, for the same reason, knowledge of the kinetic parameters of the fast reactions is no longer necessary.

Let us illustrate this method by way of two examples. First, we again consider Scheme 9. As stated above, reaction 2 can stand for phosphoglucoisomerase (PGI). This is a quasi-equilibrium enzyme, as was shown by measuring the concentration ratio *F6P/G6P in vivo,* which turned out to be nearly equal to the equilibrium constant, $q_{\mathrm{PGI}} = 0.4$ (cf. Reich and Selkov, 1981). The fast subsystem has the conservation relation $S_1 + S_2 = \mathrm{const}$. The pool variable is therefore $Y_1^{\mathrm{s}} = S_1 + S_2$ [cf. also Eq. (4.35)], which can readily be interpreted as pool of the hexose monophosphates in glycolysis. More formally, this pool variable can be deduced from the partitioned stoichiometry matrix

$$(\mathbf{N}^{\mathrm{s}}\ \mathbf{N}^{\mathrm{f}}) = \begin{pmatrix} 1 & 0 & \vdots & -1 \\ 0 & -1 & \vdots & 1 \end{pmatrix} \tag{4.60}$$

(with the second and third columns interchanged relative to the numbering of reactions). An admissible matrix $\mathbf{T}^{\mathrm{s}}$ fulfilling Eq. (4.43) is

$$\mathbf{T}^{\mathrm{s}} = (1\ \ 1). \tag{4.61}$$

The transformation matrix yielding the fast variables can be chosen, for example, as

$$\mathbf{T}^{\mathrm{f}} = (1\ \ 0). \tag{4.62}$$

S_1 thus enters both the slow variable and the fast variable $Y_1^{\mathrm{f}} = S_1$. The adjoint system reads

$$\frac{\mathrm{d}Y_1^{\mathrm{f}}}{\mathrm{d}t'} = -\hat{v}_2(S_1, S_2). \tag{4.63}$$

The fixpoint of this equation is given by $v_2 = 0$. In this equilibrium state, which is stable, relation (4.34) holds true. Note that the approximation is valid not only for the linear kinetics (4.31) but for any rate law comprised in the general mass-action kinetics. With the help of Eq. (4.34), the fast variable can be expressed in terms of the slow variable as

$$Y_1^{\mathrm{f}} = S_1 = \frac{Y_1^{\mathrm{s}}}{1 + q_2}. \tag{4.64}$$

One can also use the original variables S_1 and S_2 with, say, S_1 being the dependent variable, as we have done above [cf. Eq. (4.36)].

Care must be taken if the whole system involves conservation relations. Because of the decomposition of N into N^{s} and N^{f}, any conservation relation of the whole system also holds in the fast subsystem. For illustration of this case, consider a scheme of threonine synthesis in *E. coli* (cf. Gottschalk, 1986), as depicted in Figure 4.3. Aspartate semialdehyde dehydrogenase and homoserine dehydrogenase can be considered as quasi-equilibrium enzymes. To explain the method in a stepwise way, we consider first a fast subsystem which consists only of the former reaction. The fast subsystem then has the following conservation relations:

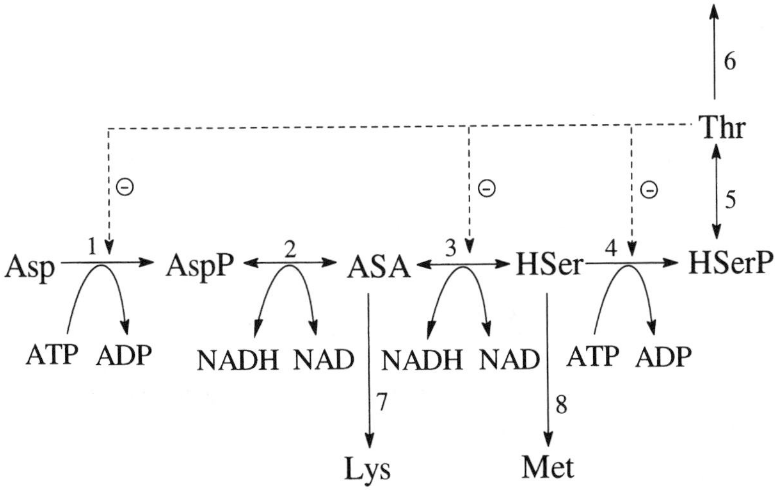

Figure 4.3 Scheme of threonine synthesis. Reactions: (1) aspartokinase I, II and III (EC 2.7.2.4); (2) aspartate semialdehyde dehydrogenase (EC .1.2.1.11); (3) homoserine dehydrogenase (EC 1.1.1.3); (4) homoserine kinase (EC 2.7.1.39); (5) threonine synthase (EC 4.2.99.2); (6) threonine consumption; (7 and 8) branches leading to lysine and methionine, respectively. Metabolites: Asp, aspartate; AspP, 4-phospho-aspartate; ASA, aspartate semialdehyde; HSer, homoserine; HSerP, homoserine phosphate; Thr, threonine; Lys, lysine; Met, methionine. The dashed arrows indicate negative feedback effects of threonine.

$$AspP + ASA = Y_1^s, \tag{4.65a}$$

$$AspP + NAD = Y_2^s, \tag{4.65b}$$

$$NADH + NAD = Y_3^s, \tag{4.65c}$$

in addition to $S_i = const.$ for those metabolites only involved in slow reactions. The conservation relation (4.65c) holds not only in the fast subsystem but also in the whole scheme. Note that relation (4.65b) cannot be interpreted in terms of conservation of moieties. If such an interpretation is desired, one could replace it by the relation $NADH + ASA = Y_2^s$, which reflects conservation of a hydrogen. However, the outlined approximation method even works if a relation with a negative coefficient is used (e.g., $AspP - NADH = Y_2^s$).

According to the conservation relations (4.65a) and (4.65b), we construct linear combinations of the governing equations,

$$\frac{dY_1^s}{dt} = \frac{d}{dt}(AspP + ASA) = v_1 - v_3 - v_7, \tag{4.66a}$$

$$\frac{dY_2^s}{dt} = \frac{d}{dt}(AspP + NAD) = v_1 + v_3, \tag{4.66b}$$

which do not involve the fast reaction rate v_2. The equilibrium condition for the aspartate semialdehyde dehydrogenase reaction (ASADH) gives

$$NADH = \frac{ASA \cdot NAD}{AspP \cdot q_{\text{ASADH}}}. \tag{4.67}$$

With the conservation relation (4.65c) and $Y_3^s = N_T$, this yields

$$NAD = \frac{N_T \cdot q_{\text{ASADH}} \cdot AspP}{ASA + q_{\text{ASADH}} \cdot AspP}. \tag{4.68}$$

Substituting this equation into Eq. (4.66b) and putting it together with Eq. (4.66a), we obtain by the chain rule of differentiation

$$\begin{pmatrix} \dfrac{dAspP}{dt} \\ \dfrac{dASA}{dt} \end{pmatrix} = \begin{pmatrix} 1 & 1 \\ 1 + \dfrac{N \cdot q_{\text{ASADH}} \cdot ASA}{Q^2} & -\dfrac{N \cdot q_{\text{ASADH}} \cdot AspP}{Q^2} \end{pmatrix}^{-1} \begin{pmatrix} v_1 - v_3 - v_7 \\ v_1 + v_3 \end{pmatrix} \tag{4.69}$$

with

$$Q = ASA + q_{\mathrm{ASADH}} \cdot AspP. \tag{4.70}$$

Now we take into account that also homoserine dehydrogenase is fast. The fast subsystem has the stoichiometry matrix

$$\mathbf{N}^{\mathrm{f}} = \begin{pmatrix} -1 & 0 \\ 1 & -1 \\ 0 & 1 \\ -1 & -1 \\ 1 & 1 \end{pmatrix}, \tag{4.71}$$

according to the following numbering of metabolites: (1) AspP, (2) ASA, (3) HSer, (4) NADH, (5) NAD. To this matrix, the conservation matrix

$$\mathbf{T}^{\mathrm{s}} = \begin{pmatrix} 1 & 1 & 1 & 0 & 0 \\ 1 & 0 & -1 & -1 & 0 \\ 0 & 0 & 0 & 1 & 1 \end{pmatrix} \tag{4.72}$$

can be attached. Accordingly, the following linear combination of governing equations does not comprise any fast reaction rates:

$$\frac{\mathrm{d}}{\mathrm{d}t}(AspP + ASA + HSer) = v_1 - v_4 - v_7, \tag{4.73a}$$

$$\frac{\mathrm{d}}{\mathrm{d}t}(AspP - HSer - NADH) = v_1 + v_4 - v_8, \tag{4.73b}$$

and an equation saying that the sum of *NAD* and *NADH* does not change in time.

Besides the equilibrium condition (4.67), now the relation

$$\frac{HSer \cdot NAD}{ASA \cdot NADH} = q_{\mathrm{HSDH}}, \tag{4.74}$$

holds true (HSDH stands for homoserine dehydrogenase). They give

$$HSer = \frac{ASA^2 \cdot q_{\mathrm{HSDH}}}{AspP \cdot q_{ASADH}}. \tag{4.75}$$

Substituting these equilibrium conditions into Eqs. (4.73a) and (4.73b), we obtain a two-dimensional equation system, which can be solved for $\mathrm{d}AspP/\mathrm{d}t$ and $\mathrm{d}ASA/\mathrm{d}t$.

In biochemical and biophysical modeling, the situation may occur that fast processes lead to side constraints other than chemical equilibria (e.g., quasi-electroneutrality conditions or osmotic equilibria). It may then occur that the fast variables cannot be eliminated analytically from these side constraints. A method to circumvent this difficulty is by differentiating the side conditions with respect to time. This gives additional differential equations, which can be integrated together with the system equations (2.8), by use of appropriate initial conditions fulfilling the side conditions. These conditions are then automatically fulfilled for any subsequent point in time. This method was used, for example, by Brumen and Heinrich (1984) to include an osmotic constraint into a model of erythrocyte metabolism.

4.4. MODAL ANALYSIS

A useful method for analyzing the behavior of cellular biological systems is provided by modal analysis. The central idea of this technique is to linearize the governing differential equations and to perform a linear transformation of the component variables, so that the equations become uncoupled of each other and move on separate time scales (Palsson *et al.,* 1984, 1985; Liao and Lightfoot Jr., 1987, 1988a, 1988b). Time constants are here used in the sense of definition (4.6) based on the eigenvalues of the Jacobian. The method is based on the normal mode analysis in classical mechanics.

The first step is to choose a reference state $\boldsymbol{S}_0$ for linearization of the system equations (2.8). $\boldsymbol{S}_0$ may be, for example, a steady state or the initial state of the system. For simplicity's sake, we first consider the case that $\boldsymbol{S}_0$ is a stable steady state and will discuss the general case at the end of this section. Linearization then yields Eq. (2.82) with $\delta\boldsymbol{S} = \boldsymbol{S} - \boldsymbol{S}_0$ (cf. Section 2.3.2). A similarity transformation is now applied to the Jacobian $\mathbf{M}$,

$$\mathbf{W}\mathbf{M}\mathbf{W}^{-1} = \boldsymbol{\Lambda}, \tag{4.76}$$

with $\boldsymbol{\Lambda}$ containing the eigenvalues of $\mathbf{M}$ as diagonal elements. In the case that $\mathbf{M}$ has multiple eigenvalues, it may not be diagonalizable. $\boldsymbol{\Lambda}$ is then the Jordan normal form of $\mathbf{M}$ (cf. Gantmacher, 1959). $\mathbf{W}$ is called modal matrix and is constructed by using the eigenrows of $\mathbf{M}$ (i.e., the eigenvectors of the transposed matrix $\mathbf{M}^{\mathrm{T}}$). The inverse matrix, $\mathbf{W}^{-1}$, encompasses the eigenvectors of $\mathbf{M}$ as columns.

$\mathbf{W}$ transforms the vector $\boldsymbol{S}$ into a vector $\boldsymbol{X}$,

$$\boldsymbol{X} = \mathbf{W}\boldsymbol{S}. \tag{4.77}$$

Provided that **M** is diagonalizable, the solution of Eq. (2.82) can be written, under consideration of $\delta\mathbf{S} = \boldsymbol{S} - \boldsymbol{S}_0$, as

$$\boldsymbol{S}(t) = \mathbf{W}^{-1}\exp(\Lambda t)\mathbf{W}(\boldsymbol{S}(0) - \boldsymbol{S}_0) + \boldsymbol{S}_0 \tag{4.78}$$

[cf. Eq. (2.84)]. From Eqs. (4.77) and (4.78), the solution for the components of $\boldsymbol{X}(t)$ can be written as

$$x_j(t) = (x_j(0) - x_{0,j})\exp(\lambda_j t) + x_{0,j}. \tag{4.79}$$

Accordingly, each component of the vector $\boldsymbol{X}$ changes in time according to a "pure" exponential function, with the time constant $-1/\mathrm{Re}\lambda_j$. If **M** is not diagonalizable, some functions $x_j(t)$ involve polynomial functions,

$$x_j(t) = (x_j(0) - x_{0,j})\exp(\lambda_j t)\sum_{i=0}^{p_j} a_i^{(j)} t^i + x_{0,j}, \tag{4.80}$$

where p_j are the multiplicities of the eigenvalues λ_j, and $a_i^{(j)}$ are constant factors. As the long-term behavior is determined by the exponential part of this function, one can consider each variable x_j to have the time constant $-1/\mathrm{Re}\lambda_j$, as in the case when **M** is diagonalizable.

The equation $\boldsymbol{S}(t) = \mathbf{W}^{-1}\,\mathbf{X}(t)$ [cf. Eq. (4.77)] shows that the components of the vector **X** are the time-dependent weights of the eigenvectors of **M** in the solution $\mathbf{S}(t)$. On the other hand, the components of the eigenrows of the Jacobian **M** are the weights of the concentrations $\boldsymbol{S}_i$ in the "pool" variables $x_j(t)$, which move on time scales corresponding to the eigenvalues of **M**. We will therefore call these eigenrows *weighting vectors,* $\mathbf{w}_k$.

The main purpose of modal analysis is to detect the various time scales of the system. For example, by applying this procedure to red blood cell metabolism, Liao and Lightfoot Jr. (1987, 1988b) showed that the time constants of this system cover a range from 0.9 ms to 12 h. In addition, modal analysis provides weighted sums (pools) of concentrations which move on the detected time scales. It is particularly interesting to evaluate the largest and smallest time scales relevant for the behavior of each substance involved in the system under study.

Modal analysis also provides information about well-separated time scales (temporal hierarchy). Detecting the fast modes can help to choose and apply suitable approximation methods. To this end, one first chooses a reference point $\boldsymbol{S}_0$. This step can hardly be circumvented in the case of nonlinear systems, because the classification of slow and fast reactions then depends on the concentration values. However, the approximations based on modal analysis are remarkably insensitive to the choice of the reference state. Next, the Jacobian, its eigenvalues λ_i, and eigenrows are calculated. Note that in the case of linear systems, the

Jacobian does not depend on concentrations. The eigenvalues are classified according to whether or not

$$-\frac{1}{\mathrm{Re}(\lambda_k)} < \tau_{\mathrm{m}}, \tag{4.81}$$

where τ_{m} is the minimum time constant of interest for the specific situation under study (given, for example, by the time resolution of experimental measurement). When studying the behavior of a metabolic pathway in the time range of, say, seconds to hours, one should define $\tau_{\mathrm{m}} = 1$ s. The weighting vectors w_k corresponding to those λ_k fulfilling condition (4.81) can be considered to have relaxed in a period shorter than the time scale of interest. Owing to Eqs. (4.77) and (4.79), this leads to the quasi-steady-state relation

$$\boldsymbol{w}_k \frac{\mathrm{d}\boldsymbol{S}}{\mathrm{d}t} \cong 0 \tag{4.82}$$

for times larger than the time constant $-1/\mathrm{Re}\lambda_k$ [with k being an index for which relation (4.81) is fulfilled]. This is equivalent to

$$\boldsymbol{w}_k \mathbf{N}\boldsymbol{v}(\boldsymbol{S}) \cong 0. \tag{4.83}$$

Because condition (4.81) excludes that $\boldsymbol{w}_k$ is an eigenrow of the Jacobian $\mathbf{M}$ belonging to the eigenvalue zero, we have

$$\boldsymbol{w}_k \mathbf{N} \neq \boldsymbol{0}^{\mathrm{T}}. \tag{4.84}$$

Due to this relation, the algebraic equation (4.83) couples several concentrations to each other, so that the system of governing differential equations can be reduced in dimension. This is particularly suitable for linear systems, in which the Jacobian and, hence, the modal matrix do not depend on concentrations.

For illustration, consider again the reaction system of Scheme 9 (Section 4.2) with the parameter values

$$k_1 = 1\ \mathrm{s}^{-1}, \quad k_2 = 100\ \mathrm{s}^{-1}, \quad k_{-2} = 50\ \mathrm{s}^{-1}, \quad k_3 = 1\ \mathrm{s}^{-1}; \tag{4.85}$$

that is, reaction 2 is assumed to be fast. The eigenrows of the modal matrix then read $(-1.99, 1)$ and $(1.007, 1)$, and the corresponding eigenvalues $\lambda_1 = -150$ s^{-1} and $\lambda_2 = -0.67\ \mathrm{s}^{-1}$, respectively. When τ_{m} is fixed to be, say, 0.1 s, the first mode can be classified to be fast. Equation (4.83) reads, in this case,

$$-1.99(P_1 - 100S_1 + 50S_2) + 100S_1 - 50S_2 - S_2 = 0, \tag{4.86}$$

which gives

$$S_2 = \frac{2.99 \times 100S_1 - 1.99P_1}{2.99 \times 50 + 1} \cong 2S_1. \tag{4.87}$$

This is an algebraic relation between S_1 and S_2, which approximately holds true after an initial time span of about $-1/\lambda_2 \cong 6.7$ ms. Interestingly, Eq. (4.87) is approximately equivalent to the quasi-equilibrium relation (4.34), due to the fact that k_2 and k_{-2} are large.

Obviously, there must exist interrelations between the modal analysis and rapid-equilibrium approximation. Both methods work with linear combinations of concentrations (pool variables). In a first attempt to elucidate this interrelation, it was shown that as the reactions classified to be fast become infinitely fast, each modal matrix **W** tends to an admissible transformation matrix **T** used in rapid-equilibrium approximation (Liao and Lightfoot Jr., 1988b; R. Schuster and S. Schuster, 1991).

Finally, we consider the general case that $\boldsymbol{S}_0$ is not necessarily a stable steady state. The linearized system equation then reads

$$\frac{\mathrm{d}}{\mathrm{d}t}(\boldsymbol{S} - \boldsymbol{S}_0) = \boldsymbol{f}(\boldsymbol{S}_0) + \mathbf{M}(\boldsymbol{S} - \boldsymbol{S}_0) \tag{4.88}$$

and, in the transformed variable vector $\boldsymbol{X}$,

$$\frac{\mathrm{d}\boldsymbol{X}}{\mathrm{d}t} = \boldsymbol{\Lambda}\boldsymbol{X} + \mathbf{W}\boldsymbol{f}(\boldsymbol{S}^0) - \boldsymbol{\Lambda}\mathbf{W}\boldsymbol{S}^0. \tag{4.89}$$

The solution is found to be

$$\boldsymbol{X}(t) = \exp(\boldsymbol{\Lambda}t)\mathbf{W}[\mathbf{M}^{-1}\boldsymbol{f}(\boldsymbol{S}^0) - \boldsymbol{S}^0 + \boldsymbol{S}(0)] + \mathbf{W}[\mathbf{S}^0 - \mathbf{M}^{-1}\boldsymbol{f}(\boldsymbol{S}^0)]. \tag{4.90}$$

We see that each pool variable x_j is composed of a constant and a term moving with the time constant $-1/\mathrm{Re}(\lambda_j)$. Therefore, modal analysis applies also in the general case. Clearly, this analysis as well as the rapid-equilibrium approximation do not require that the whole system eventually settles down to be stationary.

Modes are often difficult to handle and to interpret, because they are, in general, linear combinations of concentrations with noninteger coefficients. In nonlinear systems, these coefficients even depend on the chosen reference state $\boldsymbol{S}^0$. Therefore, the approximation based on Eq. (4.83) is often cumbersome to apply. The question of whether the modal matrix can be simplified to a matrix with easily interpretable, preferably integer entries deserves to be studied in the future.

5

Metabolic Control Analysis

From the biological point of view it is an important task to characterize the role of the particular reactions proceeding in the living cell in determining the various dynamic modes of metabolism. Due to the high number of variables and the strong stoichiometric as well as regulatory interrelations, it seems to be impossible to gain such insight by qualitative considerations only. A theoretical framework, named *metabolic control analysis,* has been developed to elucidate in quantitative terms to what extent the various reactions of metabolic pathways determine the fluxes and metabolite concentrations. The theory is based on two types of coefficients, the *control coefficients* characterizing the systemic response of the system variables (fluxes, concentrations, etc.) after parameter perturbations and the *elasticity coefficients* which quantify the changes of reaction rates after perturbations of substrate concentrations or kinetic parameters under isolated conditions. In the early papers of metabolic control analysis a partly different terminology has been used, *control strengths* (Higgins, 1965; Heinrich and Rapoport, 1973, 1974a) and *sensitivities* (Kacser and Burns, 1973) for the systemic coefficients and *effector strengths* (Heinrich and Rapoport, 1974a) for the perturbations of isolated reactions. In this chapter it will be shown that metabolic control analysis yields a number of general rules which allow one to understand the systemic behavior of metabolic networks on the basis of the kinetic properties of their enzymes.

Up to now a comprehensive theory has only been developed for the control of stationary states [for recent reviews, see Fell (1992), S. Schuster and Heinrich (1992), Liao and Delgado (1993) and Cornish-Bowden (1995)]. The fundamentals of this theory are outlined in the following sections. There are, however, several attempts to extend metabolic control analysis to time-dependent processes (relaxation processes or oscillations; see Sections 5.5, 5.8.4 and 5.8.5).

Traditional metabolic control analysis is a linear theory which considers only the effect of infinitesimally small parameter perturbations in the vicinity of a reference state where the systemic behavior is governed by linear approximations of the system equations (2.8). Recently, several attempts have been made to extend the theory to finite parameter perturbations (see Section 5.9).

It will become clear in the following that in the present theory the term *control* is used in a very special sense which should be clearly distinguished from the term *regulation.* Whereas the former merely points to the effect of a change of arbitrary parameters on a system variable, the latter is closely related to the biological function of metabolic pathways. In Section 5.10 it is shown, however, that metabolic control analysis may also be useful for quantifying metabolic regulation.

5.1. BASIC DEFINITIONS

Originally, metabolic control analysis was designed to quantify the concept of rate limitation in complex enzymic systems. Kacser and Burns (1973) drew attention to the fact that the steady-state fluxes J_j in a metabolic system depend on the values of the total concentrations E_k of the enzymes acting as catalysts of the individual reactions. Correspondingly, they defined *flux control coefficients* as follows

$$C_{E_k}^{J_j} = \left(\frac{E_k}{J_j}\frac{\Delta J_j}{\Delta E_k}\right)_{\Delta E_k \to 0} = \frac{E_k}{J_j}\frac{\partial J_j}{\partial E_k}, \tag{5.1}$$

which relate the fractional changes in the steady-state fluxes to the fractional changes in the total enzyme concentrations.

Taking into account that kinetic parameters other than enzyme concentrations may affect reaction rates v_k and, therefore, steady-state fluxes, Heinrich and Rapoport (1973, 1974a) proposed using the following definition for flux control coefficients

$$C_{v_k}^{J_j} = \left(\frac{v_k}{J_j}\frac{\Delta J_j}{\Delta v_k}\right)_{\Delta v_k \to 0} = \frac{v_k}{J_j}\frac{\partial J_j}{\partial v_k}, \tag{5.2}$$

where Δv_k denotes the change in the activity of a reaction k due to the influence of a modifier or a change of an enzyme-kinetic parameter, not necessarily the enzyme concentration, while all other parameters and concentrations are kept constant. This means that Δv_k refers to a change in the enzyme rate under isolated conditions.

Because mathematically the fluxes J_j cannot be directly expressed as functions of the rates v_k, Eq. (5.2) has to be regarded as an abbreviated notation of

$$C_{v_k}^{J_j} = \frac{v_k}{J_j}\frac{\partial J_j/\partial p_k}{\partial v_k/\partial p_k}, \tag{5.3}$$

where p_k is a kinetic parameter which affects only reaction k directly, that is,

$$\frac{\partial v_k}{\partial p_k} \neq 0, \qquad \frac{\partial v_j}{\partial p_k} = 0 \quad \text{for any } j \neq k. \tag{5.4}$$

In Sections 5.2 and 5.6 it is shown under what conditions the control coefficients calculated on the basis of formula (5.3) are fully independent of the special choice of the parameter p_k. The coefficients defined in Eq. (5.3) can then be interpreted as the extent to which reaction k (rather than some parameter) controls a given steady-state flux.

The concept of control coefficients has been extended to quantify the response of steady-state concentrations (Heinrich and Rapoport, 1973, 1974a) by introducing *concentration control coefficients*

$$C_{v_k}^{S_i} = \left(\frac{v_k}{S_i}\frac{\Delta S_i}{\Delta v_k}\right)_{\Delta v_k \to 0} = \frac{v_k}{S_i}\frac{\partial S_i/\partial p_k}{\partial v_k/\partial p_k} \tag{5.5}$$

subject to condition (5.4). In Section 5.8 we will show how metabolic control analysis may be generalized by considering variables other than concentrations and fluxes.

Concerning definitions (5.3) and (5.5), we will use in the following a somewhat modified notation which reflects that the control coefficients may be considered as elements of control matrices

$$\mathbf{C}^J = (C_{jk}^J), \tag{5.6a}$$

$$\mathbf{C}^S = (C_{ik}^S) \tag{5.6b}$$

with the first subscript (i or j) and second subscript (k) referring to the rows and columns, respectively, of the matrices. Note that $C_{jk}^J = C_{v_k}^{J_j}$ and $C_{ik}^S = C_{v_k}^{S_i}$.

Besides the coefficients defined in Eqs. (5.3) and (5.5) non-normalized (unscaled) expressions have been introduced:

$$C_{jk}^J = \frac{\partial J_j/\partial p_k}{\partial v_k/\partial p_k}, \tag{5.7a}$$

$$C_{ik}^{S} = \frac{\partial S_i/\partial p_k}{\partial v_k/\partial p_k} \tag{5.7b}$$

(see Sections 5.2 and 5.7).

The parameters p_k can be of different types. Besides enzyme concentration, which enters definition (5.1), the Michaelis constant, the elementary rate constants, the turnover number, and the concentrations of external metabolites are admissible. In the framework of metabolic control analysis, those parameters that can be changed easily in experiment [e.g., the concentrations of enzyme-specific inhibitors (Groen *et al.*, 1982)] are of special importance.

5.2. A SYSTEMATIC APPROACH

Let us consider a reaction network described by a system of ordinary differential equations of the type (2.8), as discussed in Section 2.1. The response of the steady-state concentrations S_i and the steady-state fluxes J_k toward small parameter perturbations can be systematically analyzed in the following way. The steady-state equation $\mathbf{N}\boldsymbol{v}(\boldsymbol{S},\boldsymbol{p}) = \boldsymbol{0}$ [cf. Eq. (2.9)] defines in an implicit manner the parameter dependence of the concentrations and fluxes; that is, the functions

$$\boldsymbol{S} = \boldsymbol{S}(\boldsymbol{p}), \tag{5.8a}$$

$$\boldsymbol{J} = \boldsymbol{J}(\boldsymbol{p}) = \boldsymbol{v}(\boldsymbol{S}(\boldsymbol{p}),\boldsymbol{p})\,. \tag{5.8b}$$

In the neighborhood of a stable reference state with kinetic parameters $\boldsymbol{p} = \boldsymbol{p}^0$ the effect of parameter perturbations can be evaluated using a Taylor expansion:

$$\Delta Y = \sum_k \frac{\partial Y}{\partial p_k}\Delta p_k + \frac{1}{2}\sum_{k,l}\frac{\partial^2 Y}{\partial p_k\,\partial p_l}\Delta p_k \Delta p_l + \cdots\,. \tag{5.9}$$

In this equation, Y represents the variables S_i or J_j. $\Delta p_k = p_k - p_k^0$ denotes the parameter changes, and $\Delta Y = Y(\boldsymbol{p}) - Y(\boldsymbol{p}^0)$. In the following, the first and second partial derivatives of the variables with respect to the kinetic parameters which enter the right-hand side of Eq. (5.9) are named first-order and second-order response coefficients, respectively. For metabolic systems, the steady-state equations (2.9) are generally highly nonlinear in the variables $\boldsymbol{S}$ and it is impossible to express the functions (5.8) in an explicit manner. However, restriction of the analysis to the linear terms in the Taylor expansion (5.9) enables us to derive simple expressions for ΔS_i and ΔJ_j using the following procedure (for the second-order terms, cf. Section 5.9).

Implicit differentiation of Eq. (2.9) with respect to $\boldsymbol{p}$ yields, under consideration of Eq. (5.8a),

$$\mathbf{N}\frac{\partial \boldsymbol{v}}{\partial \boldsymbol{S}}\frac{\partial \boldsymbol{S}}{\partial \boldsymbol{p}} + \mathbf{N}\frac{\partial \boldsymbol{v}}{\partial \boldsymbol{p}} = \mathbf{0}. \tag{5.10}$$

In the case that the system does not contain conservation quantities and the steady state is asymptotically stable, the Jacobian $\mathbf{M} = \mathbf{N}\partial \boldsymbol{v}/\partial \boldsymbol{S}$ is nonsingular (see Section 2.3.2). Therefore, for the first-order response of metabolite concentrations, one derives from Eq. (5.10)

$$\frac{\partial \boldsymbol{S}}{\partial \boldsymbol{p}} = -\left(\mathbf{N}\frac{\partial \boldsymbol{v}}{\partial \boldsymbol{S}}\right)^{-1}\mathbf{N}\frac{\partial \boldsymbol{v}}{\partial \boldsymbol{p}} = -\mathbf{M}^{-1}\mathbf{N}\frac{\partial \boldsymbol{v}}{\partial \boldsymbol{p}} \tag{5.11}$$

and for the response of steady-state fluxes, using Eqs. (5.8b) and (5.11),

$$\frac{\partial \boldsymbol{J}}{\partial \boldsymbol{p}} = \frac{\partial \boldsymbol{v}}{\partial \boldsymbol{p}} + \frac{\partial \boldsymbol{v}}{\partial \boldsymbol{S}}\frac{\partial \boldsymbol{S}}{\partial \boldsymbol{p}} = \left[\mathbf{I} - \frac{\partial \boldsymbol{v}}{\partial \boldsymbol{S}}\left(\mathbf{N}\frac{\partial \boldsymbol{v}}{\partial \boldsymbol{S}}\right)^{-1}\mathbf{N}\right]\frac{\partial \boldsymbol{v}}{\partial \boldsymbol{p}}. \tag{5.12}$$

It is seen that for the metabolite concentrations as well as for the fluxes, the response coefficients can be split into two terms. The terms

$$\mathbf{C}^S = -\left(\mathbf{N}\frac{\partial \boldsymbol{v}}{\partial \boldsymbol{S}}\right)^{-1}\mathbf{N} \tag{5.13}$$

and

$$\mathbf{C}^J = \mathbf{I} - \frac{\partial \boldsymbol{v}}{\partial \boldsymbol{S}}\left(\mathbf{N}\frac{\partial \boldsymbol{v}}{\partial \boldsymbol{S}}\right)^{-1}\mathbf{N} = \mathbf{I} + \frac{\partial \boldsymbol{v}}{\partial \boldsymbol{S}}\mathbf{C}^S \tag{5.14}$$

depend via the stoichiometric coefficients on the systemic properties of the network but are independent of the special choice of the perturbation parameters. In contrast, the term $\partial \boldsymbol{v}/\partial \boldsymbol{p}$ is independent of the systemic properties of the network and characterizes the effect of parameter changes on the individual reactions at fixed concentrations of the metabolites. If the parameter perturbations are infinitesimally small ($\Delta \boldsymbol{p} = \delta \boldsymbol{p}$) one may use linear approximations for $\Delta \boldsymbol{v}$ as well as for $\Delta \boldsymbol{S}$ and $\Delta \boldsymbol{J}$, that is,

$$\Delta \boldsymbol{v} \cong \delta \boldsymbol{v} = \left(\frac{\partial \boldsymbol{p}}{\partial \boldsymbol{v}}\right)\delta \boldsymbol{p}, \qquad \Delta \boldsymbol{S} \cong \delta \boldsymbol{S} = \left(\frac{\partial \boldsymbol{p}}{\partial \boldsymbol{S}}\right)\delta \boldsymbol{p}, \qquad \Delta \mathbf{J} \cong \delta \mathbf{J} = \left(\frac{\partial \boldsymbol{p}}{\partial \boldsymbol{J}}\right)\delta \boldsymbol{p}. \tag{5.15}$$

By definition, the elements of the vector $\delta \boldsymbol{v}$ are the immediate changes of the reaction rates after parameter perturbation at $t = t_0$, whereas the vectors $\delta \boldsymbol{S}$ and $\delta \boldsymbol{J}$ contain the final changes of the concentrations which are attained after adjustment of the system to the parameter perturbations for $t \to \infty$. With (5.15), it follows from Eqs. (5.11)–(5.14) that

$$\delta \boldsymbol{S} = \mathbf{C}^S \delta \boldsymbol{v}, \tag{5.16a}$$

$$\delta \boldsymbol{J} = \mathbf{C}^J \delta \boldsymbol{v}. \tag{5.16b}$$

The matrices of control coefficients $\mathbf{C}^S$ and $\mathbf{C}^J$ transform the vector $\delta \boldsymbol{v}$ into the vectors $\delta \boldsymbol{S}$ and $\delta \boldsymbol{J}$, respectively. Choosing the perturbation parameters in such a way that the matrix $\partial \boldsymbol{v}/\partial \boldsymbol{p}$ is nonsingular, the matrices of unscaled control coefficients can be rewritten as follows:

$$\mathbf{C}^S = \frac{\partial \boldsymbol{S}}{\partial \boldsymbol{p}} \left(\frac{\partial \boldsymbol{v}}{\partial \boldsymbol{p}} \right)^{-1}, \tag{5.17}$$

$$\mathbf{C}^J = \frac{\partial \boldsymbol{J}}{\partial \boldsymbol{p}} \left(\frac{\partial \boldsymbol{v}}{\partial \boldsymbol{p}} \right)^{-1}. \tag{5.18}$$

These equations can be used as definitions for control coefficients, which are more general than definitions (5.7a) and (5.7b), because the parameters p_k need not be reaction-specific. Using Eqs. (5.17) and (5.18), the set of admissible parameters can be considerably enlarged. For example, concentrations of enzymes catalyzing more than one reaction, concentrations of unspecific inhibitors, pH, or temperature can be used.

Some simplifications result if there are reaction-specific perturbation parameters fulfilling relation (5.4). Then Eqs. (5.17) and (5.18) can be specified to give Eqs. (5.7b) and (5.7a), respectively.

The partial derivatives of reaction rates with respect to substrate concentrations or kinetic parameters are called (unscaled) elasticity coefficients. We use the following notation:

$$\varepsilon_{ij} = \frac{\partial v_i}{\partial S_j}\text{: } \varepsilon\text{-elasticities,} \tag{5.19}$$

$$\pi_{jk} = \frac{\partial v_j}{\partial p_k}\text{: } \pi\text{-elasticities.} \tag{5.20}$$

Elasticity coefficients characterize the kinetic properties of the individual enzymes in isolation, in a close neighborhood of a reference state (Burns *et al.,* 1985).

Reaction systems with conservation equations: Formulas (5.13) and (5.14) for the calculation of control coefficients have to be modified if metabolic systems with conservation equations are considered (Reder, 1988). In this case, the stoichiometric matrix does not have full rank (see Section 3.1) and the Jacobian $\mathbf{M} = \mathbf{N}\partial \boldsymbol{v}/\partial \boldsymbol{S}$ is, therefore, singular. Implicit differentiation of the steady-state equation (2.9) with respect to the kinetic parameters yields

$$\mathbf{N}^0 \frac{\partial \boldsymbol{v}}{\partial \boldsymbol{S}_a} \frac{\partial \boldsymbol{S}_a}{\partial \boldsymbol{p}} + \mathbf{N}^0 \frac{\partial \boldsymbol{v}}{\partial \boldsymbol{S}_b} \frac{\partial \boldsymbol{S}_b}{\partial \boldsymbol{S}_a} \frac{\partial \boldsymbol{S}_a}{\partial \boldsymbol{p}} + \mathbf{N}^0 \frac{\partial \boldsymbol{v}}{\partial \boldsymbol{p}} = \mathbf{0}. \tag{5.21}$$

under consideration of the relation between independent concentrations $\boldsymbol{S}_a$ and dependent concentrations $\boldsymbol{S}_b$ [cf. Eq. (3.10)]. Due to $\mathbf{L}' = \partial \boldsymbol{S}_b/\partial \boldsymbol{S}_a$ and Eq. (3.7),

$$\mathbf{N}^0 \frac{\partial \boldsymbol{v}}{\partial \boldsymbol{S}} \mathbf{L} \frac{\partial \boldsymbol{S}_a}{\partial \boldsymbol{p}} + \mathbf{N}^0 \frac{\partial \boldsymbol{v}}{\partial \boldsymbol{p}} = \mathbf{0} \tag{5.22}$$

with $\partial \boldsymbol{v}/\partial \boldsymbol{S} = (\partial \boldsymbol{v}/\partial \boldsymbol{S}_a, \partial \boldsymbol{v}/\partial \boldsymbol{S}_b)$. $\mathbf{L}$ stands again for the link matrix. From Eq. (5.22) one gets

$$\frac{\partial \boldsymbol{S}_a}{\partial \boldsymbol{p}} = -(\mathbf{M}^0)^{-1} \mathbf{N}^0 \frac{\partial \boldsymbol{v}}{\partial \boldsymbol{p}}, \tag{5.23}$$

where

$$\mathbf{M}^0 = \mathbf{N}^0 \frac{\partial \boldsymbol{v}}{\partial \boldsymbol{S}} \mathbf{L} \tag{5.24}$$

is the Jacobian of the reduced system, in which the dependent concentrations $\boldsymbol{S}_b$ have been eliminated by use of the conservation relations. $\mathbf{M}^0$ is a nonsingular matrix because the steady state is assumed to be asymptotically stable. Taking into account Eqs. (3.10) and (5.23) one obtains

$$\frac{\partial \boldsymbol{S}}{\partial \boldsymbol{p}} = \mathbf{C}^S \frac{\partial \boldsymbol{v}}{\partial \boldsymbol{p}}, \tag{5.25a}$$

$$\mathbf{C}^S = -\mathbf{L}(\mathbf{M}^0)^{-1}\mathbf{N}^0 \tag{5.25b}$$

for the parameter dependence of the concentrations and

$$\frac{\partial \boldsymbol{J}}{\partial \boldsymbol{p}} = \mathbf{C}^J \frac{\partial \boldsymbol{v}}{\partial \boldsymbol{p}}, \tag{5.26a}$$

$$\mathbf{C}^J = \mathbf{I} - \frac{\partial \boldsymbol{v}}{\partial \boldsymbol{S}} \mathbf{L}(\mathbf{M}^0)^{-1}\mathbf{N}^0 \tag{5.26b}$$

for the parameter dependence of the steady-state fluxes. For systems without conservation equations ($\mathbf{L} = \mathbf{I}$), Eqs. (5.25b) and (5.26b) simplify to Eqs. (5.13) and (5.14), respectively.

Using Eqs. (5.25b) and (5.26b) and the definition (5.24) of the Jacobian of the reduced system, one can easily prove the relationships

$$\mathbf{C}^J\mathbf{C}^J = \mathbf{C}^J, \tag{5.27a}$$

$$\mathbf{C}^S\boldsymbol{\varepsilon}\mathbf{C}^S = -\mathbf{C}^S, \tag{5.27b}$$

which are valid for any metabolic system. Relation (5.27a) means that $\mathbf{C}^J$ is an idempotent matrix (Gantmacher, 1959). Obviously, this equation implies that the matrix $\mathbf{C}^J$ raised to any integer power (greater than zero) equals this matrix itself. For a further discussion of relations (5.27a) and (5.27b), see Sections 5.3.4 and 5.5.

Response coefficients: Using definition (5.20), Eqs. (5.25a) and (5.25b) may be written as follows:

$$R_{ik}^S = \sum_j C_{ij}^S \pi_{jk} \quad \text{or} \quad \mathbf{R}^S = \mathbf{C}^S\boldsymbol{\pi}, \tag{5.28}$$

and

$$R_{jk}^J = \sum_i C_{ji}^J \pi_{ik} \quad \text{or} \quad \mathbf{R}^J = \mathbf{C}^J\boldsymbol{\pi}, \tag{5.29}$$

where R_{ik}^S and R_{jk}^J denote response coefficients (Kacser and Burns, 1973; Hofmeyr et al., 1986). These relations show that the effect of a perturbation of a parameter p_k on a state variable S_i or J_j may be described as a sum of individual terms

$${}^jR_{ik}^S = C_{ij}^S \pi_{jk}, \qquad {}^iR_{jk}^J = C_{ji}^J \pi_{ik}, \tag{5.30}$$

which have been called partial response coefficients (Kholodenko, 1990).

With Eqs. (5.28) and (5.29) the response of concentrations and fluxes to simultaneous perturbations of several parameters can be written in the following form:

$$\delta S_i = \sum_k R_{ik}^S \delta p_k \quad \text{or} \quad \delta \boldsymbol{S} = \mathbf{R}^S \delta \boldsymbol{p} \tag{5.31}$$

and

$$\delta J_j = \sum_k R_{jk}^J \delta p_k \quad \text{or} \quad \delta \boldsymbol{J} = \mathbf{R}^J \delta \boldsymbol{p}. \tag{5.32}$$

Normalized coefficients: It is often useful to transform the unscaled coefficients into a normalized form. This gives, for the elasticities,

$$(\mathrm{dg}\,\boldsymbol{J})^{-1}\,\boldsymbol{\varepsilon}(\mathrm{dg}\,\boldsymbol{S}) \rightarrow \boldsymbol{\varepsilon}, \tag{5.33a}$$

$$(\mathrm{dg}\,\boldsymbol{v})^{-1}\,\boldsymbol{\pi}(\mathrm{dg}\,\boldsymbol{p}) \rightarrow \boldsymbol{\pi}, \tag{5.33b}$$

and for the control coefficients,

$$(\mathrm{dg}\,\boldsymbol{S})^{-1}\,\mathbf{C}^S\,(\mathrm{dg}\,\boldsymbol{J}) \rightarrow \mathbf{C}^S, \tag{5.34a}$$

$$(\mathrm{dg}\,\boldsymbol{J})^{-1}\,\mathbf{C}^J\,(\mathrm{dg}\,\boldsymbol{J}) \rightarrow \mathbf{C}^J. \tag{5.34b}$$

The reaction rates ($\boldsymbol{v} = \boldsymbol{J}$) and substrate concentrations $\boldsymbol{S}$ in the reference state are used for normalization. (dg$\boldsymbol{Y}$) signifies a diagonal matrix with the components of the vector $\boldsymbol{Y}$ standing in its principal diagonal. Note that premultiplication by a diagonal matrix implies that all entries of one and the same row of a matrix are multiplied by the same factor, whereas postmultiplication has a similar effect on the columns. Accordingly, the transformation rules (5.34a) and (5.34b) give the matrices defined earlier in Eqs. (5.3), (5.5) and (5.6).

The normalized matrices $\boldsymbol{\varepsilon}$ and $\boldsymbol{\pi}$ contain the elements

$$\varepsilon_{ji} = \frac{S_i}{v_j}\frac{\partial v_j}{\partial S_i}, \tag{5.35a}$$

$$\pi_{jk} = \frac{p_k}{v_j}\frac{\partial v_j}{\partial p_k}, \tag{5.35b}$$

respectively, whereas the normalized matrices $\mathbf{C}^J$ and $\mathbf{C}^S$ contain, as elements, the control coefficients defined in Eqs. (5.3) and (5.5), respectively.

The normalized coefficients can be written as logarithmic derivatives. This gives, for the elasticities,

$$\varepsilon_{ji} = \frac{\partial \ln v_j}{\partial \ln S_i}, \tag{5.36a}$$

$$\pi_{jk} = \frac{\partial \ln v_j}{\partial \ln p_k}, \tag{5.36b}$$

and for the control coefficients,

$$C_{ik}^S = \frac{\partial \ln S_i/\partial \ln p_k}{\partial \ln v_k/\partial \ln p_k}, \tag{5.37a}$$

$$C_{jk}^{J} = \frac{\partial \ln J_j / \partial \ln p_k}{\partial \ln v_k / \partial \ln p_k}. \tag{5.37b}$$

To avoid undefined values of the logarithms, we use the convention that the reaction rates are counted in such a direction that they are positive. Using normalized coefficients, Eqs. (5.13) and (5.14) are replaced by

$$\mathbf{C}^S = -[\mathbf{N}(\mathrm{dg}\,\boldsymbol{J})\boldsymbol{\varepsilon}]^{-1}[\mathbf{N}(\mathrm{dg}\,\boldsymbol{J})] \tag{5.38}$$

and

$$\mathbf{C}^J = \mathbf{I} + \boldsymbol{\varepsilon}\mathbf{C}^S, \tag{5.39}$$

respectively.

Response coefficients can also be defined in normalized form. They fulfill Eqs. (5.28) and (5.29) with normalized elasticities and control coefficients.

Using, for an enzymatic network, the total enzyme concentrations E_k as perturbation parameters, one obtains for the normalized control coefficients

$$C_{ik}^{S} = \frac{E_k}{S_i}\frac{\partial S_i}{\partial E_k}, \tag{5.40a}$$

$$C_{jk}^{J} = \frac{E_k}{J_j}\frac{\partial J_j}{\partial E_k}, \tag{5.40b}$$

under the assumption that the reaction rates are linearly dependent on enzyme concentrations, that is,

$$\frac{E_k}{v_k}\frac{\partial v_k}{\partial E_k} = \frac{\partial \ln v_k}{\partial \ln E_k} = 1. \tag{5.41}$$

Equation (5.40) is the definition of control coefficients originally proposed by Kacser and Burns (1973) [cf. Eq. (5.1)]. In deriving Eq. (5.40), a one-to-one correspondence of enzymes and reactions has been assumed (i.e., $\partial v_j/\partial E_k = 0$ for $j \neq k$). When the enzyme concentrations are not explicitly treated as variables, they belong to the system parameters. It is then more appropriate to denote the (normalized or non-normalized) partial derivatives of the system variables with respect to the enzyme concentrations as special response coefficients, $R_{Ek}^{S_i}$ and $R_{Ek}^{J_j}$, rather than to consider them as control coefficients. These response coefficients are meaningful quantities also when condition (5.41) is not satisfied.

In the above calculations, v normally stands for the rate of the overall enzyme-catalyzed reaction. It is important to note that the mathematical treatment formally remains valid when the system is described at a more detailed level of enzyme catalysis. In this case, v may play the role of the rates of the elementary steps and, accordingly, the calculated control coefficients would refer to the control exerted by these steps rather than by the whole enzyme (see Sections 5.14 and 5.15). In such a treatment, a number of additional conservation relations arise, owing to the fixed total enzyme concentrations.

5.3. THEOREMS OF METABOLIC CONTROL ANALYSIS

5.3.1. Summation Theorems

The various control coefficients are not fully independent of each other. Two types of relationships between concentration control coefficients as well as flux control coefficients can be derived which are generally valid irrespective of the complexity of the considered reaction network. Some of the relationships, called *summation theorems,* reflect the structural properties of the reaction network and are independent of the kinetic parameters of the individual enzymes. In contrast to that, the *connectivity theorems* presented in Section 5.3.2 relate the properties of the single enzymes to the systemic behavior.

We first consider the normalized control coefficients given in Eqs. (5.38) and (5.39). Postmultiplication of these equations with the r-dimensional vector $\boldsymbol{1} = (1, \ldots, 1)^{\mathrm{T}}$ yields under consideration of the steady-state condition $\mathbf{N}\boldsymbol{J} = \boldsymbol{0}$

$$\boldsymbol{C}^S\boldsymbol{1} = \boldsymbol{0} \quad \text{or} \quad \sum_{k=1}^{r} C^S_{ik} = 0 \tag{5.42}$$

and

$$\boldsymbol{C}^J\boldsymbol{1} = \boldsymbol{1} \quad \text{or} \quad \sum_{k=1}^{r} C^J_{jk} = 1; \tag{5.43}$$

that is, for each metabolic compound, the sum of the concentration control coefficients is equal to zero, whereas the control coefficients of a given steady-state flux sum up to unity. Relation (5.42) represents the summation theorem for the concentration control coefficients (Heinrich and Rapoport, 1974a), and relation (5.43) the summation theorem for the flux control coefficients (Kacser and Burns, 1973; Heinrich and Rapoport, 1974a).

It was shown by Reder (1988) that relations (5.42) and (5.43) are special cases of generalized summation theorems. This may be seen best by using the matrices of unscaled control coefficients $\boldsymbol{C}^S$ and $\boldsymbol{C}^J$ which fulfill Eqs. (5.25b) and (5.26b), respectively. Postmultiplication of these equations by the null-space matrix $\mathbf{K}$ [cf. Eq. (3.44)] yields

$$\mathbf{C}^S\mathbf{K} = \mathbf{0}, \tag{5.44a}$$

$$\mathbf{C}^J\mathbf{K} = \mathbf{K}. \tag{5.44b}$$

For the scaled control coefficients, these relations read

$$\boldsymbol{C}^S(\mathrm{dg}\,\boldsymbol{J})^{-1}\mathbf{K} = \mathbf{0}, \tag{5.45a}$$

$$\mathbf{C}^J(\mathrm{dg}\,\boldsymbol{J})^{-1}\mathbf{K} = (\mathrm{dg}\,\boldsymbol{J})^{-1}\mathbf{K}. \tag{5.45b}$$

The number of generalized summation relations equals the number of linearly independent $\boldsymbol{k}$ vectors. In the case that the stoichiometry matrix is of full rank, this number is equal to $r - n$. If the metabolic system contains conservation quantities, the matrix $\mathbf{K}$ has $r - \mathrm{rank}(\mathbf{N})$ linearly independent columns. It is worth mentioning that the structure of the generalized summation theorems does not depend on the conservation relations, as can be seen by the fact that the link matrix $\mathbf{L}$ does not enter these theorems in an explicit manner. Because the vector of the steady-state fluxes is contained in the null-space of $\mathbf{N}$, Eqs. (5.44a) and (5.44b) are also fulfilled if one uses the vector $\boldsymbol{J}$ instead of $\mathbf{K}$. The resulting equations

$$\boldsymbol{C}^S\boldsymbol{J} = \boldsymbol{0}, \tag{5.46a}$$

$$\mathbf{C}^J\boldsymbol{J} = \boldsymbol{J} \tag{5.46b}$$

for the unscaled coefficients are equivalent to the special summation relations given in Eqs. (5.42) and (5.43) for the normalized coefficients.

As outlined in Section 5.2, the control coefficients may be expressed in the form of partial derivatives of the concentrations or fluxes with respect to the enzyme concentrations in the case that the latter enter the rate laws in a linear manner. If this condition is fulfilled, the summation theorems (5.42) and (5.43) can also be derived on the basis of the following argument (Giersch, 1988). From Eqs. (2.9) and (5.8) it follows that the steady-state concentrations and steady-state fluxes are homogeneous functions of the enzyme concentrations of order 0 and 1, respectively, which means

$$S_i(\lambda E_1, \ldots, \lambda E_r) = S_i(E_1, \ldots, E_r) \tag{5.47}$$

and

$$J_k(\lambda E_1, \ldots, \lambda E_r) = \lambda J_k(E_1, \ldots, E_r). \tag{5.48}$$

Differentiation of Eqs. (5.47) and (5.48) with respect to λ yields

$$\sum_{k=1}^{r} \frac{\partial S_i}{\partial(\lambda E_k)} E_k = 0 \tag{5.49a}$$

and

$$\sum_{k=1}^{r} \frac{\partial J_j}{\partial(\lambda E_k)} E_k = J_j, \tag{5.49b}$$

respectively. For $\lambda = 1$, this gives

$$\sum_{k=1}^{r} \frac{E_k}{S_i} \frac{\partial S_i}{\partial E_k} = 0 \tag{5.50a}$$

and

$$\sum_{k=1}^{r} \frac{E_k}{J_j} \frac{\partial J_j}{\partial E_k} = 1, \tag{5.50b}$$

which correspond to Eqs. (5.42) and (5.43), respectively.

5.3.2. Connectivity Theorems

Postmultiplication of Eqs. (5.25b) and (5.26b) by $(\partial \boldsymbol{v}/\partial \boldsymbol{S})\mathbf{L}$ yields, under consideration of Eq. (5.24),

$$\mathbf{C}^S \frac{\partial \boldsymbol{v}}{\partial \boldsymbol{S}} \mathbf{L} = -\mathbf{L}, \tag{5.51a}$$

$$\mathbf{C}^J \frac{\partial \boldsymbol{v}}{\partial \boldsymbol{S}} \mathbf{L} = \mathbf{0}, \tag{5.51b}$$

respectively, which are the connectivity theorems of metabolic control analysis. With Eqs. (5.33) and (5.34) they may be rewritten using normalized control coefficients and elasticity coefficients,

$$\mathbf{C}^S \boldsymbol{\varepsilon} (\mathrm{dg}\ \boldsymbol{S})^{-1} \mathbf{L} = -(\mathrm{dg}\ \boldsymbol{S})^{-1} \mathbf{L}, \tag{5.52a}$$

$$\mathbf{C}^J \boldsymbol{\varepsilon} (\mathrm{dg}\ \boldsymbol{S})^{-1} \mathbf{L} = \mathbf{0}. \tag{5.52b}$$

A physical interpretation of the connectivity theorems related to the relaxation of fluctuations of system variables will be given in Section 5.10.3.

For systems without conservation equations ($\mathbf{L} = \mathbf{I}$), Eqs. (5.52a) and (5.52b) simplify to the relationships originally proposed by Westerhoff and Chen (1984) and Kacser and Burns (1973), respectively,

$$\mathbf{C}^S\boldsymbol{\varepsilon} = -\mathbf{I} \quad \text{or} \quad \sum_{k=1}^{r} C_{ik}^S \varepsilon_{kj} = -\delta_{ij} \tag{5.53a}$$

and

$$\mathbf{C}^J\boldsymbol{\varepsilon} = \mathbf{0} \quad \text{or} \quad \sum_{k=1}^{r} C_{jk}^J \varepsilon_{ki} = 0. \tag{5.53b}$$

Introducing $\mathbf{L} = \mathbf{I}$ in Eqs. (5.51a) and (5.51b) and comparing the resulting equations with Eqs. (5.53a) and (5.53b) it is seen that for systems without conservation quantities, the form of the connectivity theorems is invariant to scaling of control coefficients and elasticity coefficients.

The theorems derived above are also valid when the system is described at the level of elementary enzymic steps, provided the control coefficients as well as the elasticities refer to these steps.

5.3.3. Calculation of Control Coefficients Using the Theorems

In unbranched reaction chains, but not in branched systems, the traditional summation theorem (5.43) and the connectivity theorem are sufficient in number for calculating the flux control coefficients in terms of the elasticities (and similarly the concentration control coefficients) (Heinrich and Rapoport, 1975; Sauro *et al.*, 1987; Westerhoff and Kell, 1987; see Section 5.4.3.1). Attempts were made to complete the set of equations by branch-point relationships (Fell and Sauro, 1985; see Section 5.4.3.2). Later, it became clear that for arbitrary stoichiometries, the control coefficients are completely determined by the theorems if instead of the traditional summation relationships (5.42) and (5.43), the generalized summation theorems (5.44a) and (5.44b) are taken into account, because the branch-point relationships are special cases of these.

The summation and connectivity theorems [Eqs. (5.44) and (5.51), respectively] can be combined into the compact formula

$$\begin{pmatrix} \mathbf{C}^J \\ \mathbf{C}^S \end{pmatrix} (\mathbf{K} \quad \boldsymbol{\varepsilon}\mathbf{L}) = \begin{pmatrix} \mathbf{K} & \mathbf{0} \\ \mathbf{0} & -\mathbf{L} \end{pmatrix}, \tag{5.54}$$

which is a central equation in metabolic control analysis. In the following, we show the equivalence of this equation with Eqs. (5.25b) and (5.26b), which give explicit expressions for $\mathbf{C}^J$ and $\mathbf{C}^S$. Note that the stoichiometric properties of the metabolic pathway enter Eqs. (5.25b) and (5.26b) in the form of the stoichiometry matrix and Eq. (5.54) via the link matrix and null-space matrix.

We first prove that the matrix

$$\begin{pmatrix} \mathbf{A} \\ \mathbf{B} \end{pmatrix} = \begin{pmatrix} (\mathbf{K}^{\mathrm{T}}\mathbf{K})^{-1}\mathbf{K}^{\mathrm{T}}(\mathbf{I} - \boldsymbol{\varepsilon}\mathbf{L}(\mathbf{M}^0)^{-1}\mathbf{N}^0) \\ (\mathbf{M}^0)^{-1}\mathbf{N}^0 \end{pmatrix} \tag{5.55}$$

is the inverse of the matrix $(\mathbf{K}\ \boldsymbol{\varepsilon}\mathbf{L})$ entering the left-hand side of Eq. (5.54).

Because $\mathbf{K}$ contains linearly independent columns only, the matrix product $(\mathbf{K}^{\mathrm{T}}\mathbf{K})$ is nonsingular and, hence, invertible. Moreover, we have

$$(\mathbf{K}\quad \boldsymbol{\varepsilon}\mathbf{L})\begin{pmatrix} \mathbf{A} \\ \mathbf{B} \end{pmatrix} = \mathbf{K}(\mathbf{K}^{\mathrm{T}}\mathbf{K})^{-1}\mathbf{K}^{\mathrm{T}}(\mathbf{I} - \boldsymbol{\varepsilon}\mathbf{L}(\mathbf{M}^0)^{-1}\mathbf{N}^0) + \boldsymbol{\varepsilon}\mathbf{L}(\mathbf{M}^0)^{-1}\mathbf{N}^0. \tag{5.56}$$

Because all columns of $\mathbf{K}$ are orthogonal to all rows of $\mathbf{N}^0$, a matrix composed of these two submatrices is nonsingular. Therefore, we can premultiply the right-hand side of Eq. (5.56) by

$$\mathbf{I} = \begin{pmatrix} \mathbf{N}^0 \\ \mathbf{K}^{\mathrm{T}} \end{pmatrix}^{-1} \begin{pmatrix} \mathbf{N}^0 \\ \mathbf{K}^{\mathrm{T}} \end{pmatrix}. \tag{5.57}$$

This gives, due to $\mathbf{N}^0\mathbf{K} = \mathbf{0}$,

$$\begin{aligned}(\mathbf{K}\quad \boldsymbol{\varepsilon}\mathbf{L})\begin{pmatrix} \mathbf{A} \\ \mathbf{B} \end{pmatrix} &= \begin{pmatrix} \mathbf{N}^0 \\ \mathbf{K}^{\mathrm{T}} \end{pmatrix}^{-1} \begin{pmatrix} \mathbf{N}^0 \\ \mathbf{K}^{\mathrm{T}}(\mathbf{I} - \boldsymbol{\varepsilon}\mathbf{L}(\mathbf{M}^0)^{-1}\mathbf{N}^0) + \mathbf{K}^{\mathrm{T}}\boldsymbol{\varepsilon}\mathbf{L}(\mathbf{M}^0)^{-1}\mathbf{N}^0 \end{pmatrix} \\ &= \begin{pmatrix} \mathbf{N}^0 \\ \mathbf{K}^{\mathrm{T}} \end{pmatrix}^{-1} \begin{pmatrix} \mathbf{N}^0 \\ \mathbf{K}^{\mathrm{T}} \end{pmatrix} = \mathbf{I},\end{aligned} \tag{5.58}$$

which completes the proof.

Now we postmultiply Eq. (5.54) by the matrix given in Eq. (5.55) and obtain

$$\mathbf{C}^J = \mathbf{K}\mathbf{A} = \mathbf{I} - \boldsymbol{\varepsilon}\mathbf{L}\mathbf{B} = \mathbf{I} - \boldsymbol{\varepsilon}\mathbf{L}(\mathbf{M}^0)^{-1}\mathbf{N}^0 \tag{5.59}$$

(where the relation $\mathbf{K}\mathbf{A} + \boldsymbol{\varepsilon}\mathbf{L}\mathbf{B} = \mathbf{I}$ following from Eq. (5.58) has been used) and

$$\mathbf{C}^S = -\mathbf{L}(\mathbf{M}^0)^{-1}\mathbf{N}^0. \tag{5.60}$$

Eqs. (5.59) and (5.60) are equivalent to Eqs. (5.26b) and (5.25b), respectively. Thus, the connectivity and generalized summation theorems can be used to calculate the flux and concentration control coefficients in terms of elasticities and stoichiometry.

5.3.4. Geometrical Interpretation

Mazat *et al.* (1990) have shown that geometrical considerations may lead to a deeper insight into the theorems of metabolic control analysis. We will demonstrate this by analyzing the special example of an unbranched pathway consisting of only two reactions

$$P_1 \underset{}{\overset{v_1}{\rightleftharpoons}} S_1 \underset{}{\overset{v_2}{\rightleftharpoons}} P_2 \qquad \text{Scheme 10}$$

with P_1 and P_2 being outer metabolites. Reaction scheme 10 is characterized by the stoichiometric matrix $\mathbf{N} = (1 \ -1)$. The concentration S_1 which is the only internal variable affects the rate v_1 as a product of reaction 1, and the rate v_2 as a substrate of reaction 2, that is,

$$v_1 = v_1(S_1, p_1), \tag{5.61a}$$

$$v_2 = v_2(S_1, p_2), \tag{5.61b}$$

where p_1 and p_2 are kinetic parameters which are assumed to act specifically on reactions 1 and 2, respectively. Using Eq. (5.61a) to express S_1 as a function of v_1, that is, $S_1 = S_1(v_1, p_1)$, one obtains from Eq. (5.61b)

$$v_2 = v_2(S_1(v_1, p_1), p_2) = v_2^*(v_1, p_1, p_2), \tag{5.62}$$

which describes, at given values of the kinetic parameters, a curve containing all possible combinations of the reaction rates v_1 and v_2 in the space of the reaction rates (see Figure 5.1).

Due to the condition $v_1 = v_2$, the steady-state reaction rate is determined by the intersection of the curve (5.62) with the straight line which is located in the direction of the vector

$$\boldsymbol{k} = \begin{pmatrix} 1 \\ 1 \end{pmatrix}. \tag{5.63}$$

$\boldsymbol{k}$ represents the basis for the one-dimensional null-space of $\mathbf{N} = (1 \ -1)$. The tangent to the curve $v_2 = v_2^*(v_1)$ given by Eq. (5.62) at the steady-state point is given by the direction of the vector

$$\boldsymbol{t} = \begin{pmatrix} \dfrac{\partial v_1}{\partial S_1} \\ \\ \dfrac{\partial v_2}{\partial S_1} \end{pmatrix}_{S_1 = \bar{S}_1} \tag{5.64}$$

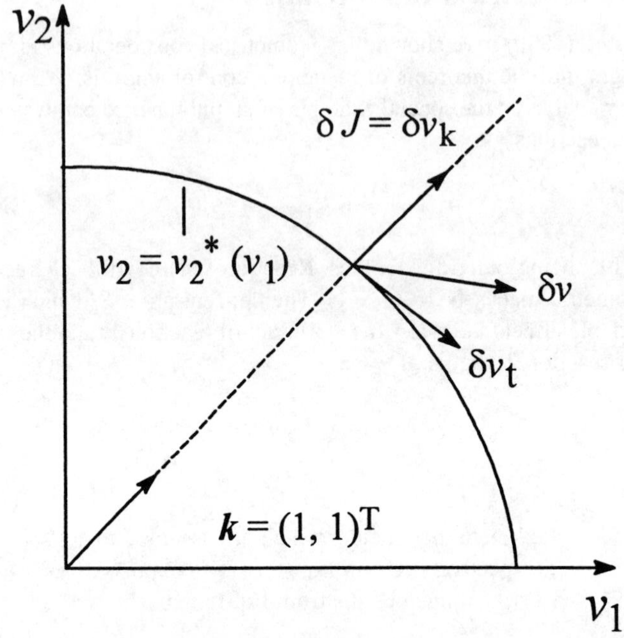

Figure 5.1 Geometrical interpretation of flux-control control coefficients for the reaction system depicted in Scheme 10.

which contains the unscaled elasticity coefficients ε_{11} and ε_{21}. $\bar{S}_1$ denotes the steady-state concentration of S_1. The vectors $\boldsymbol{k}$ and $\boldsymbol{t}$ are not parallel because, otherwise, the only element of the Jacobian matrix, $\partial v_1/\partial S_1 - \partial v_2/\partial S_1$, would be zero, so that the condition that the real parts of all eigenvalues of this matrix are negative would not be fulfilled. Therefore, initial perturbations $\delta \boldsymbol{v} = (\delta v_1, \delta v_2)^T$ of the reaction rates can be decomposed as a sum of two vectors, $\delta \boldsymbol{v}_k$ and $\delta \boldsymbol{v}_t$, in the direction of $\boldsymbol{k}$ and $\boldsymbol{t}$, respectively; that is,

$$\delta \boldsymbol{v} = \delta \boldsymbol{v}_k + \delta \boldsymbol{v}_t = a_1 \frac{\boldsymbol{k}}{|\boldsymbol{k}|} + a_2 \frac{\boldsymbol{t}}{|\boldsymbol{t}|} \tag{5.65}$$

with scalar coefficients a_1 and a_2. Thus, Eq. (5.16b) assumes the form

$$\begin{pmatrix} \delta J_1 \\ \delta J_2 \end{pmatrix} = \begin{pmatrix} C_{11} & C_{12} \\ C_{21} & C_{22} \end{pmatrix} \left(\frac{a_1}{|\boldsymbol{k}|} \begin{pmatrix} k_1 \\ k_2 \end{pmatrix} + \frac{a_2}{|\boldsymbol{t}|} \begin{pmatrix} t_1 \\ t_2 \end{pmatrix} \right), \tag{5.66}$$

which gives, under consideration of the summation and connectivity theorems (5.44b) and (5.51b),

$$\begin{pmatrix} \delta J_1 \\ \delta J_2 \end{pmatrix} = \frac{\alpha_1}{|\boldsymbol{k}|} \begin{pmatrix} k_1 \\ k_2 \end{pmatrix} = \delta \boldsymbol{v}_k. \tag{5.67}$$

Accordingly, the perturbation of the steady-state fluxes $\delta \boldsymbol{J}$ is the projection of the initial perturbation $\delta \boldsymbol{v}$ on the null-space vector $\boldsymbol{k}$ in the direction of the tangent $\boldsymbol{t}$ defined by the elasticity coefficients (see Figure 5.1).

From Eq. (5.66), it follows that $\mathbf{C}^J$ is a projection matrix, which represents the mapping of the initial perturbation of reaction rates into the null-space to give the final change in steady-state fluxes. The direction of this projection is not normally orthogonal. This reasoning can be generalized to dimensions larger than 2. The space of reaction rates can be conceived of as spanned by the column vectors, $\boldsymbol{k}_i$, of the null-space matrix, $\mathbf{K}$, and the vectors $(\varepsilon_{1i}, \ldots, \varepsilon_{ri})^{\mathrm{T}}$, $i = 1, \ldots, \mathrm{rank}\,(\mathbf{N})$, of unscaled elasticities with respect to the independent metabolites. Any perturbation $\delta \boldsymbol{v}(\delta \boldsymbol{p})$ can then be decomposed as a linear combination of these r vectors. The resulting flux change $\delta \boldsymbol{J}$ is a projection of $\delta \boldsymbol{v}$ onto the subspace spanned by the vectors $\boldsymbol{k}_i$ in the direction of the subspace spanned by the vectors of elasticities mentioned above. The property of $\mathbf{C}^J$ to be a projection matrix is also reflected in the relation $\mathbf{C}^J\mathbf{C}^J = \mathbf{C}^J$ [Eq. (5.27a)].

5.4. CONTROL ANALYSIS OF VARIOUS SYSTEMS

5.4.1. General Remarks

In the following paragraphs various applications of metabolic control analysis are presented. We start with the calculation of elasticity coefficients on the basis of enzyme-kinetic equations and consider, thereafter, control coefficients of hypothetical and real metabolic pathways.

When the rate equation of an enzymic reaction is known, the elasticity coefficients with respect to substrates, products, and effectors as well as the elasticities with respect to kinetic parameters can be calculated by differentiation (Section 5.4.2). The calculation of control coefficients is more difficult. Due to nonlinearities in the steady-state equations, it is in most cases impossible to derive explicit expressions for the parameter dependence of the steady-state concentrations and steady-state fluxes. Therefore, the direct application of formulas (5.7) or (5.37) (i.e., determination of the control coefficients by explicit differentiation) is not possible. Different methods may be envisaged. Control coefficients can be calculated by Eqs. (5.25b) and (5.26b), which have been derived by implicit differentiation of the steady-state equations. An alternative, equivalent procedure is the calculation of scaled or unscaled control coefficients on the basis of the summation and connectivity theorems as outlined in Section 5.3.3. It requires the following information about the system: (1) a complete set of basis vectors of the null-space of the stoichiometry matrix, (2) conservation relationships as expressed by the link matrix, and (3) the ε-elasticity coefficients in a reference steady state. In addition, the calculation of scaled control coefficients necessitates the knowledge of the quantities used for normalization, that is, the concentrations and fluxes in the reference steady state.

Whereas the basis vectors of the null-space and the link matrix may be easily obtained by analysis of the stoichiometry matrix, the ε-elasticities as well as concentrations and fluxes in the reference state require experimental determination or calculation on the basis of a model for the given metabolic pathway. In Section 5.4.4, we will study various models of glycolysis, which allow one to carry out a control analysis. Sometimes, conclusions concerning the control properties of metabolic pathways may also be drawn on the basis of incomplete knowledge of the stoichiometric structure and the kinetic properties of enzymes (see Sections 5.12 and 5.13). The examples in this section are chosen so that analytical and simple numerical treatment is feasible. For more complex networks, special computer programs such as those mentioned in Section 5.17 are necessary.

5.4.2. Elasticity Coefficients for Specific Rate Laws

Let us consider elasticity coefficients for several well-known rate laws used in enzyme kinetics, which were considered in Section 2.2.

(a) *Michaelis–Menten equation:* From the rate equation (2.24) one derives for the normalized ε-elasticity

$$\varepsilon_S = \frac{S}{v}\frac{\partial v}{\partial S} = \frac{K_{mS}}{K_{mS} + S}. \tag{5.68}$$

ε_S decreases monotonically with increasing substrate concentration. For very low substrate concentrations, where the Michaelis–Menten equation may be approximated by the linear equation $v \cong (V_m/K_{mS})S$, the ε-elasticity tends to unity while at saturating substrate concentrations, ε_S becomes vanishingly small. For the π-elasticity of the Michaelis constant one derives

$$\pi_{K_{mS}} = \frac{K_{mS}}{v}\frac{\partial v}{\partial K_{mS}} = -\frac{K_{mS}}{K_{mS} + S} = -\varepsilon_S. \tag{5.69}$$

From the Michaelis–Menten equation (2.20) for reversible reactions with $S = S_1$, $P = S_2$, $K_{mP} = K_{m1}$, and $K_{mS} = K_{m2}$, one gets the elasticity coefficients

$$\varepsilon_S = \left(V_m^+ + \frac{P}{K_{mP}}(V_m^+ + V_m^-)\right)\frac{S}{K_{mS}} \left[\left(1 + \frac{S}{K_{mS}} + \frac{P}{K_{mP}}\right)\left(V_m^+\frac{S}{K_{mS}} - V_m^-\frac{P}{K_{mP}}\right)\right]^{-1} \tag{5.70a}$$

and

$$\varepsilon_P = \left(V_m^- + \frac{S}{K_{mS}}(V_m^+ + V_m^-)\right)\frac{P}{K_{mP}} \left[\left(1 + \frac{S}{K_{mS}} + \frac{P}{K_{mP}}\right)\left(V_m^+\frac{S}{K_{mS}} - V_m^-\frac{P}{K_{mP}}\right)\right]^{-1}. \tag{5.70b}$$

For $v > 0$, the ε-elasticity for the substrate S is positive, and for the product P, it is negative.

Rearranging terms in Eq. (5.70a) gives

$$\varepsilon_S = V_m^+\frac{S}{K_{mS}}\left(V_m^+\frac{S}{K_{mS}} - V_m^-\frac{P}{K_{mP}}\right)^{-1} - \frac{S}{K_{mS}}\left(1 + \frac{S}{K_{mS}} + \frac{P}{K_{mP}}\right)^{-1}. \tag{5.71}$$

Owing to the Haldane relation (2.26) and with the rate of the forward reaction

$$v_f = V_m^+\frac{S}{K_{mS}}\left(1 + \frac{S}{K_{mS}} + \frac{P}{K_{mP}}\right)^{-1}, \tag{5.72}$$

Eq. (5.71) can be written as

$$\varepsilon_S = \frac{1}{1 - \dfrac{P}{qS}} - \frac{v_f}{V_m^+}. \tag{5.73}$$

The term P/qS characterizes the displacement from equilibrium (cf. Fell, 1992) and may be written in terms of the reaction affinity, $P/qS = \exp(-A/RT)$ [cf. Eq. (2.16)]. The term v_f/V_m^+ represents the fractional saturation of the enzyme with substrate.

For a near-equilibrium enzyme, the first term on the right-hand side of Eq. (5.73) is much higher than the second term (which is bounded above by 1). In this situation, the normalized elasticity only depends on the displacement from equilibrium,

$$\varepsilon_S \cong \frac{1}{1 - \dfrac{P}{qS}} = \frac{1}{1 - \exp\left(-\dfrac{A}{RT}\right)} \cong \frac{RT}{A}. \tag{5.74a}$$

Similarly,

$$\varepsilon_P \cong \frac{1}{1 - qS/P} = \frac{1}{1 - \exp\left(\dfrac{A}{RT}\right)} \cong -\frac{RT}{A}. \tag{5.74b}$$

The approximations on the right-hand sides of Eqs. (5.74a) and (5.74b) have been obtained by Taylor expansion of the exponential function up to the linear term. This approximation is justified because we assumed the enzyme to operate near equilibrium, so the affinity is small. It can be seen by comparison of Eqs. (5.70) and (5.74) that near equilibrium the elasticities become independent of the kinetic parameters and only depend on the thermodynamic properties (affinity). Clearly ε_S and ε_P tend to infinity as the reaction reaches equilibrium.

(b) *Hill equation:* For the rate equation (2.40), the normalized elasticity coefficient

$$\varepsilon_S = \frac{n_H}{1 + \left(\dfrac{S}{K_S}\right)^{n_H}} \tag{5.75}$$

obtains, which implies $\varepsilon_S \to n_H$ for $S \to 0$ and $\varepsilon_S \to 0$ for $S \to \infty$.

(c) *Model of Monod, Wyman, and Changeux:* With the rate equation (2.43a) one derives

$$\varepsilon_S = \left[1 + L\left(1 + \frac{nS}{K_S}\right)\left(\frac{1 + I/K_I}{1 + S/K_S}\right)^n\right]\left[\left(1 + \frac{S}{K_S}\right)\left(1 + L\left(\frac{1 + I/K_I}{1 + S/K_S}\right)^n\right)\right]^{-1} \tag{5.76}$$

where the possible effect of activators is included in the allosteric constant L [cf. Eq. (2.43b)]. Considering the inhibitor concentration as a parameter of the kinetic equation, differentiation of Eq. (2.43a) with respect to I gives a π-elasticity

$$\pi_I = \frac{I}{v}\frac{\partial v}{\partial I} = -\left(\frac{I}{K_I + I}\right) nL\left(\frac{1 + I/K_I}{1 + S/K_S}\right)^n\left[1 + L\left(\frac{1 + I/K_I}{1 + S/K_S}\right)^n\right]^{-1}. \tag{5.77}$$

Note that the term in the first parentheses on the right-hand side of Eq. (5.77) gives the π-elasticity of an irreversible Michaelis–Menten kinetics with noncompetitive inhibition with respect to the inhibitor concentration.

In Figures 5.2A and 5.2B, ε_S and π_I are plotted as functions of S and I, respectively, for the rate equation (2.43).

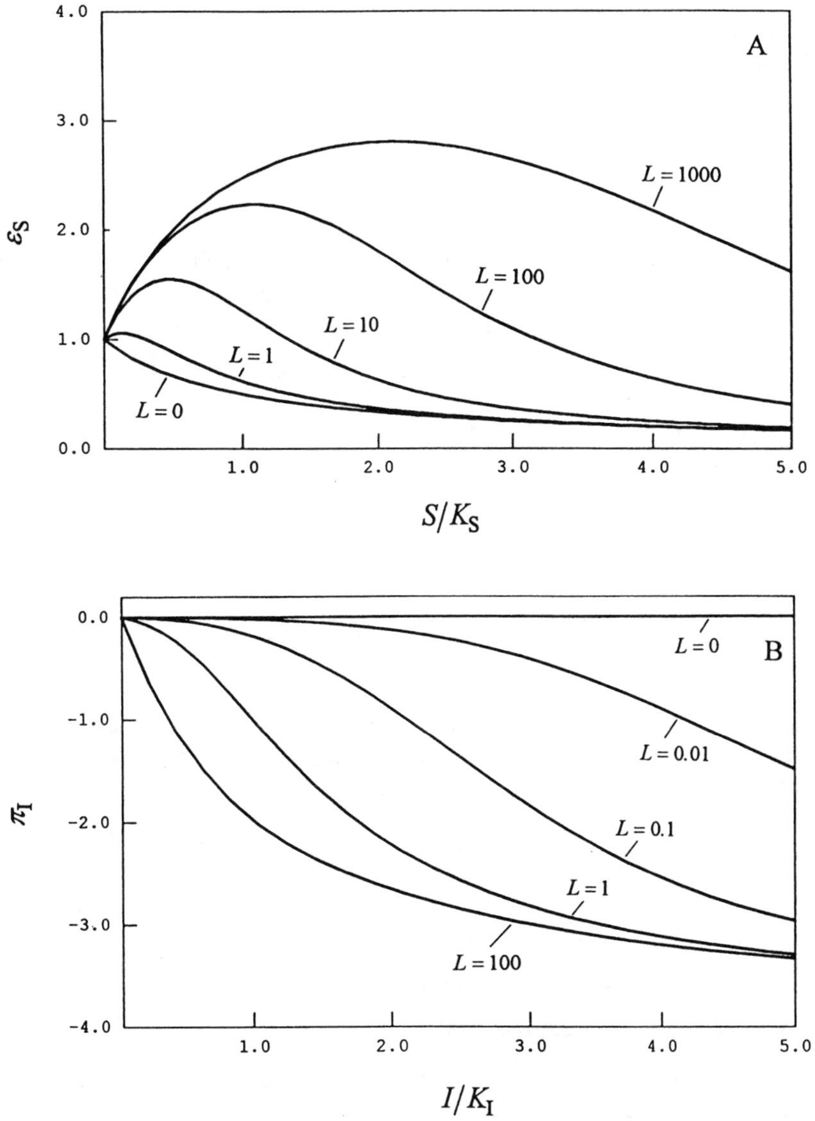

Figure 5.2 Substrate elasticity ε_S (A) and parameter elasticity π_I (B) as functions of the substrate and inhibitor concentrations for the Monod model according to Eqs. (5.76) and (5.77), respectively. Parameter values: (A) $n = 4$, $I/K_I = 0$; (B) $n = 4$, $S/K_S = 1$.

(d) *Generalized mass-action kinetics:* The above calculations can be generalized for any type of enzyme kinetics, by considering the generalized mass-action kinetics given in Eq. (2.15). One then obtains the following elasticities:

$$\varepsilon_{ji} = \frac{S_i}{v_j}\frac{\partial v_j}{\partial S_i} = \frac{S_i}{F_j}\frac{\partial F_j}{\partial S_i} + \frac{k_j^+ n_{ij}^- \prod_l S_l^{n_{lj}^-} - k_j^- n_{ij}^+ \prod_l S_l^{n_{lj}^+}}{k_j^+ \prod_l S_l^{n_{lj}^-} - k_j^- \prod_l S_l^{n_{lj}^+}}. \tag{5.78}$$

When the enzyme operates near equilibrium, the expression for ε_j simplifies considerably because the denominator of the second term on the right-hand side of Eq. (5.78) is nearly zero, so that the first term of the sum can be neglected in comparison with the second term. Furthermore, from Eqs. (2.11a) and (2.11b), it follows that

$$n_{ij} = n_{ij}^+ - n_{ij}^- . \tag{5.79}$$

Therefore, Eq. (5.78) can be simplified to

$$\varepsilon_{ji} \cong \frac{n_{ij}^- - n_{ij}^+ \prod_l S_l^{n_{lj}}/q_j}{1 - \prod_l S_l^{n_{lj}}/q_j} = \frac{n_{ij}^- - n_{ij}^+ \exp(-A_j/RT)}{1 - \exp(-A_j/RT)} \cong n_{ij}^+ - n_{ij}\frac{RT}{A}. \tag{5.80}$$

Again, the elasticity becomes, near equilibrium, independent of the kinetic properties of the enzymes, which are expressed by the function $F_j(\boldsymbol{S}, \boldsymbol{p})$. In Eq. (5.80), either n_{ij}^- or n_{ij}^+ is zero [cf. Eqs. (2.11a) and (2.11b)]. Note that Eq. (5.74a) is a special case of Eq. (5.80) with $n_{ij}^- = 1$ and $n_{ij}^+ = 0$, whereas Eq. (5.74b) is obtained with $n_{ij}^- = 0$ and $n_{ij}^+ = 1$.

5.4.3. Control Coefficients for Simple Hypothetical Pathways

5.4.3.1. Unbranched Chains

Many biochemical pathways (e.g., amino acid synthesis or glycolysis) can be modeled, in an idealized way, as unbranched reaction chains consisting of monomolecular reactions, provided that the concentrations of cofactors are kept constant. Because of their simple structure, these reaction chains, with or without feedback loops, have often been the subject of mathematical modeling (Kacser and Burns, 1973; Heinrich and Rapoport, 1974a; Savageau, 1976; Hofmeyr, 1989; Palsson *et al.*, 1985).

Unbranched reaction chains such as that shown in Scheme 11 are suitable systems to which the control analysis as presented in the preceding sections can be applied.

$$P_1 \underset{}{\overset{v_1}{\rightleftharpoons}} S_1 \underset{}{\overset{v_2}{\rightleftharpoons}} S_2 \cdots S_{n-1} \underset{}{\overset{v_n}{\rightleftharpoons}} S_n \underset{}{\overset{v_{n+1}}{\rightleftharpoons}} P_2 \qquad \text{Scheme 11}$$

At steady state, all reaction rates are equal to the steady-state flux

$$v_j = J, \text{ for } j = 1, \ldots, r = n + 1. \tag{5.81}$$

If the kinetic properties of the enzymes are described by the Michaelis–Menten equation (2.20) for reversible reactions, one may derive from the steady-state condition (2.9) the following expression for the metabolite concentrations:

$$S_i = P_1 \prod_{j=1}^{i} \frac{K_j^-(V_j^+ - J)}{K_j^+(V_j^- + J)} - J \sum_{l=1}^{i} \frac{K_l^+}{V_l^+ - J} \prod_{j=l}^{i} \frac{K_j^-(V_j^+ - J)}{K_j^+(V_j^- + J)} \tag{5.82}$$

(Heinrich *et al.*, 1987). K_j^+ and K_j^- denote the Michaelis constants of the substrate and product, respectively, of reaction j. V_j^+ and V_j^- are the maximal activities of the forward and backward reactions, respectively. Writing Eq. (5.82) for $i = n + 1$, one obtains, under the assumption that the concentrations of the pathway substrate, P_1, and of the end product, $S_{n+1} = P_2$, are constant, an expression which may be rearranged into a polynomial equation of order $n + 1$ for the flux J. For example, the steady-state flux of a chain with one internal metabolite S_1 and two reactions is determined by the quadratic equation

$$a_2 J^2 + a_1 J + a_0 = 0 \tag{5.83a}$$

with

$$a_0 = P_2 K_1^+ K_2^+ V_1^- V_2^- - P_1 K_1^- K_2^- V_1^+ V_2^+, \tag{5.83b}$$

$$a_1 = K_1^+ K_2^+ (P_2 V_1^- + P_2 V_2^- + K_2^- V_1^-) + K_1^- K_2^- (P_1 V_1^+ + P_1 V_2^+ + K_1^+ V_2^+), \tag{5.83c}$$

$$a_2 = K_1^+ K_2^+ (P_2 + K_2^-) - K_1^- K_2^- (P_1 + K_1^+). \tag{5.83d}$$

The concentration S_1 is obtained by introducing the solution of Eq. (5.83) into Eq. (5.82).

Due to Eq. (5.81), the matrix of flux control coefficients has the property that all its rows are identical. Therefore, it can be reduced to a vector with the elements $C_j^J = C_{ij}^J$ for all $i = 1, \ldots, r$.

Unbranched pathways with nonsaturated enzymes: If all enzymes operate under nonsaturating conditions, that is,

$$J \ll V_j^+, V_j^-, \tag{5.84a}$$

$$S_i \ll K_i^-, K_{i+1}^+, \tag{5.84b}$$

the reaction rates are described by linear equations

$$v_j = k_j S_{j-1} - k_{-j} S_j = k_j \left(S_{j-1} - \frac{S_j}{q_j} \right), \tag{5.85}$$

where

$$k_j = \frac{V_j^+}{K_j^+}, \quad k_{-j} = \frac{V_j^-}{K_j^-}, \quad q_j = \frac{V_j^+ K_j^-}{V_j^- K_j^+} \tag{5.86}$$

[cf. the Haldane relation (2.26))] denote the first-order rate constants and thermodynamic equilibrium constants, respectively. Equation (5.82) simplifies to

$$S_i = P_1 \prod_{j=1}^{i} q_j - J \sum_{l=1}^{i} \frac{1}{k_l} \prod_{j=l}^{i} q_j. \tag{5.87}$$

From this equation one obtains, with $S_{n+1} = P_2$, an expression for the steady-state flux

$$J = \frac{P_1 \prod_{j=1}^{n+1} q_j - P_2}{\sum_{l=1}^{n+1} \frac{1}{k_l} \prod_{j=l}^{n+1} q_j}. \tag{5.88}$$

Because the steady-state flux can be written as an analytical function of the system parameters for the case of unbranched reaction chains with linear kinetics, control coefficients can be calculated in closed form in terms of these parameters.

Let us use the rate constants k_j as perturbation parameters in such a way that the equilibrium constants are not changed (i.e., k_j and k_{-j} are changed by the same fractional amount). This is realized, for example, by changes in the enzyme concentrations. One then obtains, under consideration of $\partial v_j / \partial k_j = v_j / k_j$,

$$C_j^J = \frac{v_j}{J} \frac{\partial J / \partial k_j}{\partial v_j / \partial k_j} = \frac{k_j}{J} \frac{\partial J}{\partial k_j}, \tag{5.89}$$

which with Eq. (5.88) yields

$$C_i^J = \frac{\frac{1}{k_i} \prod_{j=i}^{n+1} q_j}{\sum_{l=1}^{n+1} \frac{1}{k_l} \prod_{j=l}^{n+1} q_j} \tag{5.90}$$

(Heinrich and Rapoport, 1974a). Equation (5.90) implies

$$0 \le C_j^J \le 1, \tag{5.91a}$$

$$\sum_{j=1}^{n+1} C_j^J = 1. \tag{5.91b}$$

It is worth noting that for unbranched chains, normalization does not change the values of the flux control coefficients, because at steady state all reaction rates are equal to the steady-state flux.

Equation (5.90) shows that C_j^J bears a direct relation to $1/k_j$; that is, fast (slow) reactions generally have low (high) control coefficients. However, the values of the flux control coefficients are also strongly dependent on the thermodynamic equilibrium constants and on the position of the reaction within the chain. This may best be demonstrated by considering the ratio of two flux control coefficients

$$\frac{C_j^J}{C_i^J} = \frac{k_i}{k_j} \prod_{l=j}^{i-1} q_l \quad \text{with } i > j. \tag{5.92}$$

From this equation, one derives, for example, that the flux control coefficient of a reaction i which is located beyond an irreversible reaction s with $q_s \to \infty$ becomes vanishingly small for any finite value k_i.

One may also take the k_j as perturbation parameters in such a way that the backward rate constants, k_{-j}, are not changed. This possibility will be discussed in more detail in Section 5.6.2.

Using Eq. (5.88), flux changes may be calculated also for finite parameter perturbations with all equilibrium constants fixed. With $\Delta J = J(k_1 + \Delta k_1, \ldots, k_{n+1} + \Delta k_{n+1}) - J(k_1, \ldots, k_{n+1})$, one derives, with the help of Eqs. (5.88) and (5.90),

$$\frac{\Delta J}{J} = \sum_{j=1}^{n+1} C_j^J \left(\frac{\Delta v_j / v_j}{1 + \Delta v_j / v_j} \right) \left(\sum_{j=1}^{n+1} \frac{C_j^J}{1 + \Delta v_j / v_j} \right)^{-1}. \tag{5.93}$$

with $\Delta v_j = \Delta k_j (S_{j-1} - S_j / q_j)$ denoting the perturbations of reaction rates considered as if the reactions proceeded in isolation. Equation (5.93) means that for unbranched chains with linear rate laws, flux control by arbitrary rate perturbations Δv_j can be characterized completely by the flux control coefficients originally defined for infinitesimal perturbations. For very small values of $\Delta v_j / v_j$, Eq. (5.93) and the summation relationship (5.91b) entail the linear approximation

$$\frac{\Delta J}{J} = \sum_{j=1}^{n+1} C_j^J \frac{\Delta v_j}{v_j}. \tag{5.94}$$

In the case that only one reaction k is perturbed, Eq. (5.93) becomes (cf. Small and Kacser, 1993; Höfer and Heinrich, 1993)

$$\frac{\Delta J}{J} = \frac{C_k^J \Delta v_k / v_k}{1 + (1 - C_k^J)\Delta v_k / v_k}. \tag{5.95}$$

This equation has some importance for the question of whether control coefficients, which are defined for infinitesimal parameter changes, are helpful for estimating the effect of finite changes, as will be discussed in Section 5.9.

General treatment: If the reaction rates of the enzymes are described by *nonlinear* kinetic equations (e.g., the Michaelis–Menten kinetics), there are generally no explicit expressions for the steady-state flux or for the metabolite concentrations, and the control coefficients cannot be calculated by direct differentiation. However, much insight is gained by application of the summation and connectivity theorems which allow to express the C_i^J and C_{ij}^S as functions of the elasticities (see Section 5.3.3).

Let us first consider the unbranched two-enzyme system depicted in Scheme 10 (Section 5.3). The summation and connectivity theorems for the flux control coefficients read

$$C_1^J + C_2^J = 1, \tag{5.96a}$$

$$C_1^J \varepsilon_{11} + C_2^J \varepsilon_{21} = 0, \tag{5.96b}$$

which have the solutions

$$C_1^J = \frac{\varepsilon_{21}}{\varepsilon_{21} - \varepsilon_{11}}, \quad C_2^J = -\frac{\varepsilon_{11}}{\varepsilon_{21} - \varepsilon_{11}}. \tag{5.97}$$

For the concentration control coefficients, these theorems read

$$C_{11}^S + C_{12}^S = 0, \tag{5.98a}$$

$$C_{11}^S \varepsilon_{11} + C_{12}^S \varepsilon_{21} = -1. \tag{5.98b}$$

From these equations, one obtains

$$C_{11}^S = \frac{1}{\varepsilon_{21} - \varepsilon_{11}}, \quad C_{12}^S = -\frac{1}{\varepsilon_{21} - \varepsilon_{11}}. \tag{5.99}$$

In the typical situation that $\varepsilon_{11} < 0$ and $\varepsilon_{21} > 0$ (neither product activation nor substrate inhibition), Eq. (5.97) requires that both flux control coefficients be positive. The control coefficient of reaction 1 with respect to the intermediate concentration is also positive, whereas the concentration control coefficient of reaction 2 is negative.

For the case $n > 1$, we assume that v_i is only affected by the concentrations of its substrate (S_{i-1}) and of its product (S_i). Then the elasticity coefficients read

$$\varepsilon_{ij} = \varepsilon_{ii}\,\delta_{ij} + \varepsilon_{i,i-1}\,\delta_{i-1,j}, \tag{5.100}$$

and the connectivity theorem (5.53b) assumes the form

$$C_i^J \varepsilon_{ii} + C_{i+1}^J \varepsilon_{i+1,i} = 0. \tag{5.101}$$

We here assume all reactions to be reversible, so that all ε_{ii} are nonzero. Equation (5.101) implies

$$C_i^J = C_{n+1}^J \prod_{j=i}^{n} \left(-\frac{\varepsilon_{j+1,j}}{\varepsilon_{jj}} \right) \tag{5.102}$$

for $1 \leq i \leq n$. The coefficient C_{n+1}^J which enters Eq. (5.102) can be determined using the summation theorem (5.43). One obtains

$$C_i^J = \frac{\prod_{j=i}^{n} \left(-\frac{\varepsilon_{j+1,j}}{\varepsilon_{jj}} \right)}{1 + \sum_{j=1}^{n} \prod_{l=j}^{n} \left(-\frac{\varepsilon_{l+1,l}}{\varepsilon_{ll}} \right)}, \tag{5.103}$$

as a general expression for the flux control coefficients in an unbranched enzymic chain.

Equation (5.101) implies that the ratio of the control coefficients of two neighboring reactions equals the negative inverse ratio of the elasticities of these reactions with respect to the intermediate shared by these reactions. Because near-equilibrium enzymes have high elasticities [cf. Eqs. (5.74a) and (5.47b)], this implies that these enzymes exert almost no flux control. Furthermore, when an enzyme is nearly saturated with its substrate, the elasticity with respect to the latter is very low. Equation (5.103) then implies, in general, that the control coefficient of substrate-saturated enzymes be high. When an enzyme is saturated with its product, it follows from this equation that the control coefficient of the subsequent enzyme in the chain is, in general, very low.

In the case that

$$\varepsilon_{ii} < 0, \quad \varepsilon_{i+1,i} > 0 \quad \text{for any } i, \tag{5.104}$$

it follows from Eq. (5.103) that all flux control coefficients are non-negative.

Therefore, the summation theorem implies that they are all smaller than, or equal to unity. Using linear kinetics, one can easily transform Eq. (5.103), which expresses the flux control coefficients in terms of elasticities, into Eq. (5.90), which expresses them in terms of system parameters.

For the calculation of the *concentration control coefficients* we use the connectivity theorem (5.53a) which reads, under consideration of Eq. (5.100),

$$C^S_{ij}\,\varepsilon_{jj} + C^S_{i,j+1}\,\varepsilon_{j+1,j} = -\delta_{ij}. \tag{5.105}$$

Applying this equation for $j \neq i$, one obtains the following two recurrent formulas:

$$C^S_{ij} = -C^S_{i,j+1}\frac{\varepsilon_{j+1,j}}{\varepsilon_{jj}} \quad \text{for } 1 \le j < i \tag{5.106a}$$

and

$$C^S_{i,j+1} = -C^S_{i,j}\frac{\varepsilon_{jj}}{\varepsilon_{j+1,j}} \quad \text{for } i+1 \le j \le n. \tag{5.106b}$$

As the flux control coefficients are known [cf. Eq. (5.103)] it is appropriate to replace the ratio $\varepsilon_{j+1,j}/\varepsilon_{jj}$ in Eqs. (5.106a) and (5.106b) by the ratio $-C^J_j/C^J_{j+1}$ [cf. Eq. (5.101)]. The resulting equations may be used to express all coefficients C^S_{ij} with $j < i$ and $j > i + 1$ as functions of C^S_{ii} and $C^S_{i,i+1}$, respectively, and of flux control coefficients,

$$C^S_{ij} = C^S_{ii}\frac{C^J_j}{C^J_i} \quad \text{for } 1 \le j \le i \tag{5.107a}$$

and

$$C^S_{ij} = C^S_{i,i+1}\frac{C^J_j}{C^J_{i+1}} \quad \text{for } i+1 \le j \le n+1. \tag{5.107b}$$

With Eqs. (5.107a) and (5.107b), the summation theorem for the concentration control coefficients reads

$$\frac{C^S_{ii}}{C^J_i}\sum_{j=1}^{i} C^J_j + \frac{C^S_{i,i+1}}{C^J_{i+1}}\sum_{j=i+1}^{n+1} C^J_j = 0. \tag{5.108}$$

This equation and the connectivity theorem (5.105) applied for $i = j$ represent two linear equations which can be solved for C^S_{ii} and $C^S_{i,i+1}$. From these, in turn, all concentration control coefficients can be calculated using Eqs. (5.107a) and (5.107b). The final result is

$$C_{ij}^{S} = \frac{C_j^J}{C_{i+1}^J \varepsilon_{i+1,i}} \sum_{k=i+1}^{n+1} C_k^J \quad \text{for } 1 \leq j \leq i, \tag{5.109a}$$

$$C_{ij}^{S} = \frac{C_j^J}{C_i^J \varepsilon_{ii}} \sum_{k=1}^{i} C_k^J \quad \text{for } i+1 \leq j \leq n+1. \tag{5.109b}$$

Under the assumption that $\varepsilon_{ii} < 0$ and $\varepsilon_{i+1,i} > 0$ [cf. relation (5.104)] which implies positive flux control coefficients, one may derive from Eqs. (5.109a) and (5.109b) $C_{ij}^S > 0$ for $j \leq i$ and $C_{ij}^S < 0$ for $j > i$; that is, activation of an enzyme leads to a decrease of the concentrations of all metabolites which are located upstream, whereas all metabolite concentrations downstream are increased. This fact is also expressed by the crossover theorem [see Higgins (1965) and Section 5.10.1].

Further conclusions from Eq. (5.109) are (a) very fast enzymes which exert no flux control ($C_j^J \cong 0$) also have vanishing concentration control coefficients; (b) when all enzymes downstream a metabolite S_i or all enzymes upstream this metabolite are very fast, so that they have very low flux control coefficients, then all control coefficients with respect to the concentration of this metabolite are very small. This may be explained by the fact that in these cases all metabolites S_j are in quasi-equilibrium with the end product or with the initial substrate of the chain.

Unbranched chain with feedback inhibition: Feedback inhibition is a frequent phenomenon in biochemical pathways. The physiological role of such regulatory loops for homeostasis has intensely been discussed (Umbarger, 1956; Othmer, 1976; Dibrov *et al.,* 1982). The apparatus of metabolic control analysis can be used to quantify such homeostatic effects.

Let us consider the reaction chain shown in Scheme 6 (Section 2.4.6) under the simplifying assumption that all enzymes catalyze irreversible reactions, that is,

$$\varepsilon_{ij} \begin{cases} = 0 & \text{for } j = i \\ \neq 0 & \text{for } j = i - 1. \end{cases} \tag{5.110}$$

It is further assumed that the feedback is exerted by the metabolite S_n which acts as an inhibitor of the first reaction, which means

$$\varepsilon_{1,n} = \frac{S_n}{v_1} \frac{\partial v_1}{\partial S_n} < 0. \tag{5.111}$$

Let us first consider the *flux control coefficients.* Using Eq. (5.110), it follows immediately from Eq. (5.101) that

$$C_j^J = 0 \quad \text{for } j = 2, \ldots, n \tag{5.112}$$

and the connectivity and summation theorems for the two remaining coefficients C_1^J and C_{n+1}^J read

$$C_1^J \,\varepsilon_{1,n} + C_{n+1}^J \,\varepsilon_{n+1,n} = 0, \tag{5.113a}$$

$$C_1^J + C_{n+1}^J = 1. \tag{5.113b}$$

Equation system (5.113) has the solution

$$C_1^J = \frac{\varepsilon_{n+1,n}}{\varepsilon_{n+1,n} - \varepsilon_{1,n}}, \quad C_{n+1}^J = -\frac{\varepsilon_{1,n}}{\varepsilon_{n+1,n} - \varepsilon_{1,n}}. \tag{5.114}$$

In the absence of feedback inhibition ($\varepsilon_{1,n} = 0$) only the first enzyme exerts the flux control ($C_1^J = 1$, $C_{n+1}^J = 0$). For $\varepsilon_{1,n} \neq 0$, flux control is shared by two enzymes: the first enzyme E_1 and the enzyme E_{n+1} which degrades the inhibitor S_n. If $\varepsilon_{n+1,n} > 0$ (which is generally fulfilled because S_n is the substrate of E_{n+1}), one derives $C_1^J > 0$ and $C_{n+1}^J > 0$ from Eq. (5.114). Because

$$\frac{C_{n+1}^J}{C_1^J} = -\frac{\varepsilon_{1,n}}{\varepsilon_{n+1,n}}, \tag{5.115}$$

the flux control is shifted entirely to the end of the chain if if the feedback inhibition is very strong, that is, $|\varepsilon_{1,n}| \gg |\varepsilon_{n+1,n}|$.

In a similar way, the summation and connectivity theorems can be used to calculate the *concentration control coefficients.* One obtains for the coefficients of the first enzyme

$$C_{i,1}^S = \frac{\varepsilon_{n+1,n}/\varepsilon_{i+1,i}}{\varepsilon_{n+1,n} - \varepsilon_{1,n}} = \frac{C_1^J}{\varepsilon_{i+1,i}} > 0 \tag{5.116}$$

($i = 1, \ldots, n$) and of the last enzyme

$$C_{i,n+1}^S = \begin{cases} -\dfrac{\varepsilon_{1,n}/\varepsilon_{i+1,i}}{\varepsilon_{n+1,n} - \varepsilon_{1,n}} = \dfrac{C_{n+1}^J}{\varepsilon_{i,i+1}} > 0 & \text{for } i \neq n \quad (5.117a) \\[2ex] -\dfrac{1}{\varepsilon_{n+1,n} - \varepsilon_{1,n}} < 0 & \text{for } i = n. \quad (5.117b) \end{cases}$$

The enzymes $E_2, \ldots, E_n$ which exert no flux control [cf. Eq. (5.112)] generally have nonvanishing control coefficients with respect to the concentrations of their substrates. One obtains

$$C^S_{i,i+1} = -\frac{1}{\varepsilon_{i+1,i}} < 0 \tag{5.118}$$

($i = 2, \ldots, n - 1$). In Eqs. (5.116)–(5.118), the inequalities refer to the case $\varepsilon_{i+1,i} > 0$, and all coefficients C^S_{ij} not listed are equal to zero.

It follows from Eq. (5.116) that a very strong feedback inhibition ($|\varepsilon_{1,n}| \gg 1$) results in very low concentration control coefficients of the input reaction with respect to all metabolites ($C^S_{i,1} \ll 1$). Furthermore, Eq. (5.117b) implies that in this situation the control of the last reaction with respect to the last intermediate, S_n, is also very weak ($|C^S_{n,n+1}| \gg 1$). Both facts indicate the homeostatic effect of the negative feedback loop (see Section 5.10.1).

5.4.3.2. *A Branched System*

For the reaction system depicted in Scheme 7 (Section 3.2.4) the stoichiometry matrix reads $\mathbf{N} = (1\ -1\ -1)$. Using

$$\boldsymbol{k}_1 = (1\ 1\ 0)^{\mathrm{T}}, \quad \boldsymbol{k}_2 = (1\ 0\ 1)^{\mathrm{T}} \tag{5.119}$$

as basis vectors for the null-space of $\mathbf{N}$ (see Section 3.2), the summation and connectivity relations for the unscaled control coefficients may be subsumed into the following matrix equation:

$$\begin{pmatrix} C^J_{11} & C^J_{12} & C^J_{13} \\ C^J_{21} & C^J_{22} & C^J_{23} \\ C^J_{31} & C^J_{32} & C^J_{33} \end{pmatrix} \begin{pmatrix} 1 & 1 & \varepsilon_{11} \\ 1 & 0 & \varepsilon_{21} \\ 0 & 1 & \varepsilon_{31} \end{pmatrix} = \begin{pmatrix} 1 & 1 & 0 \\ 1 & 0 & 0 \\ 0 & 1 & 0 \end{pmatrix}. \tag{5.120}$$

Note that this equation is a special case of Eq. (5.54). Solving this linear equation system for the flux control coefficients leads to

$$\mathbf{C}^J = \frac{1}{\varepsilon_{11} - \varepsilon_{21} - \varepsilon_{31}} \begin{pmatrix} -(\varepsilon_{21} + \varepsilon_{31}) & \varepsilon_{11} & \varepsilon_{11} \\ -\varepsilon_{21} & \varepsilon_{11} - \varepsilon_{31} & \varepsilon_{21} \\ -\varepsilon_{31} & \varepsilon_{31} & \varepsilon_{11} - \varepsilon_{21} \end{pmatrix}. \tag{5.121}$$

For example, in the usual case that $\varepsilon_{11} < 0$ (product inhibition) and $\varepsilon_{21}, \varepsilon_{31} >$

0 (substrate activation) one derives immediately from Eq. (5.121) that all flux control coefficients C^J_{ij} are positive except for C^J_{23} and C^J_{32}, which are negative.

The system represented in Scheme 7 may also serve as an example to illustrate that the *branch-point relationships* introduced by Fell and Sauro (1985) are directly related to the generalized summation theorem (5.45b). Taking this relation for one column, $\boldsymbol{k}$, of the null-space matrix gives $\mathbf{C}^J(\mathrm{dg}\boldsymbol{J})^{-1}\boldsymbol{k} = (\mathrm{dg}\boldsymbol{J})^{-1}\boldsymbol{k}$ for the normalized flux control coefficients. Using for the system depicted in Scheme 7, $\boldsymbol{k} = \boldsymbol{J} = (J_1\, J_2\, J_3)^{\mathrm{T}}$, one arrives at summation relationships for the three fluxes saying that for a given flux, the sum of all control coefficients equals unity. The branch-point relationships are obtained by choosing the $\boldsymbol{k}$ vector in three other ways. With $\boldsymbol{k} = (0\ 1\ -1)^{\mathrm{T}}$ one derives from Eq. (5.45b) that

$$\frac{C^J_{12}}{J_2} - \frac{C^J_{13}}{J_3} = 0. \tag{5.122}$$

Using the following abbreviations for the flux ratios

$$\frac{J_2}{J_1} = a, \qquad \frac{J_3}{J_1} = \frac{J_1 - J_2}{J_1} = 1 - a, \tag{5.123}$$

one obtains the first branch-point relation

$$(1 - a)C^J_{12} - aC^J_{13} = 0. \tag{5.124a}$$

Similarly, with $\boldsymbol{k} = (1\ 0\ 1)^{\mathrm{T}}$, one obtains

$$(1 - a)C^J_{21} + C^J_{23} = 0 \tag{5.124b}$$

and with $\boldsymbol{k} = (1\ 1\ 0)^{\mathrm{T}}$,

$$aC^J_{31} + C^J_{32} = 0. \tag{5.124c}$$

Equations (5.124a)–(5.124c) represent the branch-point relationships for the reaction system shown in Scheme 7, which together with the three summation relationships and the three connectivity relationships are sufficient to calculate the nine flux-control coefficients as functions of the elasticities and the flux ratio a.

5.4.4. Control of Erythrocyte Energy Metabolism

5.4.4.1. The Reaction System

We consider glycolysis in erythrocytes to demonstrate how the control properties of a real pathway may be derived on the basis of a mathematical model.

The glycolytic system has attracted much attention of both experimentalists and theoreticians for many years. This concerns, for example, glycolytic oscillations (see Section 2.4.3). However, in addition to oscillatory modes, steady states are very frequently observed. The regulatory principles of these states are worth being investigated and still involve many unsolved problems, despite the fact that many of the glycolytic enzymes have been purified and characterized kinetically and that reliable flux and concentration data exist for different conditions. One reason for the difficulties encountered is the fact that in many cells the glycolytic pathway is interconnected with other pathways such as respiration, gluconeogenesis, and the pentose phosphate pathway. In order to study glycolysis *per se,* the choice of an appropriate simple biological system is therefore of great importance. In the present section we consider mature mammalian erythrocytes where the metabolism is reduced virtually to glycolysis with some contribution of the pentose phosphate pathway. However, even in the glycolytic system of the erythrocyte, a rather high number of enzymes participate which are coupled with each other. For the purpose of deducing the essential relations in metabolism, appropriate simplifications have to be introduced in setting up a model. In particular, it is taken into account that glycolysis, as most other biochemical pathways, includes slow and very fast enzymes. This allows to apply the rapid-equilibrium approximation which leads to a reduction of the number of variables and parameters (see Section 4.3). Furthermore, the models of erythrocyte metabolism presented below neglect the pentose phosphate pathway because its contribution in the consumption of glucose is only 10% at pH 7.2.

The reactions taken into account are depicted in Figure 3.1. These are (1) the reactions of the Embden-Meyerhof pathway, (2) the two reactions of the $2,3P_2G$ bypass, and (3) the nonglycolytic ATP-consuming processes which are partly coupled to the active transport of sodium and potassium across the cellular membrane. The full stoichiometry matrix of this system is given in Table 3.1.

In this section we present three different models of erythrocyte metabolism characterized by increasing complexity. Each model has its own limits of validity. Model A (Section 5.4.4.2) neglects all nonglycolytic processes, in particular the nonglycolytic ATP-consuming processes. Model B (Section 5.4.4.3) takes into account the interplay between ATP-producing and ATP-consuming processes, and Model C (Section 5.4.4.4) considers in some detail the interaction between energy metabolism and osmotic properties of erythrocytes.

5.4.4.2. *Basic Model*

This model of the glycolytic system (Model A) is based on the following assumptions (see Heinrich and Rapoport, 1973; Rapoport *et al.,* 1974):

1. Among the glycolytic reactions, one may distinguish two different groups of en-

Table 5.1 Values of Parameters and Variables of Model A (*In Vivo* Steady State)

Parameter	Value	Variable	Value
Metabolite Concentrations		*Flux Rates*	
ATP	1.2 mM	$v_{HK\text{-}PFK}$ $(=J)$	1.25 mM/h
ADP	0.22 mM	v_{P_2GM}	0.75 mM/h
Lac	1.4 mM		
Pyr	0.08 mM	*Metabolite Concentrations*	
		G6P	0.06 mM
Kinetic Constants		*F6P*	0.025 mM
k_{HK}	1.94/h	FP_2	0.007 mM
k_{PFK}	50.8/h	*GAP*	0.006 mM
k_{PK}	125.0/h	*DHAP*	0.13 mM
k_{P_2GM}	3.76×10^5/h	$1{,}3P_2G$	0.0005 mM
V_{m,P_2Gase}	0.75 mM/h	$2{,}3\ P_2G$	5.0 mM
		3PG	0.069 mM
Equilibrium Constants			
		2PG	0.12 mM
q_{PGI}	0.41	*PEP*	0.02 mM
q_{Ald}	0.102 mM	*NAD/NADH*	2500
q_{TIM}	22.0		
q_{GAPD}	0.34×10^{-3}		
q_{PGK}	742.0		
q_{PGAM}	0.17		
q_{Enol}	1.7		
q_{LDH}	4.4×10^4		

zymes. The first group encompasses enzymes which catalyze quasi-irreversible reactions with high equilibrium constants q_j $(-\Delta G_j^0 \gg RT)$. To this group belong hexokinase (HK, EC 2.7.1.1), phosphofructokinase (PFK, EC 2.7.1.11), bisphosphoglycerate mutase (P_2GM, EC 5.4.2.4), 2,3-bisphosphoglycerate phosphatase (P_2Gase, EC 3.1.3.13), and pyruvate kinase (PK, EC 2.7.1.40). Another group of enzymes catalyzes reversible reactions, for which the mass-action ratios differ little from the equilibrium constants. To this class belong phosphoglucoisomerase (PGI, EC 5.3.1.9), fructose-bisphosphate aldolase (Ald, EC 4.1.2.13), triose-phosphate isomerase (TIM, EC 5.3.1.1), glyceraldehyde-3-phosphate dehydrogenase, (GAPD, EC 1.2.1.12), phosphoglycerate kinase (PGK, EC 2.7.2.3), phosphoglycerate mutase (PGAM, EC 5.4.2.1), enolase, (Enol, EC 4.2.1.11), and lactate dehydrogenase (LDH, EC 1.1.1.27). In accordance with experimental data, it is assumed that these enzymes are fast compared to those of the former group. The rapid-equilibrium approximation (see Section 4.3) leads to equilibrium conditions for the corresponding reactions.

2. The concentration of the adenine nucleotides ADP and ATP are considered to be fixed; that is, they are parameters of the model. Their concentrations are determined not only by glycolysis but by the interplay of ATP-producing and ATP-consuming processes (see Model B).
3. Among the many regulatory couplings realized by the action of metabolites as activators or inhibitors, only the feedback inhibitions of HK by G6P and of P_2GM by $2,3P_2G$ are taken into account.
4. Simple rate laws were used to characterize the kinetic properties of the enzymes which were based on a linear relationship between enzymatic activity and substrate concentrations. The fast enzymes were characterized solely by the equilibrium constants.
5. The mathematical treatment is confined to the steady state observed under *in vivo* conditions. The model serves two purposes: (a) calculation of the glycolytic flux $J = J_{\mathrm{glyc}}$ and of the metabolite concentrations as functions of the model parameters and comparison of the results with experimental data; (b) characterization of the control of the glycolytic flux by evaluating the control coefficients C_j^J of the glycolytic enzymes.

Under steady-state conditions, the reaction rates of the enzymes HK, PFK, and PK must fulfill the following conditions:

$$J = v_{\mathrm{HK}} = v_{\mathrm{PFK}}, \tag{5.125a}$$

$$2J = v_{PK}, \tag{5.125b}$$

where J represents the steady-state flux of glycolysis (consumption of glucose). The factor 2 in Eq. (5.125b) indicates that the flux beyond the aldolase is twice that through the PFK. By use of the rate equations

$$v_{\mathrm{HK}} = \frac{k_{\mathrm{HK}} ATP}{1 + \dfrac{G6P}{K_{\mathrm{I,G6P}}}}, \quad v_{\mathrm{PFK}} = k_{\mathrm{PFK}} F6P \tag{5.126}$$

($K_{\mathrm{I},G6P}$: inhibition constant of glucose-6-phosphate) and the equilibrium relation

$$\frac{F6P}{G6P} = q_{\mathrm{PGI}}, \tag{5.127}$$

Eq. (5.125a) becomes a quadratic equation for the concentration of F6P,

$$(F6P)^2 + q_{\mathrm{PGI}} K_{\mathrm{I,G6P}} F6P - \frac{ATP \cdot k_{\mathrm{HK}} q_{\mathrm{PGI}} K_{\mathrm{I,G6P}}}{k_{\mathrm{PFK}}} = 0. \tag{5.128}$$

Under *in vivo* conditions the concentration of glucose does not enter the rate equation of HK [Eq. (5.126a)] because the intracellular concentration of glucose ($Gluc \cong 5$ mM) is much higher than its K_m value ($K_{m,Gluc} \cong 40$ µM).

Inserting the relevant solution of Eq. (5.128) into Eq. (5.126b), one gets

$$J = k_{HK}ATP\left(\frac{1}{2} + \sqrt{\frac{1}{4} + \frac{k_{HK}ATP}{q_{PGI}k_{PFK}K_{I,G6P}}}\right)^{-1} \tag{5.129}$$

for the glycolytic flux. Using the linear relation

$$v_{PK} = k_{PK}PEP \tag{5.130}$$

and taking into account the steady-state condition (5.125b), the concentration of PEP is determined by

$$PEP = \frac{2J}{k_{PK}} \tag{5.131}$$

with J given by Eq. (5.129). The concentrations of all other metabolites may be obtained from PEP by consideration of the equilibrium conditions for the fast enzymes as well as of the steady-state condition of the two enzymes of the 2,3P_2G-bypass. As enolase, phosphoglycerate mutase, and phosphoglycerate kinase are quasi-equilibrium enzymes, we have

$$\frac{PEP}{2PG} = q_{Enol}, \quad \frac{2PG}{3PG} = q_{PGAM}, \quad \frac{3PG \cdot ATP}{1{,}3P_2G \cdot ADP} = q_{PGK} \tag{5.132}$$

with $2PG$ and $3PG$ denoting the concentrations of 2-phosphoglycerate and 3-phosphoglycerate, respectively. Therefore, with formula (5.131) one gets

$$2PG = \frac{2J}{k_{PK}q_{Enol}}, \tag{5.133a}$$

$$3PG = \frac{2J}{k_{PK}q_{Enol}q_{PGAM}}, \tag{5.133b}$$

$$1{,}3P_2G = \frac{2J \cdot ATP}{k_{PK}q_{Enol}q_{PGAM}q_{PGK}ADP}. \tag{5.133c}$$

The concentration of pyruvate and lactate are maintained *in vivo* at almost constant levels by the interplay of various tissues. Therefore, the concentrations *Pyr* and *Lac* may be considered as parameters. The *NAD/NADH* ratio is, therefore, fixed by the equilibrium condition for the LDH reaction,

$$\frac{NAD}{NADH} = q_{\mathrm{LDH}} \frac{Pyr}{Lac}. \tag{5.134}$$

The concentrations of GAP, DHAP, and FP_2 are calculated on the basis of the equilibrium conditions for the enzymes GAPD, TIM, and aldolase in the following way

$$GAP = \frac{1{,}3P_2G \cdot NADH}{q_{\mathrm{GAPD}} \cdot NAD}, \tag{5.135a}$$

$$DHAP = q_{\mathrm{TIM}} \cdot GAP, \tag{5.135b}$$

$$FP_2 = \frac{GAP \cdot DHAP}{q_{\mathrm{Ald}}} \tag{5.135c}$$

with $1{,}3P_2G$ and *NAD/NADH* resulting from Eqs. (5.133c) and (5.134), respectively.

The calculation of the $2{,}3P_2G$ concentration requires the steady-state condition

$$v_{\mathrm{P_2GM}} = v_{\mathrm{P_2Gase}}. \tag{5.136}$$

Under *in vivo* conditions the enzyme P_2Gase is saturated by its substrate, because the concentration of $2{,}3P_2G$ ($\cong$5 mM) is about 500 times higher than the corresponding K_m value. Considering the inhibition of the P_2GM by its product $2{,}3P_2G$, one may use the kinetic equations

$$v_{\mathrm{P_2GM}} = \frac{k_{\mathrm{P_2GM}} \cdot 1{,}3P_2G}{(1 + 2{,}3P_2G/K_{\mathrm{I;2,3P_2G}})}, \tag{5.137a}$$

$$v_{\mathrm{P_2Gase}} = V_{m,\mathrm{P_2Gase}}. \tag{5.137b}$$

With these equations one obtains from Eq. (5.136)

$$2{,}3P_2G = K_{\mathrm{I;2,3P_2G}}\left(\frac{k_{\mathrm{P_2GM}} \cdot 1{,}3P_2G}{V_{m,\mathrm{P_2Gase}}} - 1\right) \tag{5.138}$$

for the $2{,}3P_2G$ concentration, where for $1{,}3P_2G$ one has to insert formula (5.133c).

The values of the parameters and variables of Model A are listed in Table 5.1. The values approximate the experimental data obtained by kinetic characterization of the isolated enzymes as well as by determination of the flux rates and metabolite concentrations under *in vivo* conditions.

Inspection of formula (5.129) shows that under the given model assumptions, the glycolytic flux is only dependent on the kinetic parameters of the enzymes HK and PFK and on the equilibrium constant of PGI. It does not depend on the kinetic properties of the PK and the fast equilibrium enzymes. Therefore, the summation theorem for the normalized flux control coefficients assumes the form

$$C^J_{HK} + C^J_{PFK} = 1. \tag{5.139}$$

Because the parameters k_{HK} and k_{PFK} enter the rate laws of HK and PFK, respectively, in a linear manner the coefficients C^J_{HK} and C^J_{PFK} may be calculated as follows:

$$C^J_{HK} = \frac{\partial \ln J}{\partial \ln k_{HK}}, \quad C^J_{PFK} = \frac{\partial \ln J}{\partial \ln k_{PFK}}. \tag{5.140}$$

By use of Eq. (5.129) one gets

$$C^J_{HK} = 1 - C^J_{PFK} = 1 - \frac{a/2}{(1/2 + \sqrt{1/4 + a})\sqrt{1/4 + a}} \tag{5.141a}$$

with

$$a = \frac{k_{HK}\, ATP}{k_{PFK}\, q_{PGI}\, K_{I,G6P}}. \tag{5.141b}$$

One may easily see that

$$C^J_{HK} \geq C^J_{PFK} \tag{5.142}$$

always; that is, flux control is exerted mainly by the hexokinase, the first enzyme of the glycolytic pathway. Using the parameter values listed in Table 5.1 one obtains $C^J_{HK} = 0.69$ and $C^J_{PFK} = 0.31$. The participation of phosphofructokinase in flux control results from the feedback inhibition of hexokinase by G6P. An inhibition of the PFK, for example, would lead to an increase of its substrate F6P as well as of G6P, which would diminish the glycolytic flux by inhibition of the hexokinase. Elimination of the feedback inhibition of G6P ($K_{I,G6P} \to \infty$) would result in $C^J_{PFK} \to 0$ [cf. Eq. (5.141)].

Despite the higher control coefficient of the HK, the enzyme PFK may play an important role in the regulation of glycolysis owing to the high elasticity coefficients of a great number of internal and external effectors for this enzyme (see Otto *et al.*, 1974, 1977).

5.4.4.3. Interplay of ATP Production and ATP Consumption

Selkov (1975a, 1975b) proposed a skeleton model of glycolysis which, in contrast to Model A presented in Section 5.4.4.2, focuses on the production and degradation of ATP. It is described by the reaction scheme 12,

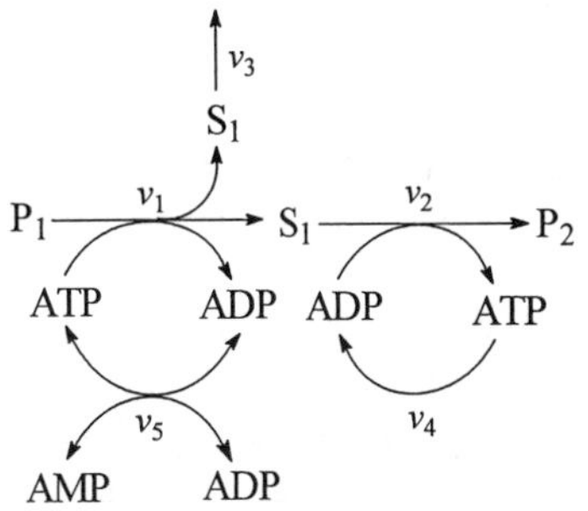

Scheme 12

where P_1 represents glucose and S_1 the pool of metabolites in the middle part of glycolysis. The ATP-consuming reactions of the upper part and the ATP-producing reactions of the lower part of glycolysis are lumped into reactions 1 and 2, respectively. v_3 represents the velocity of a side reaction without ATP production (describing, for example, the biosynthetic reactions leading to the synthesis of serine). v_4 denotes the rate of nonglycolytic ATP-consuming reactions (ATPases) and v_5 the rate of the adenylate kinase reaction (AK, EC 2.7.4.3). The model of Selkov has been modified by Heinrich and Rapoport (1975) by taking into account special features of erythrocyte glycolysis, in particular the 2,3P_2G bypass (Model B, see Figure 5.3). The reaction scheme results from that depicted in Figure 3.1

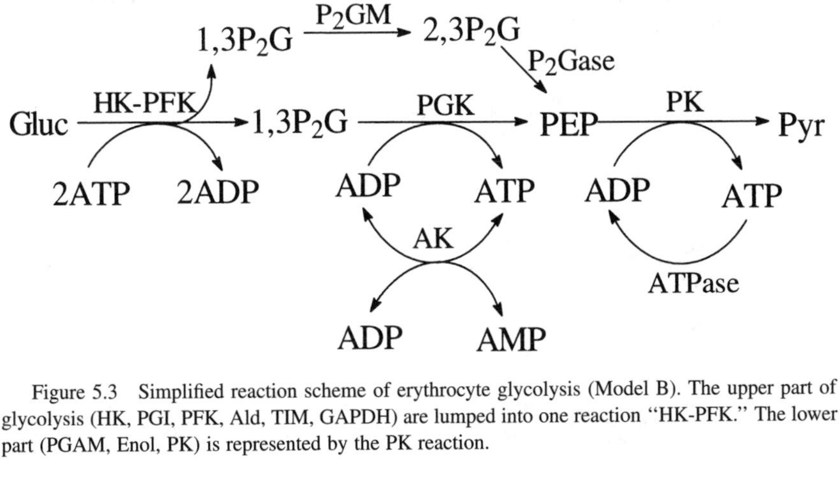

Figure 5.3 Simplified reaction scheme of erythrocyte glycolysis (Model B). The upper part of glycolysis (HK, PGI, PFK, Ald, TIM, GAPDH) are lumped into one reaction "HK-PFK." The lower part (PGAM, Enol, PK) is represented by the PK reaction.

by some simplifications. Furthermore, the substrate inhibition of phosphofructokinase by ATP was included. Because the flux through the lower part of glycolysis is twice the flux through the upper part, 4 moles of ATP are produced at the reaction steps catalyzed by the enzymes PGK and PK, whereas 2 moles of ATP are consumed by the HK-PFK system. Accordingly, the degradation of 1 mole of glucose leads to the net production of 2 moles of ATP. The actual ATP production is decreased depending on the share of the 2,3P_2G bypass which circumvents the ATP-producing PGK reaction.

As in Model A, very simple rate laws were used for all enzymes. Except for the HK-PFK system, the activities of all enzymes were characterized by linear or bilinear relationships (see Table 5.2). Furthermore, all reactions are considered to be irreversible except for the adenylate kinase reaction (AK). For the sake of simplicity, the saturation of the P_2Gase by 2,3P_2G is neglected.

The factor $f(ATP)$ in the expression $v_{\text{HK-PFK}}$ (Table 3) describes the substrate inhibition of PFK by ATP. $K_{\text{I,ATP}}$ and n_{H} are the inhibition constant and the cooperativity coefficient, respectively, of the substrate inhibition by ATP.

Figure 5.4 shows the rate of the HK-PFK system as a function of ATP according to the rate law listed in Table 5.2. Two cases are considered: $n_{\text{H}} = 1$ (no substrate inhibition) and $n_{\text{H}} = 4$ (substrate inhibition). The kinetic constant $k_{\text{HK-PFK}}$ was adjusted in such a way that in both cases a rate $v_{\text{HK-PFK}} = 1.25$ mM/h was obtained for $ATP = 1.2$ mM (*in vivo* point P in Figure 5.4).

Using the rate law given in Table 5.2 one may calculate the elasticity coefficient for the HK-PFK system with respect to the ATP concentration. One obtains:

$$\varepsilon_{\text{ATP}}^{\text{HK-PFK}} = \frac{\partial \ln v_{\text{HK-PFK}}}{\partial \ln ATP} = \begin{cases} 0.45 & (n_{\text{H}} = 1) \\ -1.70 & (n_{\text{H}} = 4). \end{cases} \tag{5.143}$$

The dynamic properties of the model depicted in Figure 5.3 are governed by the following differential equations

$$\frac{\mathrm{d}}{\mathrm{d}t} 1{,}3P_2G = 2v_{\text{HK-PFK}} - v_{\text{P2GM}} - v_{\text{PGK}}, \tag{5.144a}$$

$$\frac{\mathrm{d}}{\mathrm{d}t} 2{,}3P_2G = v_{\text{P2GM}} - v_{\text{P2Gase}}, \tag{5.144b}$$

$$\frac{\mathrm{d}}{\mathrm{d}t} PEP = v_{\text{P2Gase}} + v_{\text{PGK}} - v_{\text{PK}}, \tag{5.144c}$$

$$\frac{\mathrm{d}}{\mathrm{d}t} AMP = -v_{\text{AK}}, \tag{5.144d}$$

$$\frac{\mathrm{d}}{\mathrm{d}t} ADP = 2v_{\text{HK-PFK}} - v_{\text{PGK}} - v_{\text{PK}} + v_{\text{ATPase}} + 2v_{\text{AK}}, \tag{5.144e}$$

Table 5.2 Rate Equations of Glycolytic Enzymes Included in Model B

$v_{\text{HK-PFK}} = k_{\text{HK-PFK}} ATP \cdot f(ATP)$
$f(ATP) = \left[1 + \left(\frac{ATP}{K_{\text{I,ATP}}}\right)^{n_{\text{H}}}\right]^{-1}$
$v_{\text{P}_2\text{GM}} = k_{\text{P}_2\text{GM}} 1{,}3P_2G$
$v_{\text{P}_2\text{Gase}} = k_{\text{P}_2\text{Gase}} 2{,}3P_2G$
$v_{\text{PGK}} = k_{\text{PGK}} 1{,}3P_2G \cdot ADP$
$v_{\text{PK}} = k_{\text{PK}} PEP \cdot ADP$
$v_{\text{AK}} = k_{\text{AK}}^{+} AMP \cdot ATP - k_{\text{AK}}^{-}(ADP)^2$
$v_{\text{ATPase}} = k_{\text{ATPase}} ATP$

$$\frac{\mathrm{d}}{\mathrm{d}t} ATP = -2v_{\text{HK-PFK}} + v_{\text{PGK}} + v_{\text{PK}} - v_{\text{ATPase}} - v_{\text{AK}}. \tag{5.144f}$$

From Eqs. (5.144d)–(5.144f), it follows that

$$\frac{\mathrm{d}}{\mathrm{d}t}(AMP + ADP + ATP) = 0, \tag{5.145a}$$

$$AMP + ADP + ATP = A = \text{const.}; \tag{5.145b}$$

that is, the sum A of the concentrations of the adenine nucleotides is a conserved quantity. Because adenylate kinase is a very fast enzyme, the rapid-equilibrium approximation can be applied to Eq. (5.144d). This leads to the following equilibrium relation between the concentrations of the adenine nucleotides:

$$\frac{(ADP)^2}{AMP \cdot ATP} = q_{\text{AK}}. \tag{5.146}$$

Note that, when only steady states are considered, the adenylate kinase reaction could be considered to be at equilibrium even if it were not fast, because it represents a strictly detailed balanced reaction in the scheme given in Figure 5.3 (see Section 3.3.2). In system (5.144), the velocity v_{AK} may be eliminated by subtracting Eq. (5.144d) from Eq. (5.144f),

$$\frac{\mathrm{d}}{\mathrm{d}t}(ATP - AMP) = -2v_{\text{HK-PFK}} + v_{\text{PGK}} + v_{\text{PK}} - v_{\text{ATPase}}, \tag{5.147}$$

following the procedure explained in Section 4.3. Equations (5.145b) and (5.146) represent two algebraic conditions for the concentrations of the adenine nucleo-

Table 5.3 Values of Parameters and Steady-State Values of Variables of Erythrocyte Glycolysis (Model B)

Parameter	Value
$k_{\text{HK-PFK}}$	3.20/h
$k_{\text{P}_2\text{GM}}$	1500/h
$k_{\text{P}_2\text{Gase}}$	0.15/h
k_{PGK}	$1.57 \cdot 10^4$/mM h
k_{PK}	559/mM h
k_{ATPase}	1.46/h
n_{H}	4.0
$K_{\text{I,ATP}}$	1.0 mM
q_{AK}	2.0
A	1.5 mM
Variable	**Value**
Metabolite Concentrations (mM)	
$1{,}3P_2G$	0.0005
$2{,}3P_2G$	5.0
PEP	0.02
AMP	0.076
ADP	0.22
ATP	1.20
Metabolic Fluxes (mM/h)	
$v_{\text{HK-PFK}}$ $(=J)$	1.25
$v_{\text{P}_2\text{GM}}$	0.75
$v_{\text{P}_2\text{Gase}}$	0.75
v_{PGK}	1.75
v_{PK}	2.50
v_{ATPase}	1.75

tides. Accordingly, the concentrations of AMP and ADP may be expressed by the concentration of ATP:

$$ADP = A\left[-\frac{q_{\text{AK}}}{2} + \sqrt{\frac{q_{\text{AK}}^2}{4} + q_{\text{AK}}\frac{ATP}{A}\left(1 - \frac{ATP}{A}\right)}\right] = g_1(ATP), \qquad (5.148\text{a})$$

$$AMP = A - g_1(ATP) - ATP = g_2(ATP). \qquad (5.148\text{b})$$

The functions $ADP = g_1(ATP)$ and $AMP = g_2(ATP)$ are represented graphically in Figure 5.5.

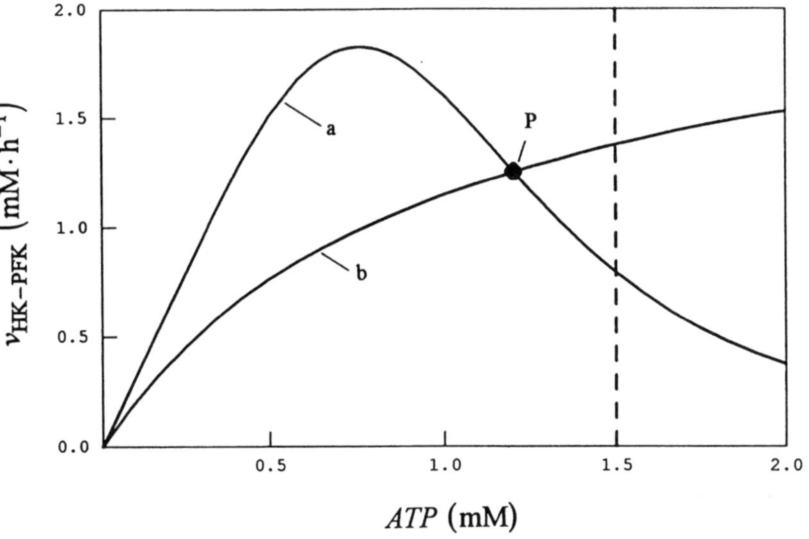

Figure 5.4 Glycolytic rate $v_{HK\text{-}PFK}$ as a function of the ATP concentration according to the rate law of the HK-PFK system given in Table 5.2. Parameter values: curve a, $n_H = 4$, $k_{HK\text{-}PFK} = 3.20$ h^{-1}; curve b, $n_H = 1$, $k_{HK\text{-}PFK} = 2.29$/h; P: *in vivo* point; broken line: $ATP = A$.

It is seen that the concentration of AMP decreases monotonically with increasing ATP concentration, whereas the function for ADP displays a maximum and becomes zero for $ATP = 0$ and $ATP = A$.

The left-hand side of Eq. (5.147) may be rewritten as follows

$$\frac{\mathrm{d}}{\mathrm{d}t}(ATP - AMP) = \left(1 - \frac{\mathrm{d}AMP}{\mathrm{d}ATP}\right)\frac{\mathrm{d}ATP}{\mathrm{d}t}. \tag{5.149}$$

From Eqs. (5.147) and (5.149), it follows that

$$\frac{\mathrm{d}}{\mathrm{d}t}ATP = \left(1 - \frac{\mathrm{d}AMP}{\mathrm{d}ATP}\right)^{-1}(-2v_{HK-PFK} + v_{PGK} + v_{PK} - v_{ATPase}). \tag{5.150}$$

The differential equation system (5.144) can now be reduced in dimension by replacement of Eqs. (5.144d)–(5.144f) by the algebraic conditions (5.148a) and (5.148b) and the differential equation (5.150).

Stationary states are defined by vanishing time derivatives of the metabolite concentrations. One obtains from Eqs. (5.144a)–(5.144c) and (5.150)

$$2v_{HK-PFK} - v_{P_2GM} - v_{PGK} = 0, \tag{5.151a}$$

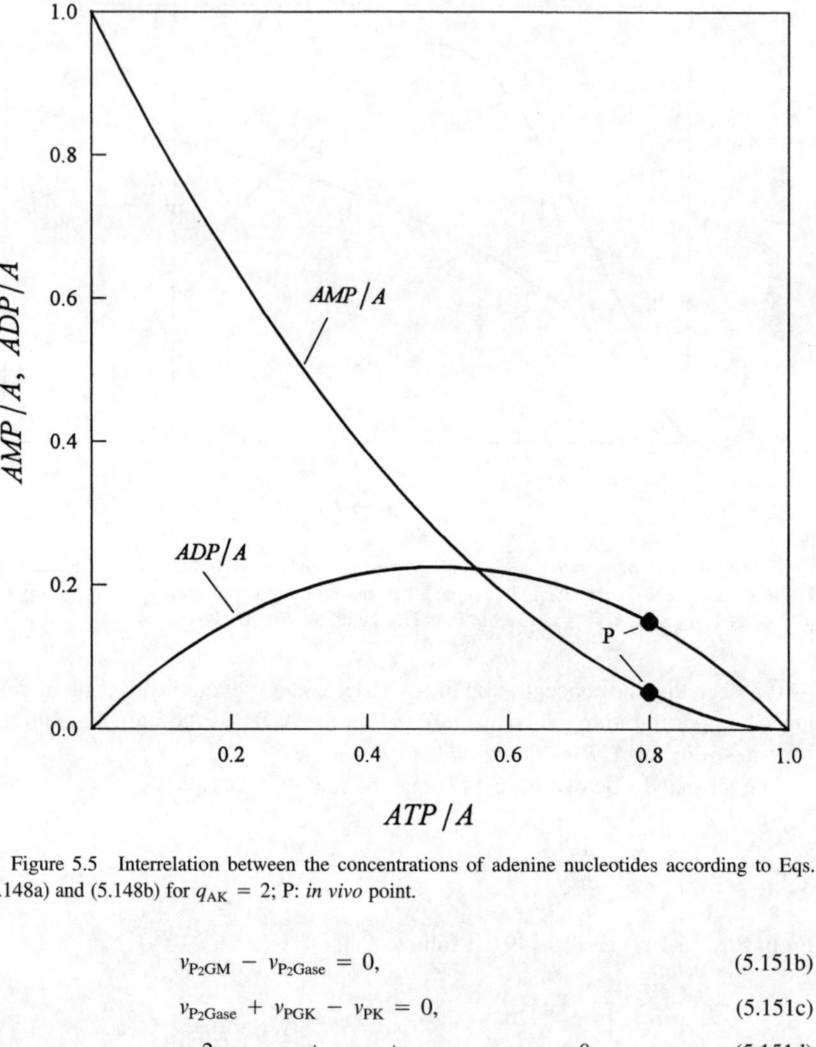

Figure 5.5 Interrelation between the concentrations of adenine nucleotides according to Eqs. (5.148a) and (5.148b) for $q_{AK} = 2$; P: *in vivo* point.

$$v_{P_2GM} - v_{P_2Gase} = 0, \tag{5.151b}$$

$$v_{P_2Gase} + v_{PGK} - v_{PK} = 0, \tag{5.151c}$$

$$-2v_{HK-PFK} + v_{PGK} + v_{PK} - v_{ATPase} = 0. \tag{5.151d}$$

Summation of Eqs. (5.151a)–(5.151d) yields

$$v_{PGK} = v_{ATPase} \tag{5.152}$$

and, by use of the kinetic equations listed in Table 5.2,

$$k_{PGK}ADP \cdot 1{,}3P_2G = k_{ATPase}ATP. \tag{5.153}$$

Because under steady-state conditions the ATP production in the PK step is compensated by the ATP consumption by the HK-PFK system ($2v_{\text{HK-PFK}} = v_{\text{PK}}$), Eqs. (5.152) and (5.153) characterize the balance between ATP-consuming and ATP-producing processes. For the calculation of the ATP concentration by use of Eq. (5.153), the concentration of $1{,}3P_2G$ is eliminated by consideration of Eq. (5.151a), which reads, in more detail,

$$2k_{\text{HK}-\text{PFK}}ATP \cdot f(ATP) - k_{\text{P}_2\text{GM}} \cdot 1{,}3P_2G - k_{\text{PGK}} \cdot 1{,}3P_2G \cdot ADP = 0. \quad (5.154)$$

This entails

$$1{,}3P_2G = \frac{2k_{\text{HK}-\text{PFK}}ATP \cdot f(ATP)}{k_{\text{P}_2\text{GM}} + k_{\text{PGK}}ADP}. \quad (5.155)$$

Inserting Eq. (5.155) into Eq. (5.153) yields

$$v_{\text{PGK}} = \frac{2k_{\text{HK}-\text{PFK}}k_{\text{PGK}}ADP \cdot ATP \cdot f(ATP)}{k_{\text{P}_2\text{GM}} + k_{\text{PGK}}ADP} = k_{\text{ATPase}}ATP = v_{\text{ATPase}}, \quad (5.156)$$

where the concentration of ADP must be considered as a function of ATP [cf. Eq. (5.148a)].

Figure 5.6 shows the net rate of the glycolytic ATP production (v_{PGK}) and the rate of the nonglycolytic ATP consumption (v_{ATPase}) as functions of the ATP concentration for various values of the rate constant of the ATPase. The values of the kinetic parameters (see Table 5.3) are close to those found in human erythrocytes. The intersection points of the curves $v_{\text{PGK}}(ATP)$ and $v_{\text{ATPase}}(ATP)$ determine the steady-state values of the ATP concentration. Evidently, the point $ATP = 0$ represents a trivial steady state which is a solution of Eq. (5.156) irrespective of the values of the kinetic parameters (state P_0). It is further seen that above a critical value, $k_{\text{ATPase}}^{\text{crit}}$ (curve a), only the trivial steady state is obtained. For $k_{\text{ATPase}} < k_{\text{ATPase}}^{\text{crit}}$, two steady states P_1 and P_2 are found, in addition to the trivial steady state. A detailed stability analysis which is based on a linearization of the equation system (5.144a)–(5.144c), (5.145), (5.148) and (5.150) and computation of the eigenvalues of the corresponding Jacobian (see Section 2.3.2), reveals that the states with low (nonvanishing) ATP concentration (states P_1) are unstable, whereas the steady states with high ATP concentration (states P_2) are stable. One may conclude that the steady state found *in vivo* corresponds to the stable high-energy state P_2.

The curves depicted in Figure 5.7 show the steady state concentration of ATP as a function of k_{ATPase} for $n_{\text{H}} = 1$ and $n_{\text{H}} = 4$. Stable and unstable states are

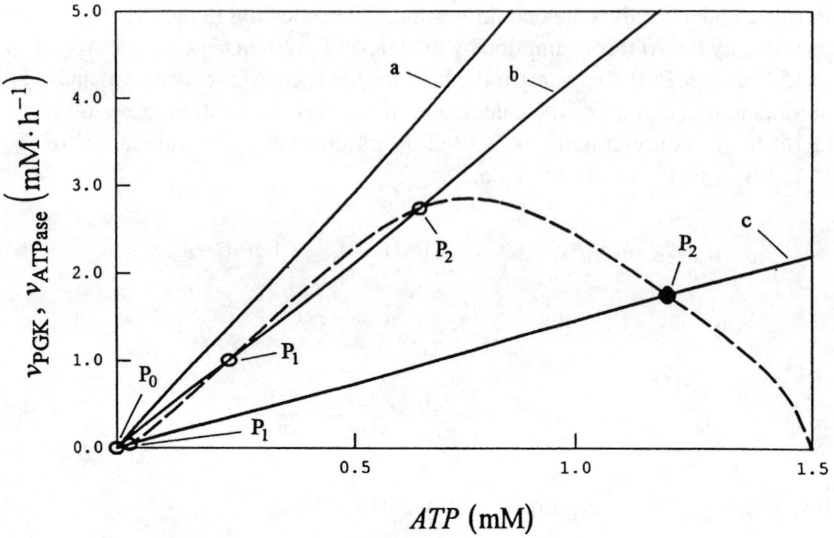

Figure 5.6 Rates of the enzymes ATPase (solid lines) and PGK (broken line) as functions of the ATP concentration according to Eq. (5.156). The intersection points P_0, P_1, and P_2 correspond to steady states. The intersection point P_2 on curve c is the *in vivo* point. Parameter values: curve a, $k_{ATPase} = 5.83$/h; curve b, $k_{ATPase} = 4.23$/h; curve c, $k_{ATPase} = 1.46$/h. The values of the other parameters correspond to those listed in Table 5.3.

characterized by solid and broken lines, respectively. It becomes clear that the critical values k_{ATPase}^{crit} represent bifurcation points which separate parameter regions with different numbers of steady states.

Figure 5.8 shows the steady-state concentration of ATP as a function of the rate of the ATPase for various values of the cooperativity coefficient (n_H) of the ATP-substrate inhibition of the HK-PFK system. The curves for high n_H values are characterized by the property that in the neighborhood of the *in vivo* state, the ATP concentration is rather insensitive against variations of the ATP-consumption rate. The regulatory property of glycolysis which leads to homeostasis of the ATP concentration in face of variations of the rate of ATP consumption was extensively studied by Selkov (1975b).

There are two reasons for ATP homeostasis. First, the share of the 2,3P_2G bypass $v_{P_2GM}/2v_{HK\text{-}PFK}$ decreases with increasing ATP-consumption rate. According to the steady-state equation (5.151a)

$$\frac{v_{PGK}}{2v_{HK-PFK}} = 1 - \frac{v_{P_2GM}}{2v_{HK-PFK}}, \tag{5.157}$$

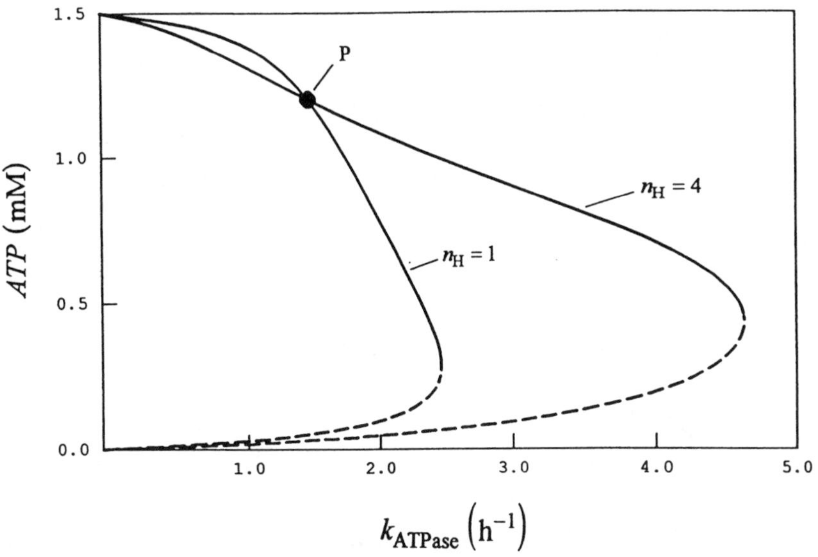

Figure 5.7 ATP concentration as a function of the rate constant k_{ATPase} for two different values of the cooperativity coefficient of the ATP inhibition of PFK ($n_H = 1$ and $n_H = 4$). Solid and broken lines indicate stable and unstable steady states, respectively.

a decrease of $v_{P_2GM}/2v_{HK\text{-}PFK}$ is accompanied by an increase of the share of the ATP-producing PGK reaction which meets the higher demand on ATP. A second effect contributing even more to homeostasis is the activation of the glycolytic flux at decreasing ATP concentration, which results from a lowering of the ATP inhibition (Figure 5.9).

In Table 5.4 the control coefficients are listed for the glycolytic flux ($J = v_{HK\text{-}PFK}$) and for the concentrations of the metabolites ATP and $2,3P_2G$. It is seen that in contrast to Model A, not only the HK and PFK but also the enzymes P_2GM, P_2Gase and ATPase exhibit nonvanishing flux control coefficients. This result is due to the circumstance that the upper and lower parts of the glycolytic system are coupled by the common cofactors ATP and ADP. Nevertheless the HK-PFK system is mainly responsible for flux control, such as in Model A. The calculations were performed for the *in vivo* state under the assumption $n_H = 1$ (no substrate inhibition of HK-PFK by ATP) and $n_H = 4$ (substrate inhibition of HK-PFK by ATP). It is seen that for $n_H = 1$, the flux control coefficient of ATPase is negative because the decrease of ATP after activation of ATPase leads to a diminution of the rate $v_{HK\text{-}PFK}$. In the more realistic case ($n_H = 4$), a decrease of ATP will activate glycolysis so that the flux control coefficient of ATPase becomes positive. The control coefficients for ATP may be considered as a quantitative

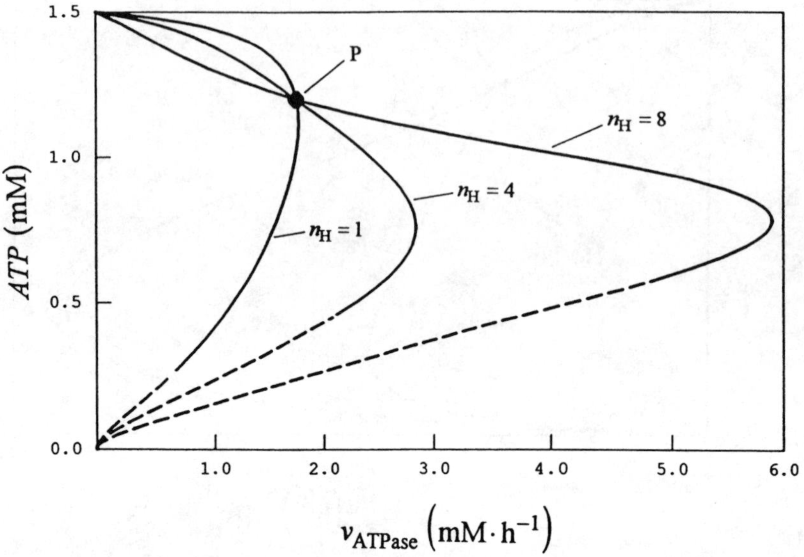

Figure 5.8 ATP concentration as a function of the rate $v_{ATPase} = k_{ATPase}ATP$ of ATP-consuming processes for different values of n_H and $k_{HK\text{-}PFK}$. Parameter values: $k_{HK\text{-}PFK} = 2.29/h$ ($n_H = 1$); $k_{HK\text{-}PFK} = 3.20/h$ ($n_H = 4$); $k_{HK\text{-}PFK} = 5.52/h$ ($n_H = 8$). Broken lines indicate unstable steady states. P: *in vivo* point.

measure for the ATP homeostasis, as already discussed. The homeostatic effect of the substrate inhibition is expressed by the fact that for $n_H = 4$ the coefficients $C^{ATP}_{HK\text{-}PFK}$ and C^{ATP}_{ATPase} are small compared to those obtained for $n_H = 1$. It is seen that the substrate inhibition of HK-PFK by ATP results not only in a homeostasis of *ATP* but also of 2,3P_2G.

The flux control coefficients of the enzymes P_2GM and PGK are of opposite sign. The negative value of C^J_{PGK} for $n_H = 4$ is easily understood by consideration of the fact that activation of PGK results in diminution of the 2,3P_2G bypass and, in this way, to an increase in ATP concentration.

Under the assumptions of this model, the pyruvate kinase reaction neither controls the concentrations of ATP and 2,3P_2G nor the glycolytic flux. This results from the simplifying assumption that the PGK reaction is irreversible. The control coefficients for the glycolytic flux and for the metabolite concentrations listed in Table 5.4 sum up to unity and zero, respectively, that is, they fulfill the summation theorems.

An extension of Model B of glycolysis was set up to study the influence of pyruvate kinase deficiency on the energy metabolism of human erythrocytes (Holzhütter *et al.*, 1985b). A new comprehensive kinetic model of the pyruvate kinase of human erythrocytes was included and account was taken of the magnesium-

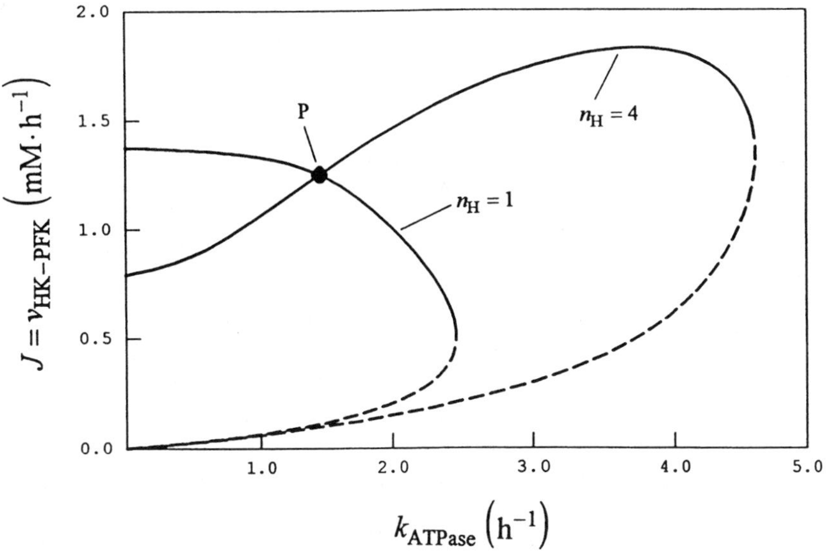

Figure 5.9 Glycolytic rate $J = v_{HK\text{-}PFK}$ as a function of the rate constant k_{ATPase} for two different values of n_H. Parameter values: $k_{HK\text{-}PFK} = 2.29\ h^{-1}$ ($n_H = 1$); $k_{HK\text{-}PFK} = 3.20\ h^{-1}$ ($n_H = 4$). P: *in vivo* point.

Table 5.4 Control Coefficients of Enzymes for the Glycolytic Flux and Metabolite Concentrations (Model B)

	Variable					
	$v_{HK\text{-}PFK}(=J)$		ATP		$2{,}3P_2G$	
Enzyme	$n_H = 1$	$n_H = 4$	$n_H = 1$	$n_H = 4$	$n_H = 1$	$n_H = 4$
HK-PFK	1.32	0.52	0.72	0.28	2.74	1.07
P_2GM	−0.10	0.14	−0.22	−0.08	0.18	0.68
P_2Gase	0.00	0.00	0.00	0.00	−1.00	−1.00
PGK	0.10	−0.14	0.22	0.08	−0.18	−0.68
ATPase	−0.32	0.48	−0.72	−0.28	−1.74	−0.07
$\sum_j C_j$	1.00	1.00	0.00	0.00	0.00	0.00

complex formation of the adenine nucleotides and 2,3-biphosphoglycerate. The analysis of individual cases with pyruvate kinase mutations permitted estimates and classification of the degree of disorder of the glycolytic pathway, which were in accord with clinical and other experimental assessments.

Other extensions consider the coupling of the glycolytic pathway with reactions responsible for the synthesis and breakdown of adenine nucleotides, in particular the 5′-nucleotidase (EC 3.1.3.5), AMP deaminase (EC 3.5.4.6), adenosine kinase (EC 2.7.1.20), adenine phosphoribosyltransferase (EC 2.4.2.7) and the uptake of adenosine across the erythrocyte membrane (Schauer *et al.* 1981a, 1981b). The main effect of including these reactions is that the total sum of the adenine nucleotides is no longer a conserved quantity. This model allows one to simulate the breakdown of adenine nucleotides after glucose depletion.

R. Schuster *et al.* (1988) developed a model of erythrocyte metabolism which comprises, in addition to glycolysis, the pentose phosphate pathway. Special attention is drawn to the fact that in erythrocytes the main function of the pentose phosphate shunt is to form NADPH. The NADPH produced by the two dehydrogenases (glucose-6-phosphate dehydrogenase, G6PD, EC 1.1.1.49, and 6-phospogluconate dehydrogenase, 6PGD, 1.1.1.43) is mainly utilized by the glutathione reductase (EC 1.6.4.2) catalyzing the reaction: GSSG + NADPH → 2GSH + NADP. Furthermore, an NADPH-dependent lactate dehydrogenase (I. Rapoport *et al.*, 1979) was included into the model. Steady states are calculated as functions of the rate constants k_{ATPase} and k_{Ox} representing the energetic load and the oxidative load, respectively, of the system. The calculation of flux control coefficients of the nonequilibrium reactions reveals that most of these coefficients are very small with the following main exceptions:

(a) Nonglycolytic ATP-consuming processes (ATPases) which affect strongly the glycolytic rate, $C^J_{\text{ATPase}} = 0.70$; see Model B for $n_H = 4$ (Table 5.4).

(b) 2,3-Bisphosphoglycerate phosphatase (P_2Gase) which controls the glycolytic flux and the reactions of the 2,3P_2G bypass, $C^J_{\text{P}_2\text{Gase}} = 0.22$, $C^{J_{\text{P}_2\text{GM}}}_{\text{P}_2\text{Gase}} = 0.94$.

(c) The reactions of the oxidative load affecting the reactions of the oxidative part of the pentose phosphate pathway, $C^{J_{\text{G6PD}}}_{\text{Ox}} = C^{J_{\text{6PGD}}}_{\text{Ox}} = 0.47$. It has been concluded that in the *in vivo* state of erythrocyte glycolysis, the 2,3P_2G bypass and the pentose phosphate pathway are almost independently controlled by the reactions consuming those metabolites which are produced by the corresponding pathways. The model was used for predicting the effect of glucose-6-phosphate dehydrogenase deficiencies (R. Schuster *et al.*, 1989) and was recently extended to predict the metabolic effect of large-scale enzyme activity alterations (R. Schuster and Holzhütter, 1995).

5.4.4.4. *Glycolytic Energy Metabolism and Osmotic States*

The theoretical investigation of energy metabolism in erythrocytes has been extended by inclusion of its interaction with active and passive fluxes of ions across the cell membrane (Brumen and Heinrich, 1984; Werner and Heinrich, 1985). This model (Model C) allows one to evaluate the state of metabolism as well as osmotic and electric effects. Accordingly, control coefficients related to the volume can be calculated. (For a general treatment of the control of variables other than concentrations and fluxes, see Section 5.8.) Compared with previous models (e.g., Model B), the set of system parameters is enlarged by the quantities characterizing the electric charges and osmotic effects of hemoglobin, the permeabilities of ions, and the cell surface area.

The metabolic part of the "metabolic-osmotic model" is essentially based on the reduced reaction scheme used for Model B (Figure 5.3). Several assumptions and simplifications are used in the model:

(a) The *in vivo* state is characterized by a fixed composition of the external medium.

(b) The inhibitory actions of H^+ ions on the enzymes PFK and P_2GM are taken into account.

(c) Two nonglycolytic ATP-consuming processes are considered: the Na/K-ATPase (EC 3.6.1.37) and the non-ion transport ATPases. It has been proposed that 25–70% of the ATP produced by glycolysis is utilized by the Na/K pump (Grimes, 1980). Maretzki *et al.* (1980) and Reimann *et al.* (1981) determined a value of 30%. The non-ion transport ATPases are linked to membrane phosphorylation processes.

(d) Consideration of the transmembrane potential ($\Delta\Psi$) and of the cell water volume (V) as system variables necessitates the incorporation of detailed electric and osmotic conditions. It is assumed that the intracellular and extracellular compartments are electrically neutral and in osmotic equilibrium.

The differential equations for the concentrations of the glycolytic metabolites are easily derived from the reaction scheme (Figure 5.3). As the metabolite concentrations may also be changed by variations of the cell volume (V), one arrives at the following type of equation:

$$\frac{1}{V^0}\frac{\mathrm{d}(S_i V)}{\mathrm{d}t} = \sum_j n_{ij} v_j, \qquad (5.158)$$

where S_i denotes the concentrations of 1,3P_2G, 2,3P_2G, PEP, and ATP [cf. Eqs. (5.144a)–(5.144c) and (5.150)]. V^0 represents the cellular volume in a reference state. For the enzymatic activities v_k rate laws are used which approximate the kinetic properties of the isolated enzymes. They are essentially the same as used in Model B. An exception is the rate equation for the Na/K-ATPase,

$$v_{\mathrm{Na/K-ATPase}} = \frac{k_{\mathrm{Na/K-ATPase}} \cdot ATP \cdot Na_{\mathrm{in}}^+}{1 + ATP/K_{\mathrm{m,ATP}}}, \qquad (5.159)$$

in which the fact that the activity of this enzyme is stimulated by intracellular sodium is considered. *In vivo,* this enzyme is almost saturated by ATP [$K_{m,ATP}$ = 0.04 mM; Cavieres (1977)].

The passive transport of sodium and potassium is described by the well-known Goldman equation (Goldman, 1943). Taking into account that the action of the Na/K-ATPase leads to the transport of 3 moles of sodium outward and 2 moles of potassium inward per 1 mole of ATP degraded, the time-dependent changes of the intracellular cation concentrations are governed by the following differential equations:

$$\frac{1}{V^0}\frac{\mathrm{d}}{\mathrm{d}t}(Na^+_{in}\cdot V) = -\frac{A_c\ln(r)}{V^0}P_{Na}\frac{(Na^+_{ex} - r\cdot Na^+_{in})}{1-r} - 3v_{Na/K-ATPase}, \tag{5.160a}$$

$$\frac{1}{V^0}\frac{\mathrm{d}}{\mathrm{d}t}(K^+_{in}\cdot V) = -\frac{A_c\ln(r)}{V^0}P_{K}\frac{(K^+_{ex} - r\cdot K^+_{in})}{1-r} + 2v_{Na/K-ATPase}, \tag{5.160b}$$

with

$$r = \exp\left(\frac{F\Delta\Psi}{RT}\right) \tag{5.161}$$

($P_{Na} = 1.3 \times 10^{-12}$ m/s, $P_K = 1.1 \times 10^{-12}$ m/s: permeabilities of sodium and potassium, respectively; $A_c = 137\ \mu m^2$: cell surface area; F: Faraday constant).

The transmembrane exchange of chloride is much faster than that of sodium and potassium. Therefore, the transport equation for chloride ions is substituted by the equilibrium condition

$$Cl^-_{ex} = \frac{Cl^-_{in}}{r}. \tag{5.162}$$

The pH in the intracellular and extracellular medium are related as follows:

$$pH_{ex} = pH_{in} - \log_{10} r. \tag{5.163}$$

The system equations are completed by the conditions of osmotic equilibrium between the intracellular and extracellular compartments,

$$RT(K^+_{in} + Na^+_{in} + Cl^-_{in} + ADP + ATP + 2{,}3P_2G + g_{Hb}\cdot Hb) = \text{const.}, \tag{5.164}$$

as well as the condition of electroneutrality,

$$K^+_{\text{in}} + Na^+_{\text{in}} - Cl^-_{\text{in}} + z_{\text{ATP}} \cdot ATP + z_{\text{ADP}} \cdot ADP + z_{2,3\text{P}_2\text{G}} \cdot 2{,}3P_2G + z_{\text{Hb}} \cdot Hb = 0. \tag{5.165}$$

In Eq. (5.164), g_{Hb} denotes the osmotic coefficient of hemoglobin which is a function of hemoglobin concentration (Gary-Bobo and Solomon, 1968; Freedman and Hoffmann, 1979). Only glycolytic metabolites with high concentrations (ADP, ATP, 2,3P_2G) are considered in Eqs. (5.164) and (5.165). The coefficients z_{ATP}, z_{ADP}, $z_{2,3\text{P}_2\text{G}}$, and z_{Hb} denote the pH-dependent charges of the compounds indicated. Equations (5.160)–(5.165) constitute a complicated nonlinear system which consists of differential as well as of algebraic equations (algebro-differential equation system). This equation system was solved by numerical procedures for the steady state *in vivo* as well as for time-dependent states (Brumen and Heinrich, 1984; Werner and Heinrich, 1985).

The model allows one to calculate the control coefficients not only for metabolite concentrations and fluxes but also for the cellular volume. This coefficient can be defined as follows:

$$C_k^V = \frac{\partial \ln V/\partial p_k}{\partial \ln v_k/\partial p_k}. \tag{5.166}$$

The various control coefficients for the volume are listed in Table 5.5. It is seen that the control coefficients fulfill the summation theorem,

$$\sum_k C_k^V = 0 \tag{5.167}$$

Table 5.5 Control Coefficients of Model C for the Cell Water Volume Under *in Vivo* Conditions

Parameter	C_j^V
HK-PFK	0.63
P_2GM	0.19
P_2Gase	−0.39
PGK	−0.19
PK	−0.01
ATPase	−0.06
Na/K-ATPase	−0.21
$v_{\text{Na,pass}}$	4.81
$v_{\text{K,pass}}$	−4.83
$\sum_j C_j^V$	0.00

Source: Brumen and Heinrich (1984)

despite the fact that there are some coefficients that differ considerably from unity. The high control coefficients of the passive transport of Na^+ and K^+ reflect the fact that these cations are of overwhelming importance for the osmotic properties of the cell. The control coefficients $C^V_{\mathrm{Na,pass}}$ and $C^V_{\mathrm{K,pass}}$ are of opposite sign. This is simply explained by the fact that an increase of the permeability P_{K} leads to a loss of potassium by the cell, whereas an increase of the permeability P_{Na} will result in an increase of intracellular sodium. The control coefficients of these transport processes almost compensate each other; that is, a simultaneous change of P_{K} and P_{Na} by the same factor would have only a negligible effect on the cellular volume.

The results of Table 5.5 confirm that the sodium-potassium pump plays an important role for the regulation of cell volume. An increase of the Na/K-ATPase activity would result in an enhancement of the outflow of sodium, which according to the (2:3)-stoichiometry of the pump is not fully compensated by the inflow of potassium. The resulting decrease in cell volume corresponds to the negative control coefficient of this enzyme (see Table 5.5). A positive volume control coefficient is obtained for the HK-PFK system, which may be explained primarily by the increase of $2,3P_2G$ concentration upon activation of the glycolytic flux.

The above-mentioned models were the basis for a more complete model of erythrocyte metabolism which includes glycolysis, the $2,3P_2G$ bypass, the pentose phosphate pathway, the adenine nucleotide metabolism, and various transmembrane processes, as well as osmotic and electrostatic conditions (Joshi and Palsson, 1989a, 1989b, 1990a, 1990b). This model comprises 33 mass balance equations which contain 41 reaction velocities. Taking into account the constraints resulting from osmotic balance, electroneutrality, and cofactor preservation, the complete description encompasses 29 system variables (metabolite concentrations, concentrations of inorganic ions, cell volume, transmembrane potential, and pH). Despite the fact that a number of relevant processes have not been considered (e.g., active and passive transport of calcium, interaction of ATP and $2,3P_2G$ with hemoglobin), the model of Joshi and Palsson (1989a, 1989b, 1990a, 1990b) is up to now the most comprehensive model of erythrocyte metabolism and, apparently, for an autonomous metabolic system in general.

5.4.5. A Simple Model of Oxidative Phosphorylation

Oxidative phosphorylation (i.e., the formation of ATP from ADP and inorganic phosphate using the energy of oxidizable substrates) is a crucial process in biological energy transduction. We will here consider oxidative phosphorylation as it occurs at the mitochondrial inner membrane. The energy transformation proceeding at bacterial plasma membranes is very similar.

According to the chemiosmotic hypothesis of Mitchell (1961), the respiratory chain uses the free energy of oxidation to extrude protons out of the mitochondrion and thus generate a proton-motive force. This quantity is defined as the electrochemical potential difference of protons across the membrane,

$$\Delta\tilde{\mu}_{H^+} = RT \ln \frac{H^+_{in}}{H^+_{out}} + F\Delta\Psi, \tag{5.168}$$

where $\Delta\Psi$ is the transmembrane potential. This force serves to produce ATP via catalysis by the H^+-transporting ATP synthase (H^+-ATPase, EC 3.6.1.34). The respiratory chain is a sequence of reactions catalyzed by a multienzyme complex. In a minimal model, its control properties can be described by treating it as one overall reaction, as will be justified in the modular approach of control analysis (Section 5.12). That approach allows not only for the existence of several enzymes in one complex, but also for more than one independent flux through this multienzyme complex. The respiratory chain actually has at least two linearly independent fluxes owing to the slippage between substrate oxidation and proton transport (see Luvisetto *et al.*, 1987; Westerhoff and Van Dam, 1987). Strictly speaking, the ATPase reaction exhibits slip also and should be described by two degrees of freedom.

A more detailed model should also include the proton leak and the adenine

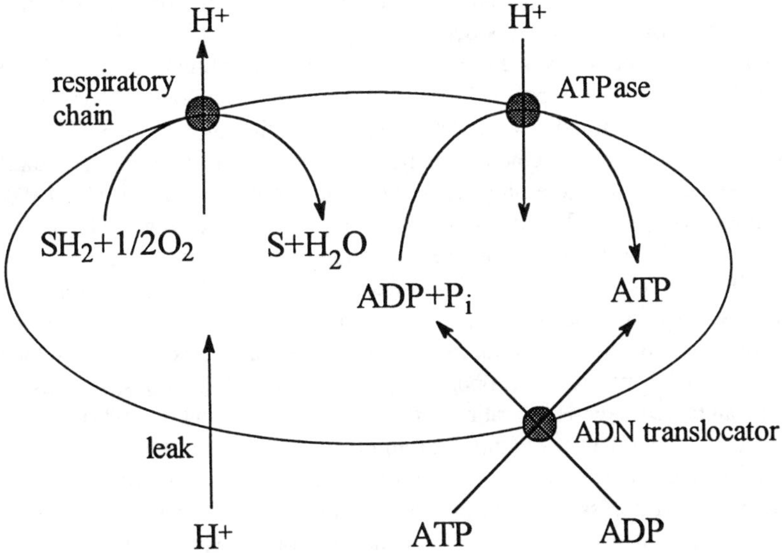

Figure 5.10 Scheme of the main processes in oxidative phosphorylation. Abbreviations: SH_2 and S, reduced and oxidized form of substrate, respectively (e.g., lactate and pyruvate); ADN translocator, adenine nucleotides translocator.

nucleotide translocator (see Figure 5.10). Although the proton leak (i.e., the passive back-flow of protons through the membrane without being used for ATP synthesis) is not catalyzed by an enzyme, control coefficients can be calculated on the basis of definitions (5.3) and (5.5). The perturbation parameter can be chosen to be the concentration of an uncoupler [e.g., FCCP (Groen *et al.,* 1982)]. Uncouplers increase the membrane permeability for protons and thus decrease the coupling between respiration and phosphorylation.

The inorganic phosphate needed for ATP synthesis enters the mitochondrion via the phosphate carrier, which transports inorganic phosphate and protons together at a ratio 1:1. When the phosphate carrier is, for simplicity's sake, neglected in the model, one must therefore consider that one extra proton is needed per produced ATP molecule than is actually utilized by the ATPase.

To calculate control coefficients, one needs knowledge of the stoichiometric proportions in the reactions, that is, the number of protons extruded per oxygen atom consumed (H^+/O ratio) and the number of protons needed to produce one molecule of ATP (H^+/P ratio). There are dissenting views in the literature about the values of these ratios. Estimates can be derived from the thermodynamic reasoning that the endergonic process cannot utilize more free energy than is produced by the exergonic process by which it is driven. Measurements of these ratios can be carried out by the oxygen pulse method and ATP pulse method (Mitchell and Moyle, 1965). The modern mainstream view (see Brand, 1994) is that $n_H^O = H^+/O = 10$ (certainly between 9 and 13) for oxidation of matrix NADH. Succinate oxidation is believed to proceed with $n_H^O = H^+/O = 6$, with minority views that the value is 8. A widely accepted estimate of $n_H^P = H^+/P$ is 4, made up of three H^+ per ATP on the ATP synthase and one H^+ on the phosphate carrier (cf. Brand, 1994). This gives P/O ratios of 2.5 for NADH-linked substrates and 1.5 for succinate oxidation. Fitton *et al.* (1994) found that the P/O ratio decreases with increasing respiration rate, from about 2.3 to about 0.9 for respiration on lactate. In a dynamic model presented by Korzeniewski and Froncisz (1991), the following values for oxidation of NADH were used: $n_H^O = H^+/O = 9$, $n_H^P = H^+/P = 3.5$. Different explanations have been given for the fact that these ratios may be noninteger. The most logical explanation is that these are average stoichiometries of complex enzymatic reaction cycles, which also include slip reactions. This view is supported by the fact that some drugs, such as almitrine, can increase the H^+/P ratio, as reported by Rigoulet et al. (1990) who gave a value of 2.7 without the addition of almitrine.

To consider that the intramitochondrial and extramitochondrial volumes are different is not necessary when the concentrations in the cytosol are assumed to be constant, as will be done here. The concentrations of S, SH_2, O_2 and H_2O will also be assumed to be constant. Furthermore, the slip in the respiratory chain and H^+-ATPase will be neglected for simplicity's sake. The system equations can then be written as

$$\frac{\mathrm{d}H^+}{\mathrm{d}t} = -n_{\mathrm{H}}^{\mathrm{O}} v_{\mathrm{O}} + n_{\mathrm{H}}^{\mathrm{P}} v_{\mathrm{P}} + v_{\mathrm{L}}, \tag{5.169}$$

$$\frac{\mathrm{d}ATP}{\mathrm{d}t} = v_{\mathrm{P}} - v_{\mathrm{A}}, \tag{5.170}$$

$$\frac{\mathrm{d}ADP}{\mathrm{d}t} = -v_{\mathrm{P}} + v_{\mathrm{A}}, \tag{5.171}$$

where v_{O}, v_{P}, v_{A} and v_{L} stand for the rates of respiration, phosphorylation, adenine nucleotide transport, and leak, respectively.

Obviously, the system involves one conservation relation, $ADP + ATP =$ const. We have

$$\mathbf{N}^0 = \begin{pmatrix} -n_{\mathrm{H}}^{\mathrm{O}} & n_{\mathrm{H}}^{\mathrm{P}} & 1 & 0 \\ 0 & 1 & 0 & -1 \end{pmatrix}, \tag{5.172}$$

$$\mathbf{L} = \begin{pmatrix} 1 & 0 \\ 0 & 1 \\ 0 & -1 \end{pmatrix}. \tag{5.173}$$

The unscaled elasticity matrix can be written as

$$\boldsymbol{\varepsilon} = \begin{pmatrix} \varepsilon_{\mathrm{H}}^{\mathrm{O}} & 0 & 0 \\ \varepsilon_{\mathrm{H}}^{\mathrm{P}} & \varepsilon_{\mathrm{ATP}}^{\mathrm{P}} & \varepsilon_{\mathrm{ADP}}^{\mathrm{P}} \\ \varepsilon_{\mathrm{H}}^{\mathrm{L}} & 0 & 0 \\ \varepsilon_{\mathrm{H}}^{\mathrm{A}} & \varepsilon_{\mathrm{ATP}}^{\mathrm{A}} & \varepsilon_{\mathrm{ADP}}^{\mathrm{A}} \end{pmatrix}. \tag{5.174}$$

The elasticity, $\varepsilon_{\mathrm{H}}^{\mathrm{A}}$, of the adenine nucleotide translocator with respect to proton concentration is not normally zero because this translocator is electrogenic. Its rate therefore depends on the transmembrane potential, $\Delta\Psi$. This potential, in turn, is linked to the inside and outside proton concentrations. Approximately, this interrelation can be written as proportionality of ΔpH and $\Delta\Psi$ (Bohnensack, 1985; Holzhütter *et al.,* 1985a). A better, quasi-linear approximation was derived by S. Schuster and Mazat (1993). In the elasticities $\varepsilon_{\mathrm{H}}^{\mathrm{O}}$, $\varepsilon_{\mathrm{H}}^{\mathrm{P}}$ and $\varepsilon_{\mathrm{H}}^{\mathrm{L}}$, the dependence of the respective processes on $\Delta\Psi$ should also be included.

Using Eqs. (5.25) and (5.26), the control coefficients for the considered model of oxidative phosphorylation can, in principle, be calculated. This method is, however, hampered by the problem that the elasticities of this system are difficult to measure, because the concentrations inside mitochondria are difficult to change specifically. Another possibility is to calculate control coefficients on the basis of a dynamic model, as was done by Korzeniewski and Froncisz (1991). They modulated enzyme activities numerically and computed the change in fluxes. Control coefficients over respiration rate with respect to substrate dehydrogenation (0.23), external ATP utilization (0.56), proton leak (0.20), and other reactions were calculated. Only some of these coefficients are in agreement with experimentally determined values. The above-mentioned problem of uncertainty in the values of elasticities corresponds in kinetic models to uncertainties in the kinetic parameters.

Control coefficients pertaining to oxidative phosphorylation can be determined in a more direct way by inhibitor titration. This has frequently been done (Groen *et al.*, 1982; Brand *et al.*, 1988; Gellerich *et al.*, 1990; Letellier *et al.*, 1993). It was found that control coefficients strongly depend on cell type and experimental conditions. For example, the control coefficients over respiration rate may vary between State 4 (no ADP supply) and State 3 (excess of ADP) from 0.9 to nearly 0 (control by the proton leak) or from 0 to 0.5 passing at 0.65 (control by phosphorylation plus ATP consumption) [see the review by Brown (1992)].

5.4.6. A Three-Step Model of Serine Biosynthesis

The system under study in this section is the pathway leading from 3-phosphoglycerate (3PG, derived from glycolysis) to serine via 3-phosphohydroxypyruvate (3PHPA) and phosphoserine (PSer) (see Figure 5.11). A control analysis of this pathway in mammalian liver (rabbit and rat) was done by Fell and Snell (1988). They used the method of calculating control coefficients from the elasticities and stoichiometric structure, as described in Sections 5.2 and 5.3.

As the flux of serine biosynthesis is small compared with the fluxes through the major pathways such as glycolysis, the three enzymes shown in Figure 5.11 have very little effect on the cellular concentrations of 3-phosphoglycerate, NAD^+, NADH, glutamate, and α-ketoglutarate. Therefore, these substances can be considered as external metabolites for the considered pathway.

3-Phosphoglycerate dehydrogenase (PGDH) and phosphoserine transaminase

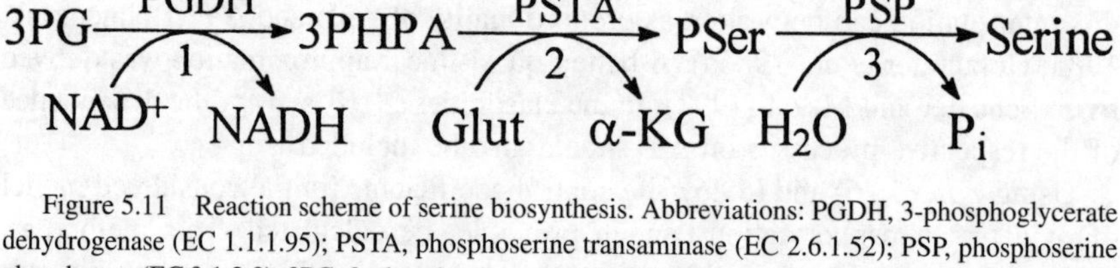

Figure 5.11 Reaction scheme of serine biosynthesis. Abbreviations: PGDH, 3-phosphoglycerate dehydrogenase (EC 1.1.1.95); PSTA, phosphoserine transaminase (EC 2.6.1.52); PSP, phosphoserine phosphatase (EC 3.1.3.3); 3PG, 3-phosphoglycerate; 3PHPA, 3-phosphohydroxypyruvate; PSer, phosphoserine.

(PSTA) are treated as a "grouped" reaction (i.e., a module in the sense defined in Section 5.12). As these enzymes operate at quasi-equilibrium, one can use the approximation formula (5.74) for the normalized elasticities of the module involving the two enzymes,

$$\pi_{3PG}^{1+2} = \frac{1}{1 - PSer/(3PG \cdot q_{\mathrm{PGDH}} \cdot q_{\mathrm{PSTA}})}, \tag{5.175}$$

$$\varepsilon_{\mathrm{PSer}}^{1+2} = \frac{1}{1 - 3PG \cdot q_{\mathrm{PGDH}} \cdot q_{\mathrm{PSTA}}/PSer}, \tag{5.176}$$

where the subscript 1 + 2 refers to the lumped process consisting of reactions PGDH and PSTA.

The quantity given in Eq. (5.175) is a π-elasticity because 3-phosphoglycerate is considered as an external metabolite here. As the average metabolite concentrations *in vitro* were measured (LaBaume *et al.,* 1987) and the equilibrium constants are known, the displacement from equilibrium can be calculated (Fell and Snell, 1988). The elasticities of phosphoserine phosphatase with respect to phosphoserine and serine were computed by numerical differentiation of the enzyme rate law

$$v = \frac{V_{\mathrm{m}} \cdot PSer(1 + Ser/K_{\mathrm{I}}')/(1 + Ser/K_{\mathrm{I}})}{PSer + K_{\mathrm{m}}(1 + Ser/K_{\mathrm{I}}')/(1 + Ser/K_{\mathrm{I}})} \tag{5.177}$$

proposed by Frieden (1964) for single-substrate enzymes in the presence of mixed-type modifiers. It is assumed that both the enzyme-substrate and enzyme-substrate-modifier complexes can yield the product and that all complex formation steps are at quasi-equilibrium. Nonlinear parameter fitting gives the parameter values $K_{\mathrm{m}} = 0.089$ mM, $K_{\mathrm{I}} = 0.60$ mM, and $K_{\mathrm{I}}' = 16.5$ mM.

The flux control coefficients can be obtained by Eq. (5.54), which implies in non-normalized form $\mathbf{C}^J = (\mathbf{K}\ \mathbf{0})(\mathbf{K}\ \boldsymbol{\varepsilon})^{-1}$ in the case of no conservation relations. For the considered system, $\mathbf{K} = (1\ 1)^{\mathrm{T}}$, so that

$$\mathbf{C}^J = \begin{pmatrix} 1 & 0 \\ 1 & 0 \end{pmatrix} \begin{pmatrix} 1 & \varepsilon_{\mathrm{PSer}}^{1+2} \\ 1 & \varepsilon_{\mathrm{PSer}}^{\mathrm{PSP}} \end{pmatrix}^{-1}, \tag{5.178}$$

where PSP stands for phosphoserine phosphatase.

The flux-response coefficients to phosphoglycerate and serine are given by

$$R_{3PG}^J = C_{1+2}^J\, \pi_{3PG}^{1+2}, \quad R_{\mathrm{Ser}}^J = C_{\mathrm{PSP}}^J\, \pi_{\mathrm{Ser}}^{\mathrm{PSP}}. \tag{5.179}$$

Fell and Snell (1988) calculated flux control coefficients for serine metabolism, using experimental values reported by LaBaume *et al.* (1987) for the metabolite concentrations in rabbit liver *in vivo* under starvation conditions (24 h-fasted animals) and 1 h after injection of glucose or ethanol or both. For starvation conditions, they obtained the normalized coefficients $C^J_{1+2} = 0.03$ and $C^J_{PSP} = 0.97$. For the situation after injection of ethanol, these control coefficients were computed to be $C^J_{1+2} = 0.22$ and $C^J_{PSP} = 0.78$ and for the situation after injection of glucose and ethanol, $C^J_{1+2} = 0.46$ and $C^J_{PSP} = 0.54$. C^J_{1+2} is very small under starvation conditions, due to the fact that PGDH and PSTA are quasi-equilibrium enzymes. The control coefficient of phosphoserine phosphatase is high in the "standard" situation because its elasticity with respect to its substrate phosphoserine is small, as this metabolite is well above the K_m value [see the discussion of Eq. (5.101)]. A possible explanation of the values in the situations after the addition of glucose and/or ethanol considering the displacement from equilibrium, redox state, and saturation was given by Fell and Snell (1988).

Although it is a widespread feature in metabolism that the first enzyme of a biosynthetic pathway exerts most flux control (see Savageau, 1976), the situation is different in the serine biosynthetic pathway, where the last enzyme is most rate-limiting. The very large control coefficient occurs, however, only under starvation conditions which are not the normal case.

5.5. TIME-DEPENDENT CONTROL COEFFICIENTS

In the preceding paragraphs, the time dependence of the system behavior after parameter perturbations was not studied. Metabolic control analysis was confined to steady states. Obviously, this restriction can be misleading upon interpretation of experimental results. In particular, it can be practically impossible to approach a steady state in reasonable times. In the present section, control analysis is extended to time-dependent states in the neighborhood of a stable steady state. From general definitions of control coefficients, we derive a calculation procedure for the time-dependent control matrices as functions of the stoichiometry of the network and of the elasticities of the reactions.

Suppose that $\boldsymbol{S}^0$ is a stable steady-state solution of equation system (2.8) for a given parameter vector $\boldsymbol{p}^0$ and let us assume that for negative times $\boldsymbol{p} = \boldsymbol{p}^0$ and $\boldsymbol{S} = \boldsymbol{S}^0$. At time zero, the parameter is perturbed and takes the value $\boldsymbol{p}$ for all positive times. For $\boldsymbol{p}$ close to $\boldsymbol{p}^0$, the solution $\boldsymbol{S}(t,\boldsymbol{p})$ of Eq. (2.8) can be approximated by

$$\boldsymbol{S}(t,\boldsymbol{p}) = \boldsymbol{S}^0 + \frac{\partial \boldsymbol{S}}{\partial \boldsymbol{p}}(t,\boldsymbol{p}^0)(\boldsymbol{p} - \boldsymbol{p}^0). \tag{5.180}$$

The time-dependent flux vector $\boldsymbol{J}(t,\boldsymbol{p})$ is defined by

$$\boldsymbol{J}(t,\boldsymbol{p}) = \boldsymbol{v}(\boldsymbol{S}(t,\boldsymbol{p}), \boldsymbol{p}). \tag{5.181}$$

For $\boldsymbol{p}$ close to $\boldsymbol{p}^0$ this flux vector can be approximated as

$$\boldsymbol{J}(t,\boldsymbol{p}) = \boldsymbol{J}^0 + \frac{\partial \boldsymbol{J}}{\partial \boldsymbol{p}}(t,\boldsymbol{p}^0)(\boldsymbol{p} - \boldsymbol{p}^0) \tag{5.182}$$

with $\boldsymbol{J}^0 = \boldsymbol{v}(\boldsymbol{S}^0,\boldsymbol{p})$. It follows from Eq. (2.8) that $\partial \boldsymbol{S}/\partial \boldsymbol{p}$ is the matrix solution of

$$\frac{\mathrm{d}}{\mathrm{d}t}\left(\frac{\partial \boldsymbol{S}}{\partial \boldsymbol{p}}\right) = \left(\mathbf{N}\frac{\partial \boldsymbol{v}}{\partial \boldsymbol{S}}\right)\frac{\partial \boldsymbol{S}}{\partial \boldsymbol{p}} + \mathbf{N}\frac{\partial \boldsymbol{v}}{\partial \boldsymbol{p}} \tag{5.183}$$

with $(\partial \boldsymbol{S}/\partial \boldsymbol{p})(t = 0,\boldsymbol{p}^0) = \mathbf{0}$. The matrices $\partial \boldsymbol{v}/\partial \boldsymbol{S}$ and $\partial \boldsymbol{v}/\partial \boldsymbol{p}$ are calculated in the reference state $(\boldsymbol{S}^0,\boldsymbol{p}^0)$. From Eq. (5.181) one derives

$$\frac{\partial \boldsymbol{J}}{\partial \boldsymbol{p}} = \frac{\partial \boldsymbol{v}}{\partial \boldsymbol{S}}\frac{\partial \boldsymbol{S}}{\partial \boldsymbol{p}} + \frac{\partial \boldsymbol{v}}{\partial \boldsymbol{p}}. \tag{5.184}$$

Let us first assume that the rows of the stoichiometry matrix $\mathbf{N}$ are linearly independent. Then there are no conservation relationships for the metabolite concentrations. Because the reference state is assumed to be stable, all the eigenvalues of the Jacobian $\mathbf{M} = \mathbf{N}(\partial \boldsymbol{v}/\partial \boldsymbol{S})$ have negative real parts, so $\mathbf{M}$ is invertible.

The formal solution of the linear differential equation system (5.183) reads

$$\frac{\partial \boldsymbol{S}}{\partial \boldsymbol{p}}(t,\boldsymbol{p}^0) = \mathbf{C}^S(t)\frac{\partial \boldsymbol{v}}{\partial \boldsymbol{p}} \tag{5.185}$$

with

$$\mathbf{C}^S(t) = [\exp(\mathbf{M}t) - \mathbf{I}]\mathbf{M}^{-1}\mathbf{N}, \tag{5.186}$$

where $\mathbf{I}$ denotes the $n \times n$ identity matrix. A possible representation of the exponential function entering this equation is

$$\exp(\mathbf{M}t) = \mathbf{B\Lambda B}^{-1}, \tag{5.187}$$

where the elements of the diagonal matrix $\boldsymbol{\Lambda}$ depend on the eigenvalues λ_i of the Jacobian $\mathbf{M}$ as follows:

$$\Lambda_{ii} = \exp(t\lambda_i). \tag{5.188}$$

The columns of $\mathbf{B}$ are the corresponding eigenvectors.

From Eqs. (5.184) and (5.185) one gets

$$\frac{\partial \boldsymbol{J}}{\partial \boldsymbol{p}}(t,\boldsymbol{p}^0) = \mathbf{C}^J(t)\frac{\partial \boldsymbol{v}}{\partial \boldsymbol{p}} \tag{5.189}$$

with

$$\mathbf{C}^J(t) = \mathbf{I} + \frac{\partial \boldsymbol{v}}{\partial \boldsymbol{S}}\mathbf{C}^S(t) = \mathbf{I} + \frac{\partial \boldsymbol{v}}{\partial \boldsymbol{S}}[\exp(\mathbf{M}t) - \mathbf{I}]\mathbf{M}^{-1}\mathbf{N}. \tag{5.190}$$

In the present case, the control matrices $\mathbf{C}^S(t)$ and $\mathbf{C}^J(t)$ are time-dependent operators which transform the initial perturbations $\delta\boldsymbol{v}$ of the reaction rates into the concentration and flux variations $\delta\boldsymbol{S}$ and $\delta\boldsymbol{J}$ at time t, that is,

$$\delta\boldsymbol{S}(t) = \mathbf{C}^S(t)\delta\boldsymbol{v}, \qquad \delta\boldsymbol{J}(t) = \mathbf{C}^J(t)\delta\boldsymbol{v}. \tag{5.191}$$

The elements $C^S_{ik}(t)$ and $C^J_{jk}(t)$ of the matrices $\mathbf{C}^S(t)$ and $\mathbf{C}^J(t)$ can be defined as the unscaled concentration control coefficients and flux control coefficients, respectively, at time t. They may be used to characterize the response of the system to parameter perturbations during the relaxation process.

Because the eigenvalues of $\mathbf{M}$ have negative real parts, the matrix $\exp(\mathbf{M}t)$ approaches zero when t tends to infinity. Equations (5.186) and (5.190) then yield the usual unscaled time-independent concentration and flux control matrices $\mathbf{C}^S$ and $\mathbf{C}^J$,

$$\mathbf{C}^S = \lim_{t\to\infty}\mathbf{C}^S(t) = -\mathbf{M}^{-1}\mathbf{N} = -\left(\mathbf{N}\frac{\partial \boldsymbol{v}}{\partial \boldsymbol{S}}\right)^{-1}\mathbf{N}, \tag{5.192}$$

$$\mathbf{C}^J = \lim_{t\to\infty}\mathbf{C}^J(t) = \mathbf{I} - \frac{\partial \boldsymbol{v}}{\partial \boldsymbol{S}}\mathbf{M}^{-1}\mathbf{N} = \mathbf{I} - \frac{\partial \boldsymbol{v}}{\partial \boldsymbol{S}}\left(\mathbf{N}\frac{\partial \boldsymbol{v}}{\partial \boldsymbol{S}}\right)^{-1}\mathbf{N} \tag{5.193}$$

[cf. Eqs. (5.13) and (5.14)]. Equations (5.186) and (5.190) show that the time-dependent control coefficients are fully determined by the stoichiometry of the reaction network and the elasticity coefficients; that is, the elements of the matrix $\boldsymbol{\varepsilon} = \partial\boldsymbol{v}/\partial\boldsymbol{S}$ calculated at the reference state.

If $\boldsymbol{p}$ contains parameters p_k acting specifically on individual reactions [cf. Eq. (5.4)], one derives from Eqs. (5.185) and (5.189)

$$C^S_{ik}(t) = \left(\frac{\partial S_i}{\partial p_k}(t,\boldsymbol{p}^0)\right)\left(\frac{\partial v_k}{\partial p_k}\right)^{-1} \tag{5.194a}$$

and

$$C^J_{jk}(t) = \left(\frac{\partial J_j}{\partial p_k}(t,\boldsymbol{p}^0)\right)\left(\frac{\partial v_k}{\partial p_k}\right)^{-1}. \tag{5.194b}$$

These formulas are directly related to the usual definitions of control coefficients (5.18), with the difference that they are time dependent.

When the network is such that some linear combinations of metabolite concentration are conserved, the rows of the stoichiometry matrix $\mathbf{N}$ are not linearly independent. In this case, one derives from formulas (3.10) and (5.190) that the time-dependent control matrices may be expressed as

$$\mathbf{C}^S(t) = \mathbf{L}[\exp(\mathbf{M}^0 t) - \mathbf{I}](\mathbf{M}^0)^{-1}\mathbf{N}^0, \tag{5.195a}$$

$$\mathbf{C}^J(t) = \mathbf{I} + \frac{\partial \boldsymbol{v}}{\partial \boldsymbol{S}}\mathbf{L}[\exp(\mathbf{M}^0 t) - \mathbf{I}](\mathbf{M}^0)^{-1}\mathbf{N}^0, \tag{5.195b}$$

with

$$\mathbf{M}^0 = \mathbf{N}^0 \frac{\partial \boldsymbol{v}}{\partial \boldsymbol{S}} \mathbf{L}, \tag{5.196}$$

where $\mathbf{L}$ and $\mathbf{N}^0$ are the link matrix and the reduced stoichiometry matrix defined in Eq. (3.7), respectively.

It follows immediately from Eqs. (5.195a) and (5.195b) that the matrices of control coefficients fulfill the following relationships:

$$\mathbf{C}^J(t_2)\mathbf{C}^J(t_1) = \mathbf{C}^J(t_1 + t_2) \tag{5.197}$$

and

$$\mathbf{C}^S(t_2)\boldsymbol{\varepsilon}\mathbf{C}^S(t_1) + \mathbf{C}^S(t_1) + \mathbf{C}^S(t_2) = \mathbf{C}^S(t_1 + t_2), \tag{5.198}$$

which generalize Eqs. (5.27a) and (5.27b) to the time-dependent case.

Equation (5.197) means that the map of an initial perturbation of fluxes by a parameter change onto the flux change after a time t_1 and the consecutive map of this flux change onto the flux change after another time span t_2, which is mediated by all reactions, is equivalent to the map of the initial perturbation onto the flux change after $t_1 + t_2$. Relationship (5.198) can be interpreted in the following way. An initial perturbation $\delta \boldsymbol{v}$ of reaction rates has a direct effect on the fluxes as expressed by the identity matrix in Eq. (5.190) and leads, after a time t_1, to a change of concentrations $\mathbf{C}^S(t_1)\delta\boldsymbol{v}$. This change also has an effect on the reaction rates, as expressed by premultiplication by $\boldsymbol{\varepsilon} = \partial \boldsymbol{v}/\partial \boldsymbol{S}$. During another time span t_2, both of these effects on the reaction rates lead to concentration changes which may be expressed by premultiplying $\mathbf{C}^S(t_2)$. These changes have to be added to the change $\mathbf{C}^S(t_1)\delta\boldsymbol{v}$ already achieved at time t_1.

Summation and connectivity theorems can also be derived for the time-dependent control coefficients. Under consideration of Eq. (3.44), one can derive from Eqs. (5.195a) and (5.195b) that

$$\mathbf{C}^S(t)\mathbf{K} = \mathbf{0}, \tag{5.199a}$$

$$\mathbf{C}^J(t)\mathbf{K} = \mathbf{K}, \tag{5.199b}$$

where **K** again denotes the null-space matrix of the stoichiometry matrix. These equalities are the summation relationships for the time-dependent control matrices $\mathbf{C}^S(t)$ and $\mathbf{C}^J(t)$. As a particular consequence,

$$\mathbf{C}^S(t)\boldsymbol{J} = \boldsymbol{0}, \tag{5.200a}$$

$$\mathbf{C}^J(t)\boldsymbol{J} = \boldsymbol{J}, \tag{5.200b}$$

because the steady-state flux $\boldsymbol{J}$ satisfies the equality $\mathbf{N}\boldsymbol{J} = \boldsymbol{0}$.

By multiplying Eqs. (5.195a) and (5.195b) by $(\partial \boldsymbol{v}/\partial \boldsymbol{S})\mathbf{L}$, one gets by the definition of $\mathbf{M}^0$

$$\mathbf{C}^S(t)\frac{\partial \boldsymbol{v}}{\partial \boldsymbol{S}}\mathbf{L} = \mathbf{L}[\exp(\mathbf{M}^0 t) - \mathbf{I}], \tag{5.201a}$$

$$\mathbf{C}^J(t)\frac{\partial \boldsymbol{v}}{\partial \boldsymbol{S}}\mathbf{L} = \frac{\partial \boldsymbol{v}}{\partial \boldsymbol{S}}\mathbf{L}\exp(\mathbf{M}^0 t). \tag{5.201b}$$

These formulas are the connectivity relationships for the time-dependent control coefficients. Equations (5.199) and (5.201) reveal the interesting fact that time enters explicitly the connectivity theorems only. This means that the summation relationships (5.199), which have the same structure as Eqs. (5.44), are satisfied during the whole relaxation process, although the control coefficients may vary considerably. As for the connectivity theorems, one derives from Eq. (5.201) the usual theorems of time-independent control analysis [cf. Eq. (5.51)] in the limit of infinite time. If, however, the steady state is unstable (i.e., if the Jacobian matrix has at least one eigenvalue with a positive real part) the exponentials on the right-hand side of Eqs. (5.201a) and (5.201b) diverge as t tends to infinity. Thus, the time-independent connectivity theorems [Eq. (5.51)] have no meaning for unstable steady states, although the derivation leading to those theorems in Section 5.3.2 applies also to such states. A more detailed analysis can be found in the work of Heinrich and Reder (1991).

It can be proved that, as in the time-independent case, the relations (5.201a) and (5.201b) are sufficient in number to calculate the control matrices from the stoichiometric and elasticity coefficients (see Section 5.3.3).

Example. We consider the branched system depicted in Scheme 7 (Section 3.2.4) with one internal metabolite and three reactions, as described by the stoichiometry matrix $\mathbf{N} = (1 \ -1 \ -1)^{\mathrm{T}}$. Admissible null-space vectors are $\boldsymbol{k}_1$ and $\boldsymbol{k}_2$ as given in Eq. (5.119). The system equation takes the form

$$\frac{dS_1}{dt} = v_1 - v_2 - v_3 . \tag{5.202}$$

We will calculate the time-dependent control coefficients $C_{ik}^S(t)$ and $C_{jk}^J(t)$ by using the summation and connectivity relationships (5.199) and (5.201), taking into account $\mathbf{L} = \mathbf{I}$ (no conservation relationships). The Jacobian $\mathbf{M}$ contains only one element

$$M_{11} = \varepsilon_{11} - \varepsilon_{21} - \varepsilon_{31} . \tag{5.203}$$

The summation relationships for the concentration control coefficients are

$$C_{11}^S(t) + C_{12}^S(t) = 0, \qquad C_{11}^S(t) + C_{13}^S(t) = 0 \tag{5.204a}$$

and for the flux control coefficients

$$\begin{aligned} C_{11}^J(t) + C_{12}^J(t) &= 1, \\ C_{11}^J(t) + C_{13}^J(t) &= 1, \\ C_{21}^J(t) + C_{22}^J(t) &= 1, \\ C_{21}^J(t) + C_{23}^J(t) &= 0, \\ C_{31}^J(t) + C_{32}^J(t) &= 0, \\ C_{31}^J(t) + C_{33}^J(t) &= 1. \end{aligned} \tag{5.204b}$$

The connectivity relationships read

$$C_{11}^S(t)\varepsilon_{11} + C_{12}^S(t)\varepsilon_{21} + C_{13}^S(t)\varepsilon_{31} = \exp[(\varepsilon_{11} - \varepsilon_{21} - \varepsilon_{31})t] - 1 \tag{5.205a}$$

and

$$C_{j1}^J(t)\varepsilon_{11} + C_{j2}^J(t)\varepsilon_{21} + C_{j3}^J(t)\varepsilon_{31} = \varepsilon_{j1} \exp[(\varepsilon_{11} - \varepsilon_{21} - \varepsilon_{31})t] \tag{5.205b}$$

with $j = 1,2,3$. Solving these equations, one obtains

$$C_{11}^S(t) = -C_{12}^S(t) = -C_{13}^S(t) = \frac{\exp[(\varepsilon_{11} - \varepsilon_{21} - \varepsilon_{31})t] - 1}{\varepsilon_{11} - \varepsilon_{21} - \varepsilon_{31}} \tag{5.206a}$$

and

$$\begin{aligned} \mathbf{C}^J = {} & \frac{1}{\varepsilon_{11} - \varepsilon_{21} - \varepsilon_{31}} \begin{pmatrix} -\varepsilon_{21} - \varepsilon_{31} & \varepsilon_{11} & \varepsilon_{11} \\ -\varepsilon_{21} & \varepsilon_{11} - \varepsilon_{31} & \varepsilon_{21} \\ -\varepsilon_{31} & \varepsilon_{31} & \varepsilon_{11} - \varepsilon_{21} \end{pmatrix} \\ & + \frac{\exp[(\varepsilon_{11} - \varepsilon_{21} - \varepsilon_{31})t]}{\varepsilon_{11} - \varepsilon_{21} - \varepsilon_{31}} \begin{pmatrix} \varepsilon_{11} & -\varepsilon_{11} & -\varepsilon_{11} \\ \varepsilon_{21} & -\varepsilon_{21} & -\varepsilon_{21} \\ \varepsilon_{31} & -\varepsilon_{31} & -\varepsilon_{31} \end{pmatrix} . \end{aligned} \tag{5.206b}$$

A sensitivity analysis of time-dependent trajectories of metabolic systems has also been developed by Kohn *et al.* (1979) and Kohn and Chiang (1982, 1983). They studied the response coefficients to a parameter perturbation and derived equations similar to Eq. (5.183). However, they did not give general definitions of control coefficients and did not, therefore, obtain results such as the summation or connectivity theorems.

5.6. ARE CONTROL COEFFICIENTS ALWAYS PARAMETER INDEPENDENT?

5.6.1. Posing the Problem

It has been shown in Section 5.2 that concentration control coefficients and flux control coefficients, defined by eqs. (5.25b) and (5.26b), respectively, are independent of the choice of the perturbation parameter. It has sometimes been questioned whether this general conclusion remains valid if the analysis of a metabolic system is based on the rates $\boldsymbol{w}$ of the elementary reactions of enzymes (instead of the rates $\boldsymbol{v}$ of the overall reactions) because the parameter dependence of the concentrations of enzyme intermediate complexes must then be taken into account in addition to that of free metabolites. It has been claimed that this problem may be of particular importance for systems with conservation equations, because at the detailed level of description, they generally include also the concentrations of enzyme-intermediate complexes (Reder, 1986; Fell and Sauro, 1990; Kholodenko *et al.,* 1992, 1993b, 1995). Here, it may be expected that perturbation parameters which affect a certain enzyme specifically but differ in their effect on the enzyme-intermediate concentrations have effects on the free-substrate concentrations or fluxes which cannot be described by one and the same control coefficient of the given enzyme. Obviously, this situation does not meet with formal difficulties if control coefficients of *elementary steps* are considered and the concentrations of free enzymes as well as of enzyme-bound species are included into the vector $\boldsymbol{S}$ of metabolite concentrations. This follows from the fact that in the general treatment presented in Section 5.2 the character of the metabolites is not specified and that no distinction is made whether the reactions are elementary or not. However, problems may arise if one is interested in control coefficients of overall enzymic steps. To clarify this problem in quantitative terms, we consider the following examples.

5.6.2. A System Without Conserved Moieties

The system depicted in Scheme 13 consists of two enzymic reactions converting the substrate P_1 into the product P_2 via the enzyme–substrate complexes

E_1S_1 and E_2S_1 and the free intermediate S_1. The concentrations P_1 and P_2 are considered to be fixed. The rates of the elementary reactions of the enzymes E_j are denoted by $w_{j,a}$ and $w_{j,b}$. In the present case the reaction rates of the isolated enzymes v_1 and v_2 are defined by fixed concentrations S_1 and by quasi-steady-state values for E_1S_1 and E_2S_1.

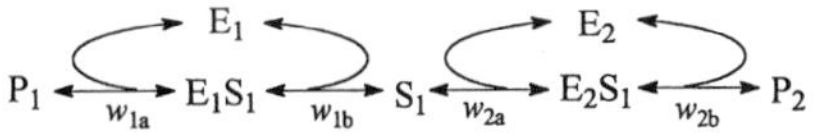

Scheme 13

First, we calculate the unscaled parameter elasticities $\partial v_j/\partial p_j$ which enter the denominator in the definition (5.7) of unscaled control coefficients. We denote by p_1 and p_2 the kinetic parameters which affect specifically the reactions catalyzed by the first and the second enzyme, respectively. The quasi-steady-state conditions for E_1S_1 and E_2S_1 read $w_{1a} = w_{1b}$ and $w_{2a} = w_{2b}$, respectively.

Taking into account the two conservation relations for the enzyme species

$$E_j + E_jS_1 = E_{T,j} \tag{5.207}$$

($j = 1,2$), one derives for the parameter elasticities

$$\frac{\partial v_j}{\partial p_j} = \frac{\partial w_{j,a}}{\partial p_j} + \left(\frac{\partial w_{j,a}}{\partial E_jS_1} - \frac{\partial w_{j,a}}{\partial E_j}\right)\frac{\partial E_jS_1}{\partial p_j}. \tag{5.208}$$

Implicit differentiation of the quasi-steady-state conditions for E_jS_1 with respect to p_1 and p_2, respectively, results in expressions for $\partial E_jS_1/\partial p_1$ which may be introduced into Eq. (5.208), to give

$$\frac{\partial v_j}{\partial p_j} = \frac{1}{a_j - b_j}\left(\frac{w_{j,b}}{\partial p_j}a_j - \frac{\partial w_{j,a}}{\partial p_j}b_j\right) \tag{5.209}$$

with

$$a_j = \frac{\partial w_{j,a}}{\partial E_j} - \frac{\partial w_{j,a}}{\partial E_jS_1}, \quad b_j = \frac{\partial w_{j,b}}{\partial E_j} - \frac{\partial w_{j,b}}{\partial E_jS_1}. \tag{5.210}$$

For the unscaled substrate elasticities $\varepsilon_{11} = \partial v_1/\partial S_1$ and $\varepsilon_{21} = \partial v_2/\partial S_1$, one obtains in a similar way

$$\varepsilon_{11} = \frac{a_1}{a_1 - b_1}\frac{\partial w_{1b}}{\partial S_1}, \quad \varepsilon_{21} = -\frac{b_2}{a_2 - b_2}\frac{\partial w_{2a}}{\partial S_1}. \tag{5.211}$$

Now we consider the steady state of the whole system where the concentration S_1 may vary depending on the kinetic parameters. This state is characterized by the steady-state condition for S_1, $w_{1b} = w_{2a}$, in addition to the steady-state conditions for E_jS_1. Implicit differentiation of the steady-state conditions with respect to p_1 and p_2 gives the following expressions for the two concentration control coefficients

$$C^S_{11} = -C^S_{12} = (a_1 - b_1)(a_2 - b_2)\left[\frac{\partial w_{2a}}{\partial S_1} b_1 b_2 + \left(\frac{\partial w_{1b}}{\partial S_1} - \frac{\partial w_{2a}}{\partial S_1}\right) a_1 b_2 - \frac{\partial w_{1b}}{\partial S_1} a_1 a_2\right]^{-1}. \quad (5.212)$$

It is seen that for the system depicted in Scheme 13 the control coefficients are independent of the special choice of the perturbation parameter and, further, that the summation theorem $C^S_{11} + C^S_{12} = 0$ is fulfilled. Taking into account relations (5.211), Eq. (5.212) may be rewritten as

$$C^S_{11} = -C^S_{12} = \frac{1}{\varepsilon_{21} - \varepsilon_{11}}, \quad (5.213)$$

which is identical to expression (5.99) derived for the concentration control coefficients for a two-enzyme system by using steady-state rate equations for the individual enzymes. This means that in the present case the general conclusions of metabolic control analysis are not affected by the level of description (i.e., whether the system is analyzed using overall rate equations or on the basis of the elementary steps).

5.6.3. A System with a Conserved Moiety

The system depicted in Scheme 14 consists of reactions converting the metabolites S_1 and S_2 in a cyclic manner.

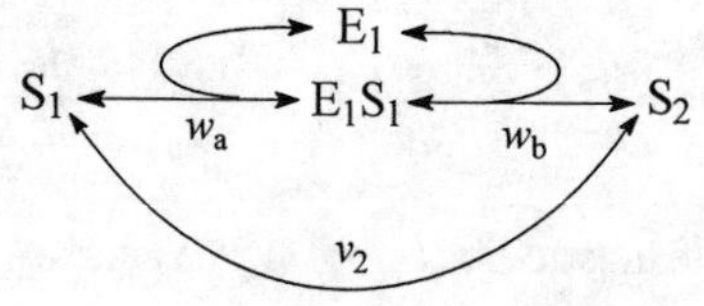

Scheme 14

For simplicity's sake it is assumed that only reaction 1 is described at the level of elementary reactions. The corresponding mechanism contains two steps described by the rates w_a and w_b. The overall velocity of reaction 2 is described by v_2. In addition to the conservation equation for the enzyme species of reaction 1, there is another one which involves the concentrations of the free intermediates and of the enzyme-intermediate complex of the first reaction. It reads

$$S_1 + S_2 + E_1S_1 = \text{const.} \tag{5.214}$$

From the quasi-steady-state condition $w_a = w_b$ for E_1S_1 one obtains

$$\frac{\partial E_1S_1}{\partial p_1} = \frac{1}{a - b}\left(\frac{\partial w_a}{\partial p_1} - \frac{\partial w_b}{\partial p_1}\right) \tag{5.215}$$

with $a = \partial w_a/\partial E_1 - \partial w_a/\partial E_1S_1$ and $b = \partial w_b/\partial E_1 - \partial w_b/\partial E_1S_1$. In a manner similar to that in Section 5.6.2, the following expression for the parameter elasticity of the enzymic reaction is derived:

$$\frac{\partial v_1}{\partial p_1} = \frac{1}{a - b}\left(\frac{\partial w_b}{\partial p_1}a - \frac{\partial w_a}{\partial p_1}b\right). \tag{5.216}$$

The steady state of the whole system is characterized by the conditions $v_2 = w_a$ and $v_2 = w_b$ and by the conservation relationships $E_1 + E_1S_1 = E_{T,1}$ and Eq. (5.214). Implicit differentiation of the steady-state conditions with respect to p_1 results in the following expressions for the concentration control coefficients with respect to reaction 1:

$$C_{11}^S = \frac{\partial S_1/\partial p_1}{\partial v_1/\partial p_1} = \left(\frac{a - b}{D}\right)\left\{\left[\frac{\partial w_a}{\partial p_1}\left(\frac{\partial v_2}{\partial S_2} - \frac{\partial w_b}{\partial S_2} - b\right) - \frac{\partial w_b}{\partial p_1}\left(\frac{\partial v_2}{\partial S_2} - a\right)\right]\left(\frac{\partial w_b}{\partial p_1}a - \frac{\partial w_a}{\partial p_1}b\right)^{-1}\right\} \tag{5.217a}$$

$$C_{21}^S = \frac{\partial S_2/\partial p_1}{\partial v_1/\partial p_1} = \left(\frac{a - b}{D}\right)\left\{\left[\frac{\partial w_b}{\partial p_1}\left(\frac{\partial v_2}{\partial S_1} - \frac{\partial w_1}{\partial S_1} - a\right) - \frac{\partial w_a}{\partial p_1}\left(\frac{\partial v_2}{\partial S_1} - b\right)\right]\left(\frac{\partial w_b}{\partial p_1}a - \frac{\partial w_a}{\partial p_1}b\right)^{-1}\right\} \tag{5.217b}$$

with

$$D = a\left(\frac{\partial v_2}{\partial S_1} - \frac{\partial v_2}{\partial S_2} + \frac{\partial w_b}{\partial S_2}\right) + b\left(\frac{\partial v_2}{\partial S_2} - \frac{\partial v_2}{\partial S_1} + \frac{\partial w_a}{\partial S_1}\right) + \frac{\partial w_a}{\partial S_1}\frac{\partial w_b}{\partial S_2} - \frac{\partial w_a}{\partial S_1}\frac{\partial v_2}{\partial S_2} - \frac{\partial w_b}{\partial S_2}\frac{\partial v_2}{\partial S_1}. \tag{5.217c}$$

Analogously, for the concentration control coefficients of reaction 2, implicit differentiation of the steady-state conditions with respect to p_2 yields

$$C_{12}^S = \frac{\partial S_1/\partial p_2}{\partial v_2/\partial p_2} = \frac{1}{D}\left(\frac{\partial w_b}{\partial S_2} + b - a\right) \tag{5.218a}$$

and

$$C_{22}^S = \frac{\partial S_2/\partial p_2}{\partial v_2/\partial p_2} = \frac{1}{D}\left(\frac{\partial w_a}{\partial S_1} - b + a\right). \tag{5.218b}$$

It is seen that the control coefficients of reaction 2 are independent of the special choice of the perturbation parameter which is in accord with the general statements made in Section 5.2. The situation is different for the control coefficients of reaction 1. Here, only the factor $(a - b)/D$ is independent of the choice of the perturbation parameter p_1, whereas, in the remaining terms, the derivatives of w_a and w_b with respect to p_1 cannot, in general, be canceled. We may conclude, therefore, that the control coefficients of enzymes with enzyme-bound metabolites involved in conserved moieties are parameter dependent. However, it is seen from Eqs. (5.217) and (5.218) that when p_1 affects w_a specifically ($\partial w_b/\partial p_1 = 0$), the term $\partial w_a/\partial p_1$ cancels in the expression of C_{11}^S and C_{21}^S. In this case, the concentration control coefficients become independent of what parameter of reaction w_a is changed (e.g., k_a or k_{-a} if the expression $w_a = k_a S_1 \cdot E_1 - k_{-a} E_1 S_1$ is used). Similarly, if p_1 affects w_b specifically, the coefficients C_{11}^S and C_{21}^S do not contain derivatives with respect to p_1 either, but they have, in general, different values from C_{11}^S and C_{21}^S in the former case.

Moreover, a description of enzyme systems at the level of elementary rates may lead to the fact that the summation theorems are violated. For example, from Eqs. (5.217a) and (5.218a) one obtains

$$C_{11}^S + C_{12}^S = \left(\frac{\partial w_a}{\partial p_1} - \frac{\partial w_b}{\partial p_1}\right)\left[D\left(\frac{\partial w_b}{\partial p_1} a - \frac{\partial w_a}{\partial p_1} b\right)\right]^{-1}\left((a - b)\frac{\partial v_2}{\partial S_2} - a\frac{\partial w_b}{\partial S_2}\right), \tag{5.219}$$

which is generally nonzero, in contrast to the summation theorem for metabolite concentrations. As a special case, one may deal with a parameter p_1 whose changes affect the rate constants of the elementary reactions by the same factor. Then one obtains $p_1 \cdot \partial w_a/\partial p_1 = w_a$ and $p_1 \cdot \partial w_b/\partial p_1 = w_b$. This implies, due to the steady-state condition $w_a = w_b$, that the summation theorems are fulfilled.

Now we use the additional assumption that the total enzyme concentration is negligibly small compared to the steady-state concentrations S_1 and S_2, that is,

$$E_{T,1} \ll S_1, S_2. \tag{5.220}$$

Using the rate laws $w_a = k_a S_1 \cdot E_1 - k_{-a} E_1 S_1$ and $w_b = k_b E_1 S_1 - k_{-b} S_2 \cdot E_1$, the solutions S_1 and S_2 resulting from the steady-state equations

$$\frac{dS_1}{dt} = v_2 - w_a = 0, \qquad \frac{dS_2}{dt} = -v_2 + w_b = 0,$$
$$\frac{dE_1S_1}{dt} = w_a - w_b = 0, \qquad \frac{dE_1}{dt} = -w_a + w_b = 0 \tag{5.221}$$

are invariant with respect to the transformation

$$k'_{\pm a} = \frac{k_{\pm a}}{\rho}, \qquad k'_{\pm b} = \frac{k_{\pm b}}{\rho}, \qquad (E_1S_1)' = \rho \cdot E_1S_1, \qquad E'_1 = \rho E_1 \tag{5.222}$$

with a scaling factor ρ. This transformation implies

$$a' = \frac{a}{\rho}, \qquad b' = \frac{b}{\rho}, \tag{5.223}$$

whereas all other quantities which enter Eqs. (5.217) and (5.218) remain unchanged. In the limit $\rho \to 0$, one obtains the control coefficients

$$C^S_{11} = -C^S_{21} = -C^S_{12} = C^S_{23} = (a - b)\left[a\left(\frac{\partial v_2}{\partial S_1} - \frac{\partial v_2}{\partial S_2} + \frac{\partial w_b}{\partial S_2}\right) + b\left(\frac{\partial v_2}{\partial S_2} - \frac{\partial v_2}{\partial S_1} + \frac{\partial w_a}{\partial S_1}\right)\right]^{-1}, \tag{5.224}$$

which no longer contain any derivative with respect to parameters. Thus, the concentration control coefficients become independent of the choice of the perturbation parameter in the case of very low enzyme concentrations. Moreover, it is seen from Eq. (5.224) that for $\rho \to 0$ the summation theorem for concentration control coefficients is fulfilled.

The general conclusions derived in the present section remain valid if the total enzyme concentrations act as perturbation parameters ($E_{T,j} = p_j$). For the system depicted in Scheme 14, for example, the quasi-steady-state condition for E_1S_1 reads

$$w_a(S_1, E_1S_1, E_1(E_1S_1, E_{T,1})) = w_b(S_2, E_1S_1, E_1(E_1S_1, E_{T,1})) \tag{5.225}$$

with $E_1(E_1S_1, E_{T,1}) = E_{T,1} - E_1S_1$. Taking the derivative with respect to $E_{T,1}$ yields

$$\frac{\partial E_1S_1}{\partial E_{T,1}} = \frac{1}{a - b}\left(\frac{\partial w_a}{\partial E_1} - \frac{\partial w_b}{\partial E_1}\right) \tag{5.226}$$

and

$$\frac{\partial v_1}{\partial E_{T,1}} = \frac{1}{a - b}\left(\frac{\partial w_b}{\partial E_1}a - \frac{\partial w_a}{\partial E_1}b\right). \tag{5.227}$$

These equations correspond to Eqs. (5.215) and (5.216), respectively, with the concentration E_1 of the *free* enzyme formally playing, on the right-hand sides of the equations, the role of the parameter. For the concentration control coefficients resulting from variations in total enzyme concentrations, one again obtains expressions (5.217a) and (5.217b), where p_1 must be replaced by E_1 on their right-hand sides.

The present results show that in systems where the total enzyme concentrations are comparable to the substrate concentrations and conservation relations involving both substrates and enzyme-substrate complexes are present, both the individual control coefficients and their sum may depend on the specific way of perturbation (cf. Kholodenko *et al.*, 1995).

The analysis in this section is closely related to the modular approach to metabolic control analysis (Section 5.12) where the subdivision of metabolic networks into functional units is studied. Accordingly, similar conclusions concerning the role of conservation relations for the control coefficients are drawn in both approaches.

5.6.4. A System Including Dynamic Channeling

In Figure 2.1, a pathway is shown in which the conversion of a substrate P_1 to a product P_2 proceeds both via a free intermediate, S_1, and a complex $E_1S_1E_2$ involving the two sequential enzymes. The scheme involves six elementary steps with net velocities $w_{j,a}$, $w_{j,b}$ and $w_{j,c}$ (see Fig. 2.1). This system is an example of a dynamically channeled pathway. The phenomenon of metabolic channeling will be commented on in more detail in Section 5.15.

We now wish to show that the concentration control coefficients of the enzymes depend on the choice of the perturbation parameter, by choosing enzyme 1 as an example. For simplicity's sake, assume the elementary step 1b in the considered pathway to be irreversible ($k_{-1b} = 0$). First, we consider a perturbation of the reaction rate, v_1, of enzyme E_1 in isolation, with the complex E_1S_1 being at quasi-steady-state. Differentiation of the Michaelis–Menten equation (2.20) yields

$$\frac{\partial v_1}{\partial k_{1a}} = \frac{P_1 E_{T,1} k_{1b}\,(k_{-1a} + k_{1b})}{(k_{1a}P_1 + k_{-1a} + k_{1b})^2}, \tag{5.228}$$

$$\frac{\partial v_1}{\partial k_{-1a}} = -\frac{P_1 E_{T,1} k_{1a} k_{1b}}{(k_{1a}P_1 + k_{-1a} + k_{-1b})^2}. \tag{5.229}$$

Now we study perturbations of the steady state of the whole system and, in particular, effects on the concentration S_1. Let $\boldsymbol{w}$ and $\boldsymbol{S}$ be the vectors of the six elementary reaction rates and of the six variable concentrations in the scheme. The response of steady-state concentrations to parameter changes can be written as

$$\frac{\partial \boldsymbol{S}}{\partial \boldsymbol{p}} = -\left(\mathbf{N}^0 \frac{\partial \boldsymbol{w}}{\partial \boldsymbol{S}} \mathbf{L}\right)^{-1} \mathbf{N}^0 \frac{\partial \boldsymbol{w}}{\partial \boldsymbol{p}} \tag{5.230}$$

[cf. Eq. (5.23)]. $\mathbf{N}^0$ and $\mathbf{L}$ are the reduced stoichiometry matrix and link matrix, respectively, of the detailed scheme consisting of elementary steps. Choosing a perturbation parameter specific to reaction 1, Eq. (5.230) implies

$$\frac{\partial S_1}{\partial p_1} = \lambda \frac{\partial w_1}{\partial p_1}, \tag{5.231}$$

because all other components of $\partial \boldsymbol{w}/\partial \boldsymbol{p}$ are zero. λ is a common factor resulting from Eqs. (5.230) and (5.231). Using $w_1 = k_{1a}P_1 \cdot E_1 - k_{-1a}E_1S_1$, one obtains from Eq. (5.231), with k_{1a} or k_{-1a} as perturbation parameters,

$$\frac{\partial S_1}{\partial k_{1a}} = \lambda \cdot P_1 \cdot E_1, \quad \frac{\partial S_1}{\partial k_{-1a}} = -\lambda \cdot E_1S_1. \tag{5.232}$$

One can calculate the non-normalized control coefficient of reaction 1 over the concentration S_1 alternatively as

$$C^S_{1,1a} = \frac{\partial S_1/\partial k_{1a}}{\partial v_1/\partial k_{1a}} = \mu \frac{P_1 \cdot E_1}{k_{-1a} + k_{1b}} \tag{5.233a}$$

or as

$$C^S_{1,-1a} = \frac{\partial S_1/\partial k_{-1a}}{\partial v_1/\partial k_{-1a}} = \mu \frac{E_1S_1}{k_{1a}} \tag{5.233b}$$

with $\mu = \lambda(k_{1a}P_1 + k_{-1a} + k_{1b})^2/(k_{1b}P_1E_{T,1})$. These two coefficients are identical if, and only if,

$$k_{1a} \cdot P_1 \cdot E_1 - (k_{-1a} + k_{1b}) \cdot E_1S_1 = 0. \tag{5.234}$$

The left-hand side of this equation equals $\mathrm{d}E_1S_1/\mathrm{d}t + w_{1c}$. As E_1S_1 is assumed to be at steady state, Eq. (5.234) holds true only if $w_{1c} = 0$ (i.e., if no channeling occurs). Consequently, if the channel is operative, the value of the concentration control coefficient of reaction 1 depends on which perturbation parameter has been chosen. This is because changing the distribution of E_1 among its subforms by altering a kinetic parameter influences the distribution of E_2 among its subforms, through the complex $E_1E_2S_1$. This effect is taken into account in the numerators of the control coefficients as given in Eq. (5.233), but not in the denominators, because the derivatives $\partial v_1/\partial k_{1a}$ and $\partial v_1/\partial k_{-1a}$ are taken for the enzyme considered in isolation. As the flux control coefficients can be calculated from the concentration control coefficients, as given in Eq. (5.14), they are not independent of the perturbation parameter in the case of channeling (cf. Kholodenko *et al.,* 1995). Note that unlike in the situation of moiety conservation considered in Section 5.6.3, in the situation of channeling even parameters of one and the same step give different results.

5.7. NORMALIZED VERSUS NON-NORMALIZED COEFFICIENTS

Upon derivation of the basic equations of metabolic control analysis in Sections 5.2 and 5.3, we have mainly used non-normalized control and elasticity coefficients, although control analysis was originally developed in terms of normalized coefficients. Generally speaking, mathematical operations with control coefficients are easier if unscaled derivatives are used (see Mazat *et al.,* 1990; Heinrich and Reder, 1991), whereas scaled coefficients are better suited for biochemical interpretation. Because the measured values of fluxes through different metabolic pathways and concentrations of different intermediates generally differ by several orders of magnitude, a quantitative measure of control should be given in terms of fractional changes. To some extent, however, it is still a matter of personal preference which type of coefficient is used. In what follows, we will discuss some advantages and drawbacks of the two methods of definition in more detail (see also Reder, 1988; Fell, 1992; S. Schuster and Heinrich, 1992). As in the present context the clear distinction between non-normalized and normalized control coefficients is essential, we denote the latter ones by C_{norm}.

Rescaling of variables and parameters: In biochemistry, the term "flux" is mostly used in the sense of steady-state velocity of the formation (or degradation) of a specified metabolite. However, this interpretation may be ambiguous in systems including reactions of higher molecularity. For example, the flux of glycolysis can be measured as the consumption rate of glucose or as the production rate of lactate (which differ by the factor 2) and the flux of Na^+/K^+-ATPase can be measured in terms of the Na^+ transport, K^+ transport or ATP consumption, all of them differing from each other. In contrast to non-normalized control coefficients, the normalized coefficients have the favorable property of being invariant with respect to rescaling of fluxes. This can be shown in the following way, for systems of any complexity. Rescaling of reaction rates can be expressed by means of a diagonal matrix (dg$\boldsymbol{A}$) constructed from an arbitrary vector $\boldsymbol{A}$ not containing a zero,

$$^{+}\boldsymbol{v} = (\text{dg}\,\boldsymbol{A})\boldsymbol{v}. \tag{5.235}$$

Similar transformation rules apply to steady-state fluxes and the derivatives of reaction rates with respect to parameters. Note the difference between rescaling of reaction rates and normalization of control coefficients. The former is related to the way the rates are registered, with those variables remaining having physical units, whereas the latter is a nondimensionalization.

Equation (5.235) implies that the transformed matrix of non-normalized control coefficients obtains as

$$^{+}\mathbf{C}^{J} = (\mathrm{dg}\,\boldsymbol{A})\mathbf{C}^{J}(\mathrm{dg}\,\boldsymbol{A})^{-1}. \tag{5.236}$$

The normalization of $^{+}\mathbf{C}^{J}$ has to be done by use of the transformed fluxes,

$$^{+}\mathbf{C}^{J}_{\mathrm{norm}} = (\mathrm{dg}\ {}^{+}\boldsymbol{J})^{-1}\ {}^{+}\mathbf{C}^{J}(\mathrm{dg}\ {}^{+}\boldsymbol{v}). \tag{5.237}$$

From these equations one obtains

$$^{+}\mathbf{C}^{J}_{\mathrm{norm}} = \mathbf{C}^{J}_{\mathrm{norm}}. \tag{5.238}$$

This identity implies, for example, that the control on the glycolytic flux is independent of whether this flux is identified with the consumption rate of glucose or with the production rate of lactate, provided that normalized coefficients are used.

It is reasonable to postulate that the control coefficients should be invariant to changes of the units of fluxes and concentrations. A change of the flux unit is expressed by Eq. (5.235) with all the components of the vector $\boldsymbol{A}$ being equal to each other. A similar equation can be written to express a rescaling of concentrations. It is easy to see that the normalized control coefficients have the favorable property to be independent of such rescaling. However, it follows from Eq. (5.236) that a change of flux units (i.e., all the components of the vector $\boldsymbol{A}$ are identical) does not affect the non-normalized flux control coefficients either (i.e., $^{+}\mathbf{C}^{J} = \mathbf{C}^{J}$). Unscaled concentration control coefficients have the same dimension as time. So they are independent of the choice of the concentration unit. Yet, they are not invariant to a different rescaling of the particular reaction rates.

Similar considerations apply if knowledge of the exact reaction mechanism of some of the reactions is incomplete, in that only the ratios of the stoichiometric coefficients are known. Any rescaling of the columns of the stoichiometry matrix must be accompanied with a reciprocal rescaling of the fluxes, as is seen from the steady-state equation (2.9). Because this rescaling of fluxes is expressed by Eq. (5.235), we again obtain Eq. (5.238) and a similar equation for concentration control coefficients. Thus, the normalized control coefficients are invariant to rescaling of the stoichiometric coefficients of the particular reactions.

Note that rescaling the perturbation parameters has neither an effect on the normalized nor on the non-normalized control coefficients, because these parameters enter both the numerator and denominator in the definition equation. π-Elasticities are invariant to rescaling of parameters in their scaled form, but in their unscaled form, they are not.

Extent of necessary knowledge: As for elasticities, it is worth mentioning that normalized coefficients are sometimes available even if the non-normalized coefficients are not. This is, in particular, the case if the substrate concentration is well below the Michaelis constant. In this situation, the normalized elasticity ε_S is virtually equal to unity [cf. Eq. (5.68)]. Similar considerations apply for the parameter elasticities. For example, the normalized elasticity coefficient of the enzyme concentration ($\pi_E = \partial \ln v / \partial \ln E$) is always unity provided that the reaction rate v depends linearly on E. Furthermore, in contrast to their unscaled counterparts, normalized elasticities have the favorable property of becoming independent of the kinetic parameters in the case of near-equilibrium reactions (see Section 5.4.1).

Interpretation: Another advantage of normalization arises from the observation that many substrates and inhibitors have *in vivo* concentrations comparable with the corresponding Michaelis constants respectively inhibition constants [Lowry and Passonneau (1964); for theoretical explanations by evolutionary arguments cf. Section 6.1 as well as the works of Crowley (1975), Cornish-Bowden (1976a), and Wilhelm *et al.* (1994)]. On the basis of Michaelis–Menten kinetics, small changes in substrate concentrations and in enzyme activities are approximately related as

$$\frac{\Delta v}{v} = \frac{1}{1 + S/K_{mS}} \frac{\Delta S}{S} . \tag{5.239}$$

Under the assumption that a certain ratio S/K_{mS} is typical for most enzymes, knowledge of the relative concentration change (which may result from the normalized concentration control coefficients) allows conclusions about the relative changes of reaction rates. In contrast, knowledge of the absolute concentration changes (which may result from the non-normalized coefficients) only allows conclusions concerning the absolute change of the reaction rate if, in addition, the reference state as well as the enzyme parameters K_{mS} and V_m are known. Similar conclusions can be drawn for the action of inhibitors or activators as far as the half-saturation constants for inhibition and activation, respectively, are comparable to the corresponding *in vivo* concentrations of effectors. One may argue, in a sense, that usage of the scaled coefficients reflects biochemical reality better than unscaled coefficients in that the evolutionary matching of average *in vivo* concentrations and the corresponding half-saturation constants are taken into account.

Singularities: As every relative quantity, the normalized control and elasticity coefficients have the drawback of having singularities if the quantity entering the definition in the denominator equals zero. As for flux control coefficients, one can distinguish the two following cases: (a) some flux is zero, but it is susceptible to control by some other reaction; (b) some flux is zero and remains zero when any parameter specific to any other reaction is changed.

An example for case (a) is provided by Scheme 7 (Section 3.2.4), for which there are special kinetic parameter values so that the steady-state flux through one reaction is zero (for example, J_3). As soon as some kinetic parameter of reaction 1 or 2 is slightly changed, J_3 is no longer zero, so that both the reactions 1 and 2 affect the flux through the third reaction. Whereas the corresponding unscaled flux control coefficients have finite values, the normalized ones are infinitely large due to division by J_3. So one may conclude that in the neighborhood of singularities, the normalized control coefficients can be very large even if the absolute flux and concentration changes are small.

An example for case (b) obtains if in Scheme 7, one of the external metabolites (e.g., P_3), is replaced by an internal metabolite, that is, by a substance with variable concentration. Reaction 3 is then detailed balanced in every steady state of the system and its flux cannot, therefore, be influenced by any reaction. The non-normalized flux control coefficients expressing the control exerted by reactions 1 or 2, which are not detailed balanced, on flux J_3 are zero, whereas the corresponding logarithmic coefficients are indeterminate.

As for concentration control coefficients, singularities need not be considered because for thermodynamic reasons, no substance participating in at least one reaction can be zero in steady states.

There may arise the misleading situation that a quantity entering the definition of some normalized coefficient in the numerator is zero, so that a nonzero unscaled coefficient can have a zero normalized counterpart. This difficulty arises, for example, for response coefficients and elasticities when artificial inhibitors with zero reference concentration are employed in experiment.

It is interesting that several favorable properties of normalized flux control coefficients are retained if normalizations other than that given by Eq. (5.34b) are used. Using, instead of the flux vector $\boldsymbol{J}$, any arbitrary vector $\boldsymbol{k}$ from the nullspace, one can define the matrix

$$\mathbf{C}^J_{\text{norm}} = (\text{dg}\,\boldsymbol{k})^{-1}\mathbf{C}^J(\text{dg}\,\boldsymbol{k}). \tag{5.240}$$

Whereas the flux vector $\boldsymbol{J}$ often is not known, a vector $\boldsymbol{k}$ can easily be computed from the stoichiometry matrix. Due to the generalized summation theorem (5.44b), the coefficients defined by Eq. (5.240) satisfy the traditional summation theorem (5.43). Furthermore, these coefficients have, like the coefficients normalized by the flux vector, the property of being invariant to a different scaling of the columns of the stoichiometry matrix, as this is accompanied with a reciprocal rescaling of the components of any vector $\boldsymbol{k}$. Normalization according to

Eq. (5.240) may be of interest in situations where the usual normalization by steady-state fluxes entails singularities.

5.8. ANALYSIS IN TERMS OF VARIABLES OTHER THAN CONCENTRATIONS AND FLUXES

5.8.1. General Analysis

Control coefficients had originally been defined to quantify the rate limitation and concentration control at steady state. Later on, the concept was extended to the control of other quantities such as transition times, cell volume, and the transmembrane potential. Sometimes concentrations and fluxes are not the only important variables for describing a biochemical system. For example, models of oxidative phosphorylation in mitochondria (Westerhoff and Van Dam, 1987; Brand *et al.*, 1988; Hafner *et al.*, 1990) often include the proton-motive force, $\Delta\tilde{\mu}_{H^+}$, rather than the proton concentrations in the cytosol and the mitochondrial matrix, as state variables. Sometimes, concentration ratios such as *ATP/ADP* (Westerhoff and Van Dam, 1987) or *acetyl-CoA/CoA* (Quant, 1993) are considered.

In the analysis presented in Sections 5.1–5.4, a tacit distinction has been made between variables which describe the state of the metabolic network and a set of variables of which the response to parameter perturbations has been studied. The former have been the set of concentrations, $\boldsymbol{S}$, and the latter the sets of concentrations and fluxes, $\boldsymbol{S}$ and $\boldsymbol{J}$. In a general treatment, a vector, $\boldsymbol{X}$, of generalized state variables, and a vector, $\boldsymbol{Y}$, of generalized response variables can be defined. Both sets may include concentrations, concentration ratios, reaction affinities, energy charge, proton-motive force, transmembrane potential, and so on. In contrast, steady-state fluxes and transient times characterizing the time necessary to reach a steady state can be taken as response variables, but not as state variables, because they depend on system parameters. As the name suggests, state variables characterize the state of the system, which need not be a stationary state. They do not directly depend on any parameters. The response variables [output variables in the terminology of Cornish-Bowden and Cárdenas (1993)] can be written as functions of the state variables and the system parameters:

$$\boldsymbol{Y} = \boldsymbol{Y}(\boldsymbol{X},\boldsymbol{p}). \tag{5.241}$$

We denote the numbers of components of $\boldsymbol{X}$ by ξ. For this analysis, we assume that the reaction rates can be written as functions of $\boldsymbol{X}$ and $\boldsymbol{p}$ [i.e., $\boldsymbol{v} = \boldsymbol{v}(\boldsymbol{X},\boldsymbol{p})$]. The state variables may be subject to m independent side constraints,

$$g_i(X_1, \ldots, X_\xi) = 0, \quad i = 1, \ldots, m. \tag{5.242}$$

Examples of such side constraints are the conservation relations (3.3) and Eq. (5.168), which links the proton-motive force to the inner and outer proton concentrations and the transmembrane potential.

Provided that the vector $\boldsymbol{X}$ contains sufficient information so that $\boldsymbol{v}$ can be expressed in terms of $\boldsymbol{X}$ and $\boldsymbol{p}$, the steady-state equation (2.9) for the concentrations requires, as only rank($\mathbf{N}$) equations in this matrix equation are linearly independent,

$$\mathbf{N}^0\boldsymbol{v}(\boldsymbol{X},\boldsymbol{p}) = \boldsymbol{0}. \tag{5.243}$$

Because this equation together with the side constraints (5.242) determines the values of the state variables $\boldsymbol{X}$ at steady state, we can conclude that $m = \xi -$ rank($\mathbf{N}$), which means that introducing an extra state variable implies imposing an extra side constraint. Total differentiation of Eq. (5.243) with respect to vector $\boldsymbol{p}$ yields

$$\mathbf{N}^0 \frac{\partial \boldsymbol{v}}{\partial \boldsymbol{p}} + \mathbf{N}^0 \frac{\partial \boldsymbol{v}}{\partial \boldsymbol{X}} \frac{\partial \boldsymbol{X}}{\partial \boldsymbol{p}} = \mathbf{0}. \tag{5.244}$$

Due to the side constraints (5.242), the vector $\boldsymbol{X}$ can be partitioned into a subvector $\boldsymbol{X}_{\mathrm{a}}$ of independent variables and a subvector $\boldsymbol{X}_{\mathrm{b}}$ of dependent variables, similar to the vector $\boldsymbol{S}$ in Eq. (3.8). These subvectors involve rank($\mathbf{N}$) components and $\xi -$ rank($\mathbf{N}$) components, respectively. Note that the dependencies between these variables need not be linear, at variance with the dependencies between $\boldsymbol{S}_{\mathrm{a}}$ and $\boldsymbol{S}_{\mathrm{b}}$ in conservation relations.

Equation (5.242) can be rewritten in vector form,

$$\boldsymbol{g}(\boldsymbol{X}_{\mathrm{a}}, \boldsymbol{X}_{\mathrm{b}}) = \boldsymbol{0}. \tag{5.245}$$

Total differentiation of Eq. (5.245) with respect to $\boldsymbol{X}_{\mathrm{a}}$ gives

$$\frac{\partial \boldsymbol{g}}{\partial \boldsymbol{X}_{\mathrm{a}}} + \frac{\partial \boldsymbol{g}}{\partial \boldsymbol{X}_{\mathrm{b}}} \frac{\partial \boldsymbol{X}_{\mathrm{b}}}{\partial \boldsymbol{X}_{\mathrm{a}}} = \mathbf{0}. \tag{5.246}$$

The matrix $\partial \boldsymbol{g}/\partial \boldsymbol{X}_{\mathrm{b}}$ is square because the number of dependent variables equals the number of constraints. Moreover, this matrix is nonsingular because the constraints are assumed to be independent. Therefore, Eq. (5.246) yields

$$\frac{\partial \boldsymbol{X}_{\mathrm{b}}}{\partial \boldsymbol{X}_{\mathrm{a}}} = -\left(\frac{\partial \boldsymbol{g}}{\partial \boldsymbol{X}_{\mathrm{b}}}\right)^{-1} \frac{\partial \boldsymbol{g}}{\partial \boldsymbol{X}_{\mathrm{a}}}. \tag{5.247}$$

So we obtain

$$\frac{\partial \boldsymbol{X}}{\partial \boldsymbol{X}_{\mathrm{a}}} = \begin{pmatrix} \mathbf{I} \\ -\left(\frac{\partial \boldsymbol{g}}{\partial \boldsymbol{X}_{\mathrm{b}}}\right)^{-1} \frac{\partial \boldsymbol{g}}{\partial \boldsymbol{X}_{\mathrm{a}}} \end{pmatrix}. \tag{5.248}$$

Equation (5.244) reads, in more detail,

$$\mathbf{N}^0 \frac{\partial \boldsymbol{v}}{\partial \boldsymbol{p}} + \mathbf{N}^0 \frac{\partial \boldsymbol{v}}{\partial \boldsymbol{X}} \frac{\partial \boldsymbol{X}}{\partial \boldsymbol{X}_{\mathrm{a}}} \frac{\partial \boldsymbol{X}_{\mathrm{a}}}{\partial \boldsymbol{p}} = \mathbf{0}, \tag{5.249}$$

which gives

$$\frac{\partial \boldsymbol{X}_{\mathrm{a}}}{\partial \boldsymbol{p}} = -\left(\mathbf{N}^0 \frac{\partial \boldsymbol{v}}{\partial \boldsymbol{X}} \frac{\partial \boldsymbol{X}}{\partial \boldsymbol{X}_{\mathrm{a}}}\right)^{-1} \mathbf{N}^0 \frac{\partial \boldsymbol{v}}{\partial \boldsymbol{p}} \tag{5.250}$$

with $\partial \boldsymbol{X}/\partial \boldsymbol{X}_{\mathrm{a}}$ given by Eq. (5.248). The derivatives $\partial \boldsymbol{v}/\partial \boldsymbol{X}$ play the role of elasticities.

One can now define a matrix of unscaled coefficients expressing the control on the variables Y_j:

$$\mathbf{C}^Y = \frac{\mathrm{d}\boldsymbol{Y}}{\mathrm{d}\boldsymbol{p}}\left(\frac{\partial \boldsymbol{v}}{\partial \boldsymbol{p}}\right)^{-1}, \tag{5.251}$$

where $\boldsymbol{p}$ is an r-dimensional subvector of the parameter vector used in Eq. (5.243) for which $(\partial \boldsymbol{v}/\partial \boldsymbol{p})$ is nonsingular. Here, the total differentiation sign $\mathrm{d}/\mathrm{d}\boldsymbol{p}$ is used because not only the direct effect of $\boldsymbol{p}$ on $\boldsymbol{Y}$ but also the indirect effect via $\boldsymbol{X}$ must be taken into account. In the definition of the flux control coefficients, the distinction between partial and total derivative is not necessary because different symbols for isolated rates, $\boldsymbol{v} = \boldsymbol{v}(\boldsymbol{S},\boldsymbol{p})$, and steady-state fluxes, $\boldsymbol{J} = \boldsymbol{J}(\boldsymbol{p})$ are used [cf. Eq. (5.8b)].

From Eqs. (5.241) and (5.250), one derives

$$\frac{\mathrm{d}\boldsymbol{Y}}{\mathrm{d}\boldsymbol{p}} = \frac{\partial \boldsymbol{Y}}{\partial \boldsymbol{p}} + \frac{\partial \boldsymbol{Y}}{\partial \boldsymbol{X}} \frac{\partial \boldsymbol{X}}{\partial \boldsymbol{X}_{\mathrm{a}}} \frac{\partial \boldsymbol{X}_{\mathrm{a}}}{\partial \boldsymbol{p}} = \frac{\partial \boldsymbol{Y}}{\partial \boldsymbol{p}} - \frac{\partial \boldsymbol{Y}}{\partial \boldsymbol{X}} \frac{\partial \boldsymbol{X}}{\partial \boldsymbol{X}_{\mathrm{a}}} (\mathbf{M}_{\mathrm{X}}^0)^{-1} \mathbf{N}^0 \frac{\partial \boldsymbol{v}}{\partial \boldsymbol{p}}, \tag{5.252}$$

where

$$\mathbf{M}_X^0 = \mathbf{N}^0 \frac{\partial \boldsymbol{v}}{\partial \boldsymbol{X}} \frac{\partial \boldsymbol{X}}{\partial \boldsymbol{X}_a} \tag{5.253}$$

is the Jacobian matrix of the vector function on the left-hand side of Eq. (5.243). Substitution of Eq. (5.252) into Eq. (5.251) gives

$$\mathbf{C}^Y = \frac{\partial \boldsymbol{Y}}{\partial \boldsymbol{p}}\left(\frac{\partial \boldsymbol{v}}{\partial \boldsymbol{p}}\right)^{-1} - \frac{\partial \boldsymbol{Y}}{\partial \boldsymbol{X}} \frac{\partial \boldsymbol{X}}{\partial \boldsymbol{X}_a} (\mathbf{M}_X^0)^{-1}\mathbf{N}^0. \tag{5.254}$$

Some particular cases deserve special mention. When interested in the control on the state variables X_j, one can simplify Eq. (5.254) to

$$\mathbf{C}^X = -\frac{\partial \boldsymbol{X}}{\partial \boldsymbol{X}_a} (\mathbf{M}_X^0)^{-1}\mathbf{N}^0, \tag{5.255}$$

as $\boldsymbol{X}$ does not depend on $\boldsymbol{p}$ directly. Often, $\boldsymbol{Y}$ can be written as a function of $\boldsymbol{X}$ and the reaction rates $\boldsymbol{v}$,

$$\boldsymbol{Y} = \boldsymbol{Y}(\boldsymbol{X}, \boldsymbol{v}(\boldsymbol{X},\boldsymbol{p})). \tag{5.256}$$

An example is provided by the transient times defined by Easterby (1981), which will be dealt with in Section 5.8.4. From Eq. (5.256), we obtain

$$\frac{d\boldsymbol{Y}}{d\boldsymbol{p}} = \frac{\partial \boldsymbol{Y}}{\partial \boldsymbol{v}} \frac{\partial \boldsymbol{v}}{\partial \boldsymbol{p}} + \left(\frac{\partial \boldsymbol{Y}}{\partial \boldsymbol{v}} \frac{\partial \boldsymbol{v}}{\partial \boldsymbol{X}} + \left.\frac{\partial \boldsymbol{Y}}{\partial \boldsymbol{X}}\right|_{\boldsymbol{v}}\right) \frac{\partial \boldsymbol{X}}{\partial \boldsymbol{X}_a} \frac{\partial \boldsymbol{X}_a}{\partial \boldsymbol{p}}, \tag{5.257}$$

where the symbol $|_{\boldsymbol{v}}$ means that the derivative is taken at constant $\boldsymbol{v}$. This gives, due to Eqs. (5.250), (5.251), and (5.253),

$$\mathbf{C}^Y = \frac{\partial \boldsymbol{Y}}{\partial \boldsymbol{v}} - \left(\frac{\partial \boldsymbol{Y}}{\partial \boldsymbol{v}} \frac{\partial \boldsymbol{v}}{\partial \boldsymbol{X}} + \left.\frac{\partial \boldsymbol{Y}}{\partial \boldsymbol{X}}\right|_{\boldsymbol{v}}\right) \frac{\partial \boldsymbol{X}}{\partial \boldsymbol{X}_a} (\mathbf{M}_X^0)^{-1}\mathbf{N}^0. \tag{5.258}$$

In particular, if $\boldsymbol{Y}$ equals the flux vector $\boldsymbol{J}$, we have

$$\mathbf{C}^J = \mathbf{I} - \frac{\partial \boldsymbol{v}}{\partial \boldsymbol{X}} \frac{\partial \boldsymbol{X}}{\partial \boldsymbol{X}_a} (\mathbf{M}_X^0)^{-1}\mathbf{N}^0. \tag{5.259}$$

If, in addition, $\boldsymbol{X} = \boldsymbol{S}$, Eq. (5.259) coincides with Eq. (5.26b).

Note that if the response variables $\boldsymbol{Y}$ depend only on the state variables, $\boldsymbol{X}$, or on the reaction rates, $\boldsymbol{v}$, the control matrix $\mathbf{C}^Y$ does not depend on the special choice of the parameter vector $\boldsymbol{p}$ [cf. Eqs. (5.255) and (5.258)]. In contrast, if $\boldsymbol{Y}$

additionally depends on $\boldsymbol{p}$ directly, the control coefficients do depend on this choice.

Unified summation and connectivity theorems: Postmultiplying Eq. (5.254) by the null-space matrix $\mathbf{K}$, we obtain, due to Eq. (3.44),

$$\mathbf{C}^{Y}\mathbf{K} = \frac{\partial \boldsymbol{Y}}{\partial \boldsymbol{p}}\left(\frac{\partial \boldsymbol{v}}{\partial \boldsymbol{p}}\right)^{-1}\mathbf{K}. \tag{5.260}$$

This equation can be regarded as a unified summation theorem. Postmultiplication of Eq. (5.254) by $(\partial \boldsymbol{v}/\partial \boldsymbol{X})(\partial \boldsymbol{X}/\partial \boldsymbol{X}_{\mathrm{a}})$ gives, owing to Eq. (5.253),

$$\mathbf{C}^{Y}\frac{\partial \boldsymbol{v}}{\partial \boldsymbol{X}}\frac{\partial \boldsymbol{X}}{\partial \boldsymbol{X}_{\mathrm{a}}} = \left[\frac{\partial \boldsymbol{Y}}{\partial \boldsymbol{p}}\left(\frac{\partial \boldsymbol{v}}{\partial \boldsymbol{p}}\right)^{-1}\frac{\partial \boldsymbol{v}}{\partial \boldsymbol{X}} - \frac{\partial \boldsymbol{Y}}{\partial \boldsymbol{X}}\right]\frac{\partial \boldsymbol{X}}{\partial \boldsymbol{X}_{\mathrm{a}}}, \tag{5.261}$$

which is a unified connectivity theorem. The notion "unified" refers to the fact that both for the summation theorem and the connectivity theorem, only one equation need be written instead of separate equations for concentration control, flux control, and possible other quantities.

If $\boldsymbol{Y} = \boldsymbol{X}$, Eq. (5.260) simplifies to

$$\mathbf{C}^{X}\mathbf{K} = \mathbf{0}. \tag{5.262}$$

This equation is a generalization of the summation theorem (5.44a) for any generalized state variable.

When $\boldsymbol{Y}$ depends on $\boldsymbol{p}$ only via $\boldsymbol{v}$, as indicated in Eq. (5.256), we have

$$\frac{\partial \boldsymbol{Y}}{\partial \boldsymbol{p}} = \frac{\partial \boldsymbol{Y}}{\partial \boldsymbol{v}}\frac{\partial \boldsymbol{v}}{\partial \boldsymbol{p}}, \tag{5.263a}$$

$$\frac{\partial \boldsymbol{Y}}{\partial \boldsymbol{X}} = \left.\frac{\partial \boldsymbol{Y}}{\partial \boldsymbol{X}}\right|_{\boldsymbol{v}} + \frac{\partial \boldsymbol{Y}}{\partial \boldsymbol{v}}\frac{\partial \boldsymbol{v}}{\partial \boldsymbol{X}}. \tag{5.263b}$$

Note that the derivative on the left-hand side of Eq. (5.263b) is taken at constant $\boldsymbol{p}$. Therefore, Eq. (5.260) yields

$$\mathbf{C}^{Y}\mathbf{K} = \frac{\partial \boldsymbol{Y}}{\partial \boldsymbol{v}}\mathbf{K}. \tag{5.264}$$

Moreover, Eq. (5.261) gives

$$\mathbf{C}^{Y}\frac{\partial \boldsymbol{v}}{\partial \boldsymbol{X}}\frac{\partial \boldsymbol{X}}{\partial \boldsymbol{X}_{\mathrm{a}}} = -\left.\frac{\partial \boldsymbol{Y}}{\partial \boldsymbol{X}}\right|_{\boldsymbol{v}}\frac{\partial \boldsymbol{X}}{\partial \boldsymbol{X}_{\mathrm{a}}}. \tag{5.265}$$

Equations (5.264) and (5.265) are generalizations of the summation and connectivity theorems, respectively, for any response variable that depends on $\boldsymbol{v}$ and $\boldsymbol{X}$ only [cf. Eq. (5.256)]. In particular, the summation and connectivity theorems for concentration control coefficients and flux control coefficients result from the above equations by obvious substitutions.

5.8.2. Concentration Ratios and Free-Energy Differences as State Variables

This section is devoted to further illustration of the general analysis presented in the previous section. Consider first the situation that in a given system two concentrations enter a conservation relation of the form $S_1 + S_2 = \text{const.}$ S_1 and S_2 may stand, for example, for *NAD* and *NADH*, respectively. In this situation, it is usual practice in biochemistry to interpret experimental results in terms of the concentration ratio $X_1 = S_1/S_2$ (see Hofmeyr *et al.*, 1986; Quant, 1993). It is then possible to replace the concentration vector $\boldsymbol{S}$ by a vector

$$\boldsymbol{X} = \left(\frac{S_1}{S_2}, S_1 + S_2, S_3, \ldots, S_n\right)^{\mathrm{T}}. \tag{5.266}$$

The elasticity $\partial v_k/\partial X_1 = \partial v_k/\partial(S_1/S_2)$ is taken at constant $S_1 + S_2$ and the elasticity $\partial v_k/\partial X_2 = \partial v_k/\partial(S_1 + S_2)$ at constant ratio S_1/S_2.

The unified connectivity theorem (5.265) implies, with $\boldsymbol{Y} = \boldsymbol{J}$,

$$\sum_{i,k} C^J_{jk} \frac{\partial v_k}{\partial X_i} \frac{\partial X_i}{\partial(S_1/S_2)} = 0. \tag{5.267}$$

for the flux control coefficients. The right-hand side is zero because the partial derivative of $\boldsymbol{J}$ with respect to $\boldsymbol{X}$ vanishes. Equation (5.267) can be simplified to

$$\sum_{k} C^J_{jk} \frac{\partial v_k}{\partial(S_1/S_2)} = 0. \tag{5.268}$$

This connectivity theorem has also been given by Hofmeyr *et al.* (1986). Equation (5.268) holds true only if the elasticities with respect to the concentration ratio are determined with the sum of the two concentrations kept constant. It also applies when X_1 is defined as $\ln(S_1/S_2)$, which is relevant for an interpretation in terms of free energy (see below).

The above calculations are of particular importance for reaction systems involving cofactor pairs such as ATP/ADP or NAD(P)/NAD(P)H. In the case that $S_1 = NAD$ and $S_2 = NADH$, it is easy to see that Eq. (5.268) implies, for the normalized coefficients, the following connectivity relation:

$$\sum_k C^J_{jk}\, \varepsilon^k_{NAD/NADH} = 0, \tag{5.269}$$

where $\varepsilon^k_{NAD/NADH}$ is the elasticity of reaction k with respect to the *NAD/NADH* ratio. To calculate this elasticity, one can represent *NAD* and *NADH* in terms of the *NAD/NADH* ratio and the conservation sum, $NAD + NADH$. One obtains

$$\varepsilon^k_{NAD/NADH} = \frac{\varepsilon^k_{NAD} \cdot NADH - \varepsilon^k_{NADH} \cdot NAD}{NAD + NADH}. \tag{5.270}$$

A similar elasticity can be calculated with respect to the *acetyl-CoA/CoA* ratio (Quant, 1993).

Related elasticities were defined with respect to molar free-energy differences of reactions (Westerhoff *et al.,* 1983; Westerhoff and Van Dam, 1987),

$$\varepsilon^k_{\Delta G_j} = \frac{\Delta G_j}{v_k} \frac{\partial v_k}{\partial \Delta G_j}. \tag{5.271}$$

A relevant example is the electrochemical potential difference for protons (proton-motive force), $\Delta\tilde{\mu}_{\mathrm{H}}$, across mitochondrial or other membranes [cf. Eq. (5.168)]. Such elasticities have been used in various studies (Brand *et al.,* 1988; Hafner *et al.,* 1990; S. Schuster *et al.,* 1993a).

In a mathematically rigorous notation, one should indicate what quantities remain constant when the derivative $\partial v_k/\partial \Delta G_j$ is calculated. This depends on what variables other than ΔG_j are included in the vector $\boldsymbol{X}$. The difficulty arises from the fact that reaction rates cannot, in general, be written as functions of the free-energy differences only (see Section 2.2.3). For example, elasticities with respect to $\Delta\tilde{\mu}_{\mathrm{H}}$ depend on what other variables are kept constant (H^+_{ex}, H^+_{in}, $\Delta\Psi$).

5.8.3. Entropy Production as a Response Variable

A thermodynamically relevant response variable is the total entropy production of metabolic pathways, σ, defined by Eq. (3.74). In analogy to Eq. (5.16) one may define unscaled control coefficients for the entropy production rate as follows:

$$\delta\sigma = \boldsymbol{C}^\sigma \delta\boldsymbol{v}, \tag{5.272}$$

where the vector $\boldsymbol{C}^\sigma$ contains the elements

$$C^\sigma_k = \frac{\partial\sigma/\partial p_k}{\partial v_k/\partial p_k}. \tag{5.273}$$

According to Eq. (3.76), the entropy production rate is a linear function of the steady-state fluxes J_j because for given external conditions, the coefficients $\ln(\tilde{q}_j)$ are fixed. This entails a direct relation between the flux control coefficients and the control coefficients for entropy production rate. For the unscaled coefficients, one obtains

$$C_l^\sigma = R \sum_{j=1}^{r} C_{jl}^J \ln(\tilde{q}_j). \tag{5.274}$$

This equation may be rewritten as follows:

$$C_l^\sigma = R \ln\left(\prod_{j=1}^{r} (\tilde{q}_j)^{C_{jl}^J} \right). \tag{5.275}$$

It shows that control of entropy production is closely related to control of all independent fluxes. Although our aim is here primarily to give an example of the general treatment presented in Section 5.8.1, we note that Eq. (5.275) may be a starting point for optimization analysis concerning thermodynamic efficiencies (Kedem and Caplan, 1965; Stucki *et al.*, 1983).

From Eqs. (3.76) and (5.44b) it follows that the control coefficients for entropy production fulfill the generalized summation theorem

$$\sum_{l=1}^{r} C_l^\sigma k_{li} = R \sum_{j=1}^{r} k_{ji} \ln(\tilde{q}_j) = R \ln\left(\prod_{j=1}^{r} \tilde{q}_j^{k_{ji}} \right), \tag{5.276}$$

where the k_{ji} denote the elements of the null-space matrix $\mathbf{K}$. Note that the term in parentheses in Eq. (5.276) also enters the generalized Wegscheider condition (3.64b).

5.8.4. Control of Transient Times

As outlined in Section 4.1, an agreed definition for transient times only exists for isolated reactions obeying first-order kinetics [Eq. (4.1)]. Another definition, which takes into account the systemic interactions, is based on the eigenvalues of the Jacobian matrix [cf. Eq. (4.6)]. There are several approaches to define average relaxation times. However, all of these definitions are applicable only under very special conditions. For control analysis two different definitions of transient times have been used primarily.

1. Easterby considered an unbranched reaction sequence (see Scheme 11, Section 5.4.3.1) where the input and output reactions are assumed to be irreversible (Eas-

terby, 1981). Initiating the reactions at $t = 0$ with initial metabolite concentrations $S_i(0) = 0$ and $P_2(0) = 0$, the condition of mass conservation leads to

$$v_1 t = P_2(t) + \sum_{i=1}^{n} S_i(t) \tag{5.277}$$

where it is assumed that the rate $v_1 = k_1 P_1$ of the input reaction is constant. Because the output reaction is irreversible, the concentrations S_i may attain a stationary state for $t \to \infty$. With $S_i = \text{const.}$ Eq. (5.277) defines in a (t,P_2) diagram a straight line which intersects the time axis at $t = \tau$ with

$$\tau = \frac{1}{J} \sum_{i=1}^{n} S_i = \sum_{i=1}^{n} \tau_i. \tag{5.278}$$

τ is called the "overall transient time" of the pathway. In Eq. (5.278), $J = v_1$ has been taken into account. According to definition (5.278), τ characterizes the time needed to generate the steady-state concentrations. $\tau_i = S_i/J$ is the transient time of intermediate S_i. A generalization of this definition for the case of unbranched reaction chains with some steps having non-unitary stoichiometries was given by Meléndez-Hevia *et al.* (1990). In eq. (5.278), the sum of concentrations has then to be replaced by a linear combination with the coefficients being products of stoichiometric coefficients.

2. A more general definition of the average transient time for a metabolite S_i is

$$\tau_i = \frac{\int_0^{\infty} t \Delta S_i(t) dt}{\int_0^{\infty} \Delta S_i(t) dt} \equiv \frac{B_i}{A_i} \tag{5.279}$$

with

$$\Delta S_i(t)|_{t=0} = \Delta S_i^0, \qquad \lim_{t \to \infty} \Delta S_i(t) = 0 \tag{5.280}$$

(Heinrich and Rapoport, 1975). If the perturbations are sufficiently small, the functions $\Delta S_i(t) = \delta S_i(t)$ are determined by the linear approximations (2.82) of the system equations (2.8). An advantage of expression (5.279) compared with definition (5.278) is that it applies not only to unbranched chains with a constant input but to the intermediate concentrations of any reaction network. It has been applied also in the theory of tracer kinetics (Gitterman and Weiss, 1994), in the analysis of metabolic channeling (Heinrich and S. Schuster, 1991) and in other fields of mathematical modelling of biological systems (Overholser *et al.*, 1994). Note that

Eq. (5.279) results in Eq. (4.1) when applied to a single monomolecular reaction, because the relaxation process may then be described by only one exponential function.

For the case that the system involves conservation relations, definition (5.279) only applies if the perturbations do not violate these relations.

For the calculation of the transient times $\boldsymbol{\tau} = (\tau_1, \ldots, \tau_n)^{\mathrm{T}}$ defined in Eq. (5.279), the following general procedure may be applied. With $\delta\boldsymbol{S} = (\delta S_1, \ldots, \delta S_n)^{\mathrm{T}}$, integration of Eq. (2.82) yields

$$\int_0^\infty \frac{\mathrm{d}\delta\boldsymbol{S}(t)}{\mathrm{d}t}\,\mathrm{d}t = \mathbf{M}\int_0^\infty \delta\boldsymbol{S}(t)\,\mathrm{d}t, \tag{5.281a}$$

and under consideration of Eq. (5.280),

$$-\delta\boldsymbol{S}^0 = \mathbf{M}\boldsymbol{A}, \tag{5.281b}$$

where the integrals on the right-hand side of Eq. (5.281a) constitute the vector $\boldsymbol{A} = (A_1, \ldots, A_n)^{\mathrm{T}}$.

Partial integration on the left-hand side of the equation

$$\int_0^\infty t\frac{\mathrm{d}\delta\boldsymbol{S}}{\mathrm{d}t}\,\mathrm{d}t = \mathbf{M}\int_0^\infty t\delta\boldsymbol{S}(t)\,\mathrm{d}t \tag{5.282a}$$

yields, due to $\lim_{t\to\infty}[t\delta\boldsymbol{S}(t)] = 0$

$$-\boldsymbol{A} = \mathbf{M}\boldsymbol{B}, \tag{5.282b}$$

where the integrals on the right-hand side of Eq. (5.282a) constitute the vector $\boldsymbol{B} = (B_1, \ldots, B_n)^{\mathrm{T}}$.

From Eqs. (5.281b) and (5.282b) the vectors $\boldsymbol{A}$ and $\boldsymbol{B}$ and, therefore, the vector of the transient times $\boldsymbol{\tau} = (B_1/A_1, \ldots, B_n/A_n)^{\mathrm{T}}$ can be calculated from the initial perturbations $\delta\boldsymbol{S}^0$ and the Jacobi matrix $\mathbf{M}$ without explicit knowledge of the relaxation function $\delta S_i(t)$. With $\mathbf{M} = \mathbf{N}\,\partial\boldsymbol{v}/\partial\boldsymbol{S}$, one obtains

$$\boldsymbol{A} = -\left(\mathbf{N}\frac{\partial\boldsymbol{v}}{\partial\boldsymbol{S}}\right)^{-1}\delta\boldsymbol{S}^0, \tag{5.283a}$$

$$\boldsymbol{B} = \left(\mathbf{N}\frac{\partial\boldsymbol{v}}{\partial\boldsymbol{S}}\right)^{-2}\delta\boldsymbol{S}^0. \tag{5.283b}$$

These equations have been derived for the case of no conservation relations. If the system does involve such relations, Eqs. (5.283a) and (5.283b) include the link matrix **L**.

Using definitions (5.278) or (5.279), control coefficients for the transient time τ_i of a metabolite S_i may be calculated in the following way:

$$C_{ik}^{\tau} = \frac{v_k}{\tau_i} \frac{\partial \tau_i / \partial p_k}{\partial v_k / \partial p_k} \equiv \frac{\partial \ln \tau_i}{\partial \ln v_k}, \tag{5.284}$$

where p_k denotes any reaction-specific perturbation parameter. The control coefficients C_k^{τ} of the "overall transient time" are defined similarly.

Using definition (5.278) the problem of transient-time control is closely related to that of fluxes and metabolite concentrations (see Meléndez-Hevia *et al.*, 1990). With $\tau_i = S_i / J$, one obtains

$$\frac{1}{\tau_i} \frac{\partial \tau_i}{\partial p_k} = \frac{1}{S_i} \frac{\partial S_i}{\partial p_k} - \frac{1}{J} \frac{\partial J}{\partial p_k}. \tag{5.285}$$

This equation implies, under consideration of Eqs. (5.37a), (5.37b) and (5.284),

$$C_{ik}^{\tau} = C_{ik}^{S} - C_k^{J}, \tag{5.286}$$

where C_{ik}^{τ} denotes the normalized transient-time control coefficients of metabolite S_i. Analogously, one gets for the control coefficients of the overall transient time

$$C_k^{\tau} = \frac{\sum_{i=1}^{n} S_i C_{ik}^{S}}{\sum_{i=1}^{n} S_i} - C_k^{J}. \tag{5.287}$$

Substitution of Eqs. (5.278) and (5.284) into expression (5.287) gives

$$C_k^{\tau} = \sum_{i=1}^{n} \frac{\tau_i}{\tau} C_{ik}^{\tau}. \tag{5.288}$$

On the basis of definition (5.279) the normalized control coefficients for the transient time of a metabolite S_i is given by the expression

$$C_{ik}^{\tau} = \frac{\partial \ln B_i / \partial p_k}{\partial \ln v_k / \partial p_k} - \frac{\partial \ln A_i / \partial p_k}{\partial \ln v_k / \partial p_k}. \tag{5.289}$$

Summation theorems: From Eqs. (5.286) and (5.287), which are derived from

Easterby's definition of transient times, one gets under, consideration of summation theorems for concentrations and fluxes [cf. Eqs. (5.42) and (5.43)],

$$\sum_{k=1}^{r} C_{ik}^{\tau} = -1, \tag{5.290a}$$

$$\sum_{k=1}^{r} C_{k}^{\tau} = -1, \tag{5.290b}$$

which are the summation theorems for the transient times of individual metabolites and for the overall transient time, respectively (Heinrich and Rapoport, 1975). These theorems, which can also be derived from the unified summation theorem (5.260), express the fact that an activation of all enzymes by the same fractional amount (which is equivalent to a division of the time scale by this factor) reduces all transient times by this factor. This fact holds true for all time-independent variables with dimension of time (Acerenza and Kacser, 1990).

Now it is shown that the control coefficients derived from the alternative definition (5.279) of transient times also fulfill the summation theorem (5.290) as long as the perturbation parameters are reaction-specific and enter the rate equation $\boldsymbol{v}(\boldsymbol{S},\boldsymbol{p})$ as multipliers, that is,

$$\frac{\partial v_i}{\partial p_k} = \frac{v_k}{p_k}\,\delta_{ik}\,. \tag{5.291}$$

Differentiation of Eq. (5.281b) with respect to the perturbation parameters yields, with the Jacobian $\mathbf{M} = \mathbf{N}(\partial \boldsymbol{v}/\partial \boldsymbol{S})$,

$$\mathbf{N}\frac{\partial^2 \boldsymbol{v}}{\partial \boldsymbol{p}\,\partial \boldsymbol{S}}\boldsymbol{A} + \left(\mathbf{N}\frac{\partial \boldsymbol{v}}{\partial \boldsymbol{S}}\right)\frac{\partial \boldsymbol{A}}{\partial \boldsymbol{p}} = \mathbf{0}. \tag{5.292}$$

In this equation, mixed second derivatives of the reaction rates with respect to metabolite concentrations and kinetic parameters appear (cf. the second-order approach to metabolic control analysis in Section 5.9). However, from Eq. (5.291), the following simplification results:

$$\frac{\partial^2 v_i}{\partial p_k\,\partial S_j} = \frac{\delta_{ik}}{p_k}\frac{\partial v_k}{\partial S_j}\,. \tag{5.293}$$

One gets from Eq. (5.292)

$$\mathbf{N}\left(\mathrm{dg}\left(\frac{\partial \boldsymbol{v}}{\partial \boldsymbol{S}}\boldsymbol{A}\right)\right)(\mathrm{dg}\,\boldsymbol{p})^{-1} + \mathbf{N}\frac{\partial \boldsymbol{v}}{\partial \boldsymbol{S}}\frac{\partial \boldsymbol{A}}{\partial \boldsymbol{p}} = \mathbf{0}. \tag{5.294}$$

Postmultiplication of this equation by the parameter vector $\boldsymbol{p}$ yields

$$\mathbf{N}\frac{\partial \boldsymbol{v}}{\partial \boldsymbol{S}}\boldsymbol{A} + \mathbf{N}\frac{\partial \boldsymbol{v}}{\partial \boldsymbol{S}}\frac{\partial \boldsymbol{A}}{\partial \boldsymbol{p}}\boldsymbol{p} = \boldsymbol{0}, \tag{5.295}$$

and because $\mathbf{M} = \mathbf{N}\,\partial \boldsymbol{v}/\partial \boldsymbol{S}$ is assumed to be invertible,

$$\frac{\partial \boldsymbol{A}}{\partial \boldsymbol{p}}\boldsymbol{p} = -\boldsymbol{A} \quad \text{or} \quad \sum_{k=1}^{r} \frac{\partial \ln A_i}{\partial \ln p_k} = -1. \tag{5.296}$$

In a similar way, one derives, by differentiation of Eq. (5.282b)

$$-\frac{\partial \boldsymbol{A}}{\partial \boldsymbol{p}}\boldsymbol{p} = \mathbf{N}\frac{\partial \boldsymbol{v}}{\partial \boldsymbol{S}}\boldsymbol{B} + \mathbf{N}\frac{\partial \boldsymbol{v}}{\partial \boldsymbol{S}}\frac{\partial \boldsymbol{B}}{\partial \boldsymbol{p}}\boldsymbol{p} \tag{5.297}$$

and with Eq. (5.296),

$$-2\boldsymbol{B} = \frac{\partial \boldsymbol{B}}{\partial \boldsymbol{p}}\boldsymbol{p} \quad \text{or} \quad \sum_{k=1}^{r} \frac{\partial \ln B_i}{\partial \ln p_k} = -2. \tag{5.298}$$

Together with Eq. (5.289), Eqs. (5.296) and (5.298) lead directly to the summation theorem (5.290).

Connectivity theorems: Using Eq. (5.287), the connectivity theorems for the fluxes and metabolite concentrations imply connectivity relationships for the transient-time control coefficients which read, for unbranched chains,

$$\sum_{k=1}^{r} C_{ik}^{\tau}\varepsilon_{kj} = -\delta_{ij}, \tag{5.299a}$$

$$\sum_{k=1}^{r} C_{k}^{\tau}\varepsilon_{kj} = -S_j\left(\sum_{i=1}^{n} S_i\right)^{-1} \tag{5.299b}$$

(Meléndez-Hevia *et al.*, 1990).

We will now show that the connectivity theorems for the transient times are special cases of the unified connectivity theorem derived in Section 5.8.1. According to Eq. (5.278), the overall transient time is expressed as a function of the metabolite concentrations and the flux in a reference steady state. As the kinetic parameters do not enter expression (5.278) in an explicit manner, the unified connectivity theorem assumes the form given in Eq. (5.265) with $Y = \tau$ and $\boldsymbol{X}$

$= \mathbf{S} = (S_1, \ldots, S_n)^{\mathrm{T}}$. From Eqs. (5.265) and (5.278), one obtains, for the unscaled coefficients,

$$\sum_{k=1}^{r} C_k^{\tau} \frac{\partial v_k}{\partial S_i} = -\frac{\partial \tau}{\partial S_i} = -\frac{1}{J}. \tag{5.300}$$

Introducing normalized coefficients (i.e., $C_k^{\tau} v_k/\tau \rightarrow C_k^{\tau}$) and by taking into account the steady-state condition $v_k = J$, relation (5.300) can easily be transformed into the connectivity theorem (5.299b). In a similar way, Eq. (5.299a) for the transient-time control coefficients of individual metabolite concentrations can be derived from the unified connectivity theorem (5.265).

For *unbranched reaction chains* with no allosteric regulations, Eq. (5.299b) simplifies to

$$C_j^{\tau}\varepsilon_{jj} + C_{j+1}^{\tau}\varepsilon_{j+1,j} = -S_j\left(\sum_{i=1}^{n} S_i\right)^{-1}, \tag{5.301}$$

because the rate of any reaction depends on its substrate and product only. With $\varepsilon_{jj} < 0$, $\varepsilon_{j+1,j} > 0$ [cf. assumption (5.104)], it follows immediately from Eq. (5.301) that:

(a) if $C_j^{\tau} > 0$, then

$$C_i^{\tau} > 0 \text{ for } i < j \tag{5.302a}$$

(b) if $C_j^{\tau} < 0$, then

$$C_i^{\tau} < 0 \text{ for } i > j \tag{5.302b}$$

This result and the summation theorem (5.290b) imply that the transient-time control coefficient of the last enzyme will always be negative (Meléndez-Hevia *et al.,* 1990).

Let us consider the most simple case that in Scheme 11 (Section 5.4.3.1) all reactions are irreversible and may be described by first-order kinetic equations. Then, the steady-state concentrations and the steady-state flux are determined by

$$S_i = \frac{J}{k_{i+1}}, \qquad J = k_1 P_1 \tag{5.303}$$

and the concentration control coefficients and flux control coefficients turn out to be

$$C^S_{i,j} = \begin{cases} 1 & \text{for } j = 1 \\ -\delta_{i+1,j} & \text{for } j \geq 2, \end{cases} \tag{5.304a}$$

$$C^J_j = \begin{cases} 1 & \text{for } j = 1 \\ 0 & \text{for } j \geq 2. \end{cases} \tag{5.304b}$$

Introducing this into Eq. (5.287) yields

$$C^\tau_1 = 0, \qquad C^\tau_j = -\frac{1/k_j}{\sum_{i=2}^{n+1} (1/k_i)} \quad \text{for } j \geq 2. \tag{5.305}$$

A comparison of Eqs. (5.304b) and (5.305) shows that reactions which have a low or even vanishing influence on steady-state fluxes may exert a strong control on transient times and *vice versa.* Furthermore, it is easily verified that the coefficients given in Eq. (5.305) fulfill the summation relationship (5.290b).

5.8.5 Control of Oscillations

The possibility of periodic time-dependent changes of enzymic systems has always been a central point in the mathematical analysis of metabolic processes (see also Section 2.4). Despite the fact that for many systems the physiological role of the observed oscillations is still unclear, it has sometimes been argued that the cellular response toward oscillations is governed by their frequencies rather than by their amplitudes or by the mean levels of oscillating concentrations (Rapp *et al.,* 1981; Rapp, 1987; Berridge, 1989; Goldbeter and Li, 1989; Goldbeter *et al.,* 1990).

It seems to be worth generalizing the concept of control coefficients to oscillating systems in order to characterize, for example, the role of the individual reactions in determining the frequency of the observed oscillations. Such a generalization, however, meets with several difficulties, outlined below. The theory of control of oscillations, in the sense of metabolic control analysis, is far from being elaborate, and we will give some basic ideas only.

As a first step toward a control analysis of oscillating processes, one might, therefore, define control coefficients characterizing the effect of changes in enzyme activities on the frequency f or the period T of oscillations. A direct application of the usual definitions (5.3) or (5.5) of control coefficients is hindered by the problem that there is no well-defined time-independent reference state for the activity v_j of step j which enters the denominators of these equations. In contrast, this problem does not arise for response coefficients which may be defined in the following way:

$$R^T_j = \frac{p_j}{T}\frac{\partial T}{\partial p_j}, \tag{5.306a}$$

$$R_j^f = -R_j^T = \frac{p_j}{f}\frac{\partial f}{\partial p_j} \tag{5.306b}$$

(Markus and Hess, 1990; Baconnier *et al.*, 1993; Westerhoff *et al.*, 1996), where p_j denotes a parameter, for example the enzyme concentration of reaction j. In principle, it is also possible to define response coefficients for the amplitudes A_i of oscillating variables, notably concentrations $S_i(t)$, in the following way (Baconnier *et al.*, 1993):

$$R_{ij}^A = \frac{p_j}{A_i}\frac{\partial A_i}{\partial p_j}. \tag{5.307}$$

In the case that the kinetic parameters p_j enter the rate equations in a linear manner, the normalized response coefficients for oscillations fulfill summation relationships which may be rationalized as follows. Let us compare the time-dependent changes of two oscillating systems A and B starting with the same initial conditions. Concerning the kinetic parameters, we assume that $p_j^B = \lambda p_j^A$ for $j = 1, \ldots, r$. Because the system equations depend in a linear manner on the rates v_j [cf. Eq. (2.8)], a change of the parameters p_j by a common factor λ results merely in a change of the time scale. In system B, the same motions take place as in system A but faster ($\lambda > 1$) or slower ($\lambda < 1$). In particular, one obtains, for the frequency,

$$f(\lambda p_1, \ldots, \lambda p_r) = \lambda f(p_1, \ldots, p_r) \tag{5.308}$$

and for the amplitudes of oscillating concentrations,

$$A_i(\lambda p_1, \ldots, \lambda p_r) = A_i(p_1, \ldots, p_r). \tag{5.309}$$

Differentiation of these equations with respect to λ yields, for $\lambda = 1$,

$$\sum_{j=1}^{r} \frac{\partial \ln f}{\partial \ln p_j} = \sum_{j=1}^{r} R_j^f = 1, \tag{5.310a}$$

$$\sum_{j=1}^{r} \frac{\partial \ln A_i}{\partial \ln p_j} = \sum_{j=1}^{r} R_{ij}^A = 0 \tag{5.310b}$$

[see Acerenza *et al.*, 1989; Acerenca and Kacser, 1990) as well as the derivation of the relationships (5.50a) and (5.50b) in Section 5.3.1].

As an example we consider the two-component model which has been proposed by Higgins (1964) and Selkov (1968) for the explanation of glycolytic oscillations (see Section 2.4.2). With $\gamma = 2$ and $v_1 = k_1 P_1$, the system equations (2.119a) and (2.119b) assume the form

$$\frac{dS_1}{dt} = k_1 P_1 - k_2 S_1 S_2^2, \tag{5.311a}$$

$$\frac{dS_2}{dt} = k_2 S_1 S_2^2 - k_3 S_2 \tag{5.311b}$$

with $S_0 = \text{const}$. The only steady state obtains as

$$S_1 = \frac{k_3^2}{k_2 k_1 P_1}, \qquad S_2 = \frac{k_1 P_1}{k_3}. \tag{5.312}$$

Equation (2.122) implies that this state is unstable for

$$k_2 < \frac{k_3^3}{k_1^2 P_1^2} = k_2^{\text{crit}}. \tag{5.313}$$

Let us consider oscillations obtained for parameter values at $k_2 \cong k_2^{\text{crit}}$ (i.e., near the Hopf bifurcation). There, one obtains with Eq. (2.121) the following estimate for the oscillation frequency

$$f \cong \frac{\sqrt{\Delta}}{2\pi} = \frac{k_1 P_1}{2\pi} \sqrt{\frac{k_2}{k_3}}, \tag{5.314}$$

whereas the amplitudes A_i of the oscillating concentrations S_i are vanishingly small. From Eq. (5.314), one derives, for the normalized response coefficients of the frequency,

$$\frac{\partial \ln f}{\partial \ln k_1} = 1, \qquad \frac{\partial \ln f}{\partial \ln k_2} = \frac{1}{2}, \qquad \frac{\partial \ln f}{\partial \ln k_3} = -\frac{1}{2}, \tag{5.315}$$

which means that a stimulation of reactions 1 or 2 will result in an increase, and a stimulation of reaction 3 in a decrease of oscillation frequency. It is seen that the response coefficients fulfill the summation relationship (5.310a). According to Eq. (5.306b), the response coefficients of the period T sum up to -1, which is in accord with the summation relationship for the control coefficients of transient times given in Eq. (5.290).

In a more general treatment, one may consider the frequencies of oscillations in unbranched two-component systems described by the differential equations

$$\frac{dS_1}{dt} = v_1 - v_2, \tag{5.316}$$

$$\frac{dS_2}{dt} = v_2 - v_3. \tag{5.317}$$

The determinant Δ and, therefore, the frequency f may be expressed by the unscaled elasticities in the following way:

$$\Delta = \varepsilon_{11}\varepsilon_{22} - \varepsilon_{12}\varepsilon_{21} - \varepsilon_{11}\varepsilon_{32} + \varepsilon_{12}\varepsilon_{31} + \varepsilon_{21}\varepsilon_{32} - \varepsilon_{22}\varepsilon_{31} \tag{5.318}$$

with $\varepsilon_{ij} = \partial v_i/\partial S_j$. For the system depicted in Scheme 5 (Section 2.4.2), one has $\varepsilon_{11} = \varepsilon_{12} = \varepsilon_{31} = 0$ and Eq. (5.318) simplifies to

$$\Delta = \varepsilon_{21}\varepsilon_{32} = \frac{\partial v_2}{\partial S_1}\frac{\partial v_3}{\partial S_2}. \tag{5.319}$$

Taking into account that the elasticities are dependent on the steady-state concentrations of S_1 and S_2, one derives for the response coefficients for the frequency

$$\begin{aligned} \frac{\partial \ln f}{\partial \ln k_j} &= \frac{k_j}{2\varepsilon_{21}}\left(\frac{\partial^2 v_2}{\partial S_1\,\partial k_j} + \frac{\partial^2 v_2}{\partial S_1^2}\frac{\partial S_1}{\partial k_j} + \frac{\partial^2 v_2}{\partial S_1\,\partial S_2}\frac{\partial S_2}{\partial k_j}\right) \\ &+ \frac{k_j}{2\varepsilon_{32}}\left(\frac{\partial^2 v_3}{\partial S_2\,\partial k_j} + \frac{\partial^2 v_3}{\partial S_1\,\partial S_2}\frac{\partial S_1}{\partial k_j} + \frac{\partial^2 v_3}{\partial S_2^2}\frac{\partial S_2}{\partial k_j}\right). \end{aligned} \tag{5.320}$$

From this equation, one may conclude that the response coefficients of frequencies for oscillations observed near a Hopf bifurcation may be expressed by coefficients characterizing the control of steady states; that is, first-order and second-order elasticity coefficients $\partial v_j/\partial S_i$, $\partial^2 v_j/\partial S_i\partial S_k$, and $\partial^2 v_j/\partial S_k\partial k_i$ as well as the first-order response coefficients for steady-state concentrations $(\partial S_i/\partial k_j)$. For more details concerning the second-order elasticities see Section 5.9.

For parameter combinations within the interior of the instability region oscillations with finite amplitudes for the concentrations S_i may be obtained. There, explicit solutions cannot be obtained for the frequency nor for the amplitudes, so that response coefficients should be calculated numerically.

5.9. A SECOND-ORDER APPROACH

Owing to definition (5.16), control coefficients describe the response of the system variables to infinitesimally small rate perturbations. In this sense, they characterize *local* systemic properties of a biochemical network in the vicinity of a stable steady state. Elasticities are also sometimes referred to as local properties, because they characterize single enzymes rather than the whole system. To avoid confusion, one should prefer the term *component* property in the context of elasticities (cf. Liao and Delgado, 1993).

As regulation of enzymes by effectors can cause substantial changes of their activities, it may be questioned to what extent the effects of relevant parameter perturbations can be described by the linear approximation (5.15). Apart from being of theoretical interest this problem is related to practical applications such as genetic engineering in biotechnology for which metabolic control analysis has been suggested as a useful tool (Westerhoff and Kell, 1987; Galazzo and Bailey, 1990; Fell, 1992).

The question of to what extent an enzyme controls a flux may also be analyzed in the way that one asks what happens when the enzyme is completely inhibited. Whether or not the flux under consideration is then still present can be decided by analyzing the zero and nonzero entries in the null-space matrix. It may occur that an enzyme has a high flux control coefficient, although it is not necessary for that flux because a parallel route bypassing the enzyme exists. Such an analysis can, however, only provide qualitative assertions.

In the present section we wish to analyze how metabolic control analysis can be extended to give more accurate predictions for the changes of the system variables than the simple linear approximation for finite parameter perturbations. The starting point is the power series expansion (5.9). We now focus on the quadratic approximation which takes into account the second-order derivatives of the system variables with respect to the kinetic parameters. The concentration and flux changes after perturbations of the parameter vector is approximated by

$$\Delta \boldsymbol{S} = \frac{\partial \boldsymbol{S}}{\partial \boldsymbol{p}}(\boldsymbol{p})\Delta \boldsymbol{p} + \frac{1}{2}\sum_{\alpha,\beta} \frac{\partial^2 \boldsymbol{S}}{\partial p_\alpha \partial p_\beta}(\boldsymbol{p})\Delta p_\alpha \Delta p_\beta, \tag{5.321a}$$

$$\Delta \boldsymbol{J} = \frac{\partial \boldsymbol{J}}{\partial \boldsymbol{p}}(\boldsymbol{p})\Delta \boldsymbol{p} + \frac{1}{2}\sum_{\alpha,\beta} \frac{\partial^2 \boldsymbol{J}}{\partial p_\alpha\, \partial p_\beta}(\boldsymbol{p})\Delta p_\alpha \Delta p_\beta. \tag{5.321b}$$

It has been shown in Section 5.2 that the first-order terms $\partial \boldsymbol{S}/\partial \boldsymbol{p}$ and $\partial \boldsymbol{J}/\partial \boldsymbol{p}$ can be obtained by implicit differentiation of the steady-state equation (2.9) with results given in Eqs. (5.11) and (5.12). In a similar way, the second-order terms are obtained by differentiating Eqs. (2.9) and (5.8) twice with respect to the parameters. By consideration of Eqs. (5.13) and (5.14), this results, after some algebra, in the following expressions:

$$\frac{\partial^2 S_a}{\partial p_\alpha \partial p_\beta} = \sum_{i,l,m=1}^{r} \sum_{j,k=1}^{n} C_{ai}^S \frac{\partial^2 v_i}{\partial S_j \, \partial S_k} C_{jl}^S C_{km}^S \frac{\partial v_l}{\partial p_\alpha} \frac{\partial v_m}{\partial p_\beta}$$

$$+ \sum_{i,k=1}^{r} \sum_{j=1}^{n} C_{ai}^S \left(\frac{\partial^2 v_i}{\partial S_j \, \partial p_\beta} C_{jk}^S \frac{\partial v_k}{\partial p_\alpha} + \frac{\partial^2 v_i}{\partial S_j \, \partial p_\alpha} C_{jk}^S \frac{\partial v_k}{\partial p_\beta} \right) \tag{5.322}$$

$$+ \sum_{i=1}^{r} C_{ai}^S \frac{\partial^2 v_i}{\partial p_\alpha \, \partial p_\beta}$$

and

$$\frac{\partial^2 J_b}{\partial p_\alpha \, \partial p_\beta} = \sum_{i,\, l,m=1}^{r} \sum_{j,\, k=1}^{n} C_{bi}^J \frac{\partial^2 v_i}{\partial S_j \, \partial S_k} C_{jl}^S C_{km}^S \frac{\partial v_l}{\partial p_\alpha} \frac{\partial v_m}{\partial p_\beta}$$

$$+ \sum_{i,\, k=1}^{r} \sum_{j=1}^{n} C_{bi}^J \left(\frac{\partial^2 v_i}{\partial S_j \, \partial p_\beta} C_{jk}^S \frac{\partial v_k}{\partial p_\alpha} + \frac{\partial^2 v_i}{\partial S_j \partial p_\alpha} C_{jk}^S \frac{\partial v_k}{\partial p_\beta} \right) \tag{5.323}$$

$$+ \sum_{i=1}^{r} C_{bi}^J \frac{\partial^2 v_i}{\partial p_\alpha \partial p_\beta}$$

for $a = 1, \ldots, n$ and $b = 1, \ldots, r$.

If only reaction-specific parameters are considered, certain sums in Eqs. (5.322) and (5.323) can be reduced to one term by taking into account $\partial v_i / \partial p_\alpha = 0$ for $i \neq \alpha$. The second-order terms can be written in the following compact notation:

$$\delta^2 \boldsymbol{S} = \boldsymbol{C}^S(\delta_{SS}^2 \boldsymbol{v} + 2\delta_{Sp}^2 \boldsymbol{v} + \delta_{pp}^2 \boldsymbol{v}), \tag{5.324a}$$

$$\delta^2 \boldsymbol{J} = \boldsymbol{C}^J(\delta_{SS}^2 \boldsymbol{v} + 2\delta_{Sp}^2 \boldsymbol{v} + \delta_{pp}^2 \boldsymbol{v}), \tag{5.324b}$$

where the vectors $\delta^2 \boldsymbol{S}$, $\delta^2 \boldsymbol{J}$, $\delta_{SS}^2 \boldsymbol{v}$, $\delta_{pp}^2 \boldsymbol{v}$ and $\delta_{pp}^2 \boldsymbol{v}$ have the following components:

$$\delta^2 S_a = \sum_{\alpha,\beta} \frac{\partial^2 S_a}{\partial p_\alpha \, \partial p_\beta} \Delta p_\alpha \Delta p_\beta, \qquad \delta^2 J_b = \sum_{\alpha,\beta} \frac{\partial^2 J_b}{\partial p_\alpha \, \partial p_\beta} \Delta p_\alpha \Delta p_\beta,$$

$$\delta_{SS}^2 v_i = \sum_{j,k} \frac{\partial^2 v_i}{\partial S_j \, \partial S_k} \delta S_j \delta S_k, \qquad \delta_{Sp}^2 v_i = \sum_{\alpha,j} \frac{\partial^2 v_i}{\partial S_j \, \partial p_\alpha} \delta S_j \Delta p_\alpha, \tag{5.325}$$

$$\delta_{pp}^2 v_i = \sum_{\alpha,\beta} \frac{\partial^2 v_i}{\partial p_\alpha \, \partial p_\beta} \Delta p_\alpha \Delta p_\beta$$

with

$$\delta S_j = \sum_{\beta,l} C_{jl}^S \frac{\partial v_l}{\partial p_\beta} \Delta p_\beta \tag{5.326}$$

being the concentration change in the linear approximation.

In addition to the quantities of the linear theory, the second-order response coefficients (5.322) and (5.323) contain the following second derivatives of the individual rates:

$$\varepsilon_{ijk} = \frac{\partial^2 v_i}{\partial S_j \, \partial S_k}: \text{ second-order } \varepsilon\text{-elasticities,} \tag{5.327a}$$

$$\pi_{i\alpha\beta} = \frac{\partial^2 v_i}{\partial p_\alpha \, \partial p_\beta}: \text{ second-order } \pi\text{-elasticities,} \tag{5.327b}$$

and

$$(\varepsilon - \pi)_{ij\alpha} = \frac{\partial^2 v_i}{\partial S_j \, \partial p_\alpha}: \text{ mixed second-order } \varepsilon - \pi\text{-elasticities.} \tag{5.327c}$$

Hence, the local characterization of the individual rates has to be extended to the second-order elasticity coefficients in order to determine the response of the system variables to parameter perturbations in the quadratic approximation. Because of the occurrence of mixed derivatives of the reaction rates with respect to metabolite concentrations, a general definition of parameter-independent second-order control coefficients is impossible. In particular, the parameter perturbations cannot be replaced by the rate perturbations as independent variables in Eqs. (5.321a) and (5.321b). Therefore, the perturbation parameters do not merely play a technical role as in the linear theory (see Section 5.2). Another interesting feature of the second-order terms is that they contain, besides derivatives characterizing the influence of a single reaction on a steady-state variable, mixed derivatives also (e.g., $\partial^2 S_a / \partial p_\alpha \partial p_\beta$, where p_α and p_β may belong to different rate equations). The effects of simultaneous perturbations of several rates are thus not simply approximated as the sum of the individual effects as in the case of the linear theory. In this sense the nonlinear terms in the expansions (5.321a) and (5.321b) reflect a fundamental characteristic of the underlying expressions for $\boldsymbol{S}$ and $\boldsymbol{J}$; namely they are nonlinear functions of the kinetic parameters even in the simplest case of linear rate laws. [See also Eq. (5.88) for the steady-state flux of an unbranched chain under nonsaturating conditions.]

A first discussion of Eqs. (5.322) and (5.323) becomes easier when reaction-specific perturbation parameters are considered, which enter the rate laws linearly,

$$v_j = p_j v_j^*(\boldsymbol{S}). \tag{5.328}$$

It turns out that, under this condition, one can introduce the rate perturbations instead of the parameter perturbations as independent variables. For example, with Eqs. (5.327) and (5.328), the second-order response coefficients for the flux simplify to

$$\frac{\partial^2 J_b}{\partial p_a \, \partial p_\beta} = \sum_{j=1}^{r} \sum_{j,k=1}^{n} C^J_{bi}\varepsilon_{ijk}C^S_{ja}C^S_{k\beta}\frac{v_a}{p_a}\frac{v_\beta}{p_\beta} + \frac{1}{p_a p_\beta}[C^J_{b\beta}(C^J_{\beta a} - \delta_{\beta a})v_a + C^J_{ba}(C^J_{a\beta} - \delta_{a\beta})v_\beta]. \tag{5.329}$$

The following definition of second-order flux control coefficients is appropriate:

$$D^J_{ba\beta} = \frac{1}{2}\frac{\partial^2 J_b}{\partial p_a p_\beta}\left(\frac{\partial v_a}{\partial p_a}\frac{\partial v_\beta}{\partial p_\beta}\right)^{-1}. \tag{5.330}$$

With $\partial v_a / \partial p_a = v_a / p_a$ one derives from Eq. (5.329)

$$D^J_{ba\beta} = \frac{1}{2}\left[\sum_{i=1}^{r} \sum_{j,k=1}^{n} C^J_{bi}\varepsilon_{ijk}C^S_{ja}C^S_{k\beta} + \frac{1}{v_a}C^J_{ba}C^J_{a\beta} + \frac{1}{v_\beta}C^J_{b\beta}C^J_{\beta a} - \delta_{a\beta}\left(\frac{C^J_{ba}}{v_a} + \frac{C^J_{b\beta}}{v_\beta}\right)\right]. \tag{5.331}$$

Similar equations are obtained for the second-order concentration control coefficients. The coefficients (5.331) are independent of the special choice of the (linear) perturbation parameter. Hence, Eqs. (5.321a) and (5.321b) can be written in the form

$$\Delta S_i = \sum_{a=1}^{r} C^S_{ia}\Delta v_a + \sum_{a,\beta}^{r} D^S_{ia\beta}\Delta v_a \Delta v_\beta, \tag{5.332a}$$

$$\Delta J_j = \sum_{a=1}^{r} C^J_{ja}\Delta v_a + \sum_{a,\beta}^{r} D^J_{ja\beta}\Delta v_a \Delta v_\beta, \tag{5.332b}$$

where the changes of the steady-state variables are related to perturbations of the individual reaction rates rather than of parameters.

For the second-order control coefficients, *summation theorems* exist similar to those of the linear theory. Denoting by $\boldsymbol{k}_\gamma$ and $\boldsymbol{k}_\delta$ two vectors in the null-space matrix, one obtains from Eq. (5.331)

$$\sum_{a,\beta=1}^{r} D^J_{ba\beta}k_{a\gamma}k_{\beta\delta} = \sum_{j,k=1}^{n} C^J_{bi}\varepsilon_{ijk}\left(\sum_{a=1}^{r} C^S_{ja}k_{a\gamma}\right)\left(\sum_{\beta=1}^{r} C^S_{i\beta}k_{\beta\delta}\right) + \sum_{a=1}^{r}\frac{C^J_{ba}}{v_a}\left(\sum_{\beta=1}^{r}(C^J_{a\beta} - \delta_{a\beta})k_{\beta\delta}\right)k_{a\gamma} + \sum_{\beta=1}^{r}\frac{C^J_{b\beta}}{v_\beta}\left(\sum_{a=1}^{r}(C^J_{\beta a} - \delta_{\beta a})k_{a\gamma}\right)k_{\beta\delta}. \tag{5.333}$$

With the summation theorems of the linear theory [Eqs. (5.44a) and (5.44b)], it follows immediately that

$$\sum_{\alpha,\beta}^{r} D^J_{b\alpha\beta} k_{\alpha\gamma} k_{\beta\delta} = 0. \tag{5.334}$$

Extending the analysis to second-order perturbations of the metabolite concentrations, one can show that

$$\sum_{\alpha,\beta}^{r} D^S_{a\alpha\beta} k_{\alpha\gamma} k_{\beta\delta} = 0 \tag{5.335}$$

(Höfer and Heinrich, 1993). The summation relationships for the second-order control coefficients for metabolite concentrations and fluxes thus have the same form, in contrast to the coefficients of the linear theory.

Example. We investigate the second-order approximation for flux control for the unbranched metabolic chain, which has been analyzed using the linear theory in Section 5.4.3.1. With the rate equations (5.85) the second-order elasticity coefficients ε_{ijk} vanish and expression (5.331) simplifies to

$$D^J_{ijk} = \frac{1}{2}\left[\frac{C^J_{ij} C^J_{jk}}{v_j} + \frac{C^J_{ik} C^J_{kj}}{v_k} - \delta_{jk}\left(\frac{C^J_{ij}}{v_j} + \frac{C^J_{ik}}{v_k}\right)\right]. \tag{5.336}$$

The index i in Eq. (5.336) may be omitted, because in the present case there is only one steady-state flux ($J = v_j = v_k$). Furthermore, for an unbranched chain, normalized and non-normalized control coefficients are equal (see Sections 5.4.3.1 and 5.7) and Eq. (5.336) turns into

$$D^J_{jk} = \frac{1}{J} C^J_j (C^J_k - \delta_{jk}). \tag{5.337}$$

As the first-order flux control coefficients are confined to the range between zero and unity, the second-order coefficients fulfill the relations

$$D^J_{jk} \begin{cases} \leq 0 & \text{if } j = k \\ \geq 0 & \text{if } j \neq k. \end{cases} \tag{5.338}$$

For an unbranched chain with linear rate equations the flux change ΔJ resulting from a perturbation of a reaction may be fully expressed by the first-order flux control coefficients [cf. Eq. (5.95)]. In the present case it is therefore easily possible to compare the accuracy of the linear and quadratic approximation using the formula

$$\eta^J = \frac{\Delta J_{\mathrm{approx}} - \Delta J_{\mathrm{exact}}}{\Delta J_{\mathrm{exact}}}, \tag{5.339}$$

where η^J denotes the relative error of the approximations. With the linear approximation

$$\Delta J_{\mathrm{lin}} = C_k^J \Delta v_k \tag{5.340}$$

and the second-order approximation

$$\Delta J_{\mathrm{sec}} = C_k^J \Delta v_k + \frac{C_k^J}{J}(C_k - 1)\Delta v_k^2, \tag{5.341}$$

one gets with Eqs. (5.95) and (5.339)

$$\eta_{\mathrm{lin}}^J(\Delta v_k) = (1 - C_k^J)\frac{\Delta v_k}{v_k}, \tag{5.342}$$

$$\eta_{\mathrm{sec}}^J(\Delta v_k) = -(1 - C_k^J)^2\left(\frac{\Delta v_k}{v_k}\right)^2 \leq 0. \tag{5.343}$$

A large linear flux control coefficient of the perturbed reaction implies small relative errors of both approximations. [Note that in the limiting case $C_k^J = 1$, Eq. (5.95) predicts that $\Delta J = C_k^J \Delta v_k$ becomes exact for any finite perturbation Δv_j.] For $C_k^J < 1$, in a certain range of rate perturbations the quadratic approximation is more accurate than the linear one, whereas the opposite holds for large rate changes (due to the rapid divergence of the quadratic terms for large Δv_k). According to Eqs. (5.342) and (5.343), rate perturbations which lead to a given (permissible) error $\tilde{\eta}$ are related as follows:

$$\left(\frac{\Delta v_k}{v_k}\right)_{\mathrm{sec}} = \frac{1}{\sqrt{\tilde{\eta}}}\left(\frac{\Delta v_k}{v_k}\right)_{\mathrm{lin}}. \tag{5.344}$$

This equation shows that the second-order approximation is more accurate up to an error of 100%. The treatment has been applied also to metabolic chains with saturation kinetics as well as to a model of glycolysis (Höfer and Heinrich, 1993).

Dealing with the effects of large changes in enzyme activities on the fluxes Small and Kacser (1993) introduced a *deviation index* in the following way:

$$D_k^J = \frac{v_k}{J} \frac{\Delta J}{\Delta v_k}, \tag{5.345}$$

where $\Delta v_k = v_k(p_k^0 + \Delta p_k) - v_k(p_k^0)$ represents the effect of a finite parameter perturbation on the activity of an isolated reaction, and ΔJ the resulting change of the steady-state activity. For normalization the enzyme activity v_k and the flux J for the new parameter value $p_k^0 + \Delta p_k$ are used.

As shown above, in the case of unbranched reaction chains with linear kinetics, the effect of parameter perturbations may be evaluated analytically for arbitrary rate perturbations. Applying the concept of deviation index to such systems, one obtains, with formulas (5.95) and (5.345),

$$D_k^J = C_k^J. \tag{5.346}$$

This equation means that for this special case, the deviation index equals the control coefficient at the reference state for any parameter perturbation. Furthermore, with formula (5.95), one may calculate the ratio of the steady-state fluxes for the perturbed state and the reference state. If it is assumed that the rate of reaction k is changed by a factor μ [i.e., $\Delta v_k = (\mu - 1)v_k$], one obtains, for the *amplification factor,*

$$f_k^J = \frac{J}{J^0} = \frac{1}{1 - C_k^J \dfrac{(\mu - 1)}{\mu}}, \tag{5.347}$$

where J^0 denotes the flux at the reference state. The results expressed by formulas (5.346) and (5.347) underline the general conclusion made in Section 5.4.3.1 that the steady-state properties of an unbranched chain with linear kinetics are characterized completely by the first-order control coefficients in a reference state. Obviously, a similar conclusion cannot be drawn for systems with nonlinear kinetic equations. Here, the effect of finite parameter changes has to be calculated using a kinetic model. The second-order approach presented above may be useful if the finite changes are not too large.

5.10. METABOLIC REGULATION FROM THE VIEWPOINT OF CONTROL ANALYSIS

5.10.1. Coresponse Coefficients

In the introduction to Chapter 5 it has been discussed that, in the framework of metabolic control analysis, the term *control* is used merely in its descriptive sense, which means that control coefficients describe the effect of a parameter perturbation on metabolite concentrations or fluxes, irrespective of whether or not this parameter actually changes under physiological conditions. The question of whether metabolic control analysis may provide quantities specifically characterizing the *regulation* of metabolic systems arises. It has often been stressed that regulation is related to the teleonomic response of biological systems to external and internal signals. Therefore, a *regulation analysis* must take into account the biological function of a given metabolic pathway (e.g., the synthesis of ATP in glycolysis or oxidative phosphorylation, the synthesis of amino acids in the corresponding pathways, *etc.*). Furthermore, it seems to be necessary to quantify certain regulative properties of metabolic systems such as homeostasis.

A clue to the quantification of regulation may be the distinction between the effect of parameters on steady-state variables and the correlation between changes of two steady-state variables (Hofmeyr *et al.,* 1993).

Concerning the action of *external signals,* the problem can be tackled within the framework of traditional metabolic control analysis. The effect of external inhibitors or activators or the effect of changed enzyme concentrations may be quantified by response coefficients [cf. Eqs. (5.28) and (5.29)]. The problem of whether it is possible to characterize in an adequate way the effect of internal regulators, such as substances which exert feedback inhibitions arises (Hofmeyr and Cornish-Bowden, 1991, 1993; Kahn and Westerhoff, 1993a, 1993b; Hofmeyr *et al.,* 1993). Using metabolic control analysis, one is confronted with the conceptual difficulty that after perturbation of concentrations of internal metabolites

$$\boldsymbol{S}(t) = \boldsymbol{S}^0 \quad \text{for } t < 0, \qquad \boldsymbol{S}(t) = \boldsymbol{S}^0 + \delta\boldsymbol{S}(t) \quad \text{for } t \geq 0, \qquad \delta\boldsymbol{S}(0) = \delta\boldsymbol{S}^0 \tag{5.348}$$

the system will generally relax to the original steady state, which means that eventually the total effects of the perturbation on the system variables vanish as long as the considered reference steady state is asymptotically stable. We will show below that some problems of regulation (in particular, the quantification of the effect of internal regulators) may be tackled within the framework of metabolic control analysis.

To arrive at a more complete description of the response of internal variables after perturbations of parameters, Hofmeyr *et al.* (1993) introduced normalized coresponse coefficients in the following way:

$$ {}^kO_{ij}^{S,S} = \frac{R_{ik}^S}{R_{jk}^S} = \frac{\sum\limits_m C_{im}^S \pi_{mk}}{\sum\limits_m C_{jm}^S \pi_{mk}}, \tag{5.349}$$

which characterize the concomitant change in two steady-state concentrations S_i and S_j resulting from a perturbation of a parameter p_k. Analogously, coresponse coefficients can be defined for two fluxes or for one concentration and one flux or for two other steady-state variables. Experimentally, coresponse coefficients may be calculated from the slopes of the tangents in a plot of one internal variable versus another variable, obtained by variation of a parameter.

If reaction-specific perturbation parameters are considered, Eq. (5.349) simplifies to

$$ {}^{k}O_{ij}^{S,S} = \frac{C_{ik}^{S}}{C_{jk}^{S}}, \tag{5.350} $$

where the left-hand superscript k denotes the number of the perturbed reaction. Similarly, non-normalized coresponse coefficients may be defined by replacing, in Eqs. (5.349) and (5.350), the normalized π-elasticities and the control coefficients by their non-normalized counterparts.

Concerning the homeostatic property observed in many metabolic pathways, a system may be considered to be effectively regulated if strong changes of fluxes are accompanied by low variations of the metabolite concentrations, that is, if the coresponse coefficients ${}^{k}O^{S,J}$ of a reaction k whose reaction rate may change under physiological conditions have small absolute values (Hofmeyr and Cornish-Bowden, 1991).

Example. In Section 5.4.3.1. it has been shown that for the unbranched chain with feedback inhibition (Scheme 6) the effect of perturbations of the consumption rate v_{n+1} on the steady-state flux J and the end-product concentration S_n may be characterized by the following control coefficients:

$$ C_{n+1}^{J} = -\frac{\varepsilon_{1,n}}{\varepsilon_{n+1,n} - \varepsilon_{1,n}}, \tag{5.351a} $$

$$ C_{n,n+1}^{S} = -\frac{1}{\varepsilon_{n+1,n} - \varepsilon_{1,n}}, \tag{5.351b} $$

where the elasticity $\varepsilon_{1,n}$ describes the strength of the feedback inhibition. From this, one obtains for the coresponse coefficient of the end-product concentration and the steady-state flux at perturbations of reaction $n + 1$,

$$ {}^{n+1}O^{S_n,J} = \frac{C_{n,n+1}^{S}}{C_{n+1}^{J}} = \frac{1}{\varepsilon_{1,n}} < 0. \tag{5.352} $$

As expected an effective regulation, $|{}^{n+1}O^{S_n,J}| \ll 1$, results when the feedback inhibition is strong, $|\varepsilon_{1,n}| \gg 1$.

It is worth mentioning that coresponse coefficients of metabolite concentrations are related to the "crossover theorem" which dates back to the very beginning of the mathematical analysis of metabolic networks (Chance *et al.*, 1955, 1958; Holmes, 1959; Higgins, 1965) and which has been used to identify interaction sites with outer effectors. In its simplest form, this theorem can be stated in the following way: *The variations of the concentrations of the metabolites upstream and downstream an enzyme which is influenced by an effector have different signs.* Accordingly, when in an unbranched sequence, a reaction k is the target of an effector and the corresponding coresponse coefficient is negative,

$$^{k}O_{i,j}^{S,S} = \frac{C_{ik}^{S}}{C_{jk}^{S}} < 0, \tag{5.353}$$

then this reaction is located in between the metabolites S_i and S_j [cf. Eq. (5.109)]. It should be noted, however, that there are severe limitations to the crossover theorem if it is applied to more complex pathways. It has been shown that in systems with conserved quantities and in other more complex situations, the interaction with an external effector does not always produce a crossover at the affected enzyme and that "pseudo-crossovers" may also occur at unaffected enzymes (Heinrich and Rapoport, 1974b).

5.10.2. Fluctuations of Internal Variables Versus Parameter Perturbations

Perturbations of internal variables generally have a nonvanishing effect at *finite times*. We now show that the time-dependent responses $\delta S(t)$ for $0 \le t < \infty$ after perturbations defined by Eq. (234) can be mimicked by responses taking place after parameter perturbations (see Kahn and Westerhoff, 1993a).

Perturbations of concentrations: After small perturbations δS^0 in the neighborhood of a steady state, the dynamics of the system is governed by the linearized equations

$$\frac{d(\delta \boldsymbol{S})}{dt} = \left(\mathbf{N}\frac{\partial \boldsymbol{v}}{\partial \boldsymbol{S}}\right)\delta \boldsymbol{S} = \mathbf{M}\delta \boldsymbol{S}. \tag{5.354}$$

With the initial conditions given in Eq. (5.348), this has the solution

$$\delta \boldsymbol{S}(t) = \exp(\mathbf{M}t)\delta \boldsymbol{S}^0. \tag{5.355}$$

[For the definition of the exponential function for matrices, cf. Eqs. (2.85) and (5.187).] To characterize the time-dependent effect of fluctuations of internal variables, Kahn and Westerhoff (1993a) introduced the response function

$$\delta \boldsymbol{S}^{\text{resp}} = \delta \boldsymbol{S}(t) - \delta \boldsymbol{S}^0 = [\exp(\mathbf{M}t) - \mathbf{I}]\delta \boldsymbol{S}^0 \tag{5.356}$$

with

$$\delta \boldsymbol{S}^{\text{resp}}|_{t \to 0} = \boldsymbol{0}, \tag{5.357a}$$

$$\delta \boldsymbol{S}^{\text{resp}}|_{t \to \infty} = -\delta \boldsymbol{S}^0. \tag{5.357b}$$

For the time-dependent flux response $\delta \boldsymbol{J}(t)$ after small perturbations of metabolite concentrations, one obtains with $\boldsymbol{J} = \boldsymbol{v}(\boldsymbol{S}(t),\boldsymbol{p})$ in the linear approximation

$$\delta \boldsymbol{J}^{\text{resp}}(t) = \frac{\partial \boldsymbol{v}}{\partial \boldsymbol{S}} \delta \boldsymbol{S}(t) = \frac{\partial \boldsymbol{v}}{\partial \boldsymbol{S}} \exp(\mathbf{M}t)\delta \boldsymbol{S}^0, \tag{5.358}$$

where Eq. (5.355) has been taken into account. In particular,

$$\delta \boldsymbol{J}^{\text{resp}}|_{t=0} = \delta \boldsymbol{J}^0 = \frac{\partial \boldsymbol{v}}{\partial \boldsymbol{S}} \delta \boldsymbol{S}^0, \tag{5.359a}$$

$$\delta \boldsymbol{J}^{\text{resp}}|_{t \to \infty} = \boldsymbol{0}. \tag{5.359b}$$

Parameter perturbations: According to Eqs. (5.185) and (5.186), perturbations of parameters will result in changes in the metabolite concentrations described by the functions

$$\delta \boldsymbol{S}(t) = [\exp(\mathbf{M}t) - \mathbf{I}]\mathbf{M}^{-1}\mathbf{N} \frac{\partial \boldsymbol{v}}{\partial \boldsymbol{p}} \delta \boldsymbol{p}. \tag{5.360}$$

The corresponding flux changes are

$$\delta \boldsymbol{J}(t) = \left(\mathbf{I} + \frac{\partial \boldsymbol{v}}{\partial \boldsymbol{S}} [\exp(\mathbf{M}t) - \mathbf{I}]\mathbf{M}^{-1}\mathbf{N}\right) \frac{\partial \boldsymbol{v}}{\partial \boldsymbol{p}} \delta \boldsymbol{p} \tag{5.361}$$

[cf. Eqs. (5.189) and (5.190)]. Now we choose special parameter perturbations which cause the same immediate changes in the reaction rates for $t = 0$ as the perturbations $\delta \boldsymbol{S}^0$ considered in Eq. (5.355),

$$\delta \boldsymbol{v}^0 = \frac{\partial \boldsymbol{v}}{\partial \boldsymbol{p}} \delta \boldsymbol{p} = \frac{\partial \boldsymbol{v}}{\partial \boldsymbol{S}} \delta \boldsymbol{S}^0. \tag{5.362}$$

Choosing the set of parameters in such a way that the matrix $\partial \boldsymbol{v}/\partial \boldsymbol{p}$ is invertible, one obtains

$$\delta \boldsymbol{p} = \left(\frac{\partial \boldsymbol{v}}{\partial \boldsymbol{p}}\right)^{-1} \left(\frac{\partial \boldsymbol{v}}{\partial \boldsymbol{S}}\right) \delta \boldsymbol{S}^0. \tag{5.363}$$

Introducing this into Eqs. (5.360) and (5.361) gives, by consideration of $\mathbf{M} = \mathbf{N}(\partial \boldsymbol{v}/\partial \boldsymbol{S})$,

$$\delta \boldsymbol{S}(t) = [\exp(\mathbf{M}t) - \mathbf{I}]\delta \boldsymbol{S}^0, \tag{5.364a}$$

$$\delta \boldsymbol{J}(t) = \frac{\partial \boldsymbol{v}}{\partial \boldsymbol{S}} \exp(\mathbf{M}t)\delta \boldsymbol{S}^0. \tag{5.364b}$$

Comparison of Eqs. (5.364a) and (5.364b) with Eqs. (5.356) and (5.358), respectively, shows that $\delta \boldsymbol{S}(t)$ and $\delta \boldsymbol{J}(t)$ brought about by parameter perturbations equal $\delta \boldsymbol{S}^{\mathrm{resp}}(t)$ and $\delta \boldsymbol{J}^{\mathrm{resp}}(t)$, respectively, resulting from a perturbation of the internal variables.

5.10.3. Internal Response Coefficients

A further quantity which may characterize metabolic regulation is the *internal response coefficient* introduced by Hofmeyr and Cornish-Bowden (1993) as well as Kahn and Westerhoff (1993a):

$$\tilde{R}_{ij}^S = \sum_{k=1}^{r} {}^k\tilde{R}_{ij}^S = \sum_{k=1}^{r} C_{ik}^S \varepsilon_{kj} = \sum_{k=1}^{r} C_{ik}^S \frac{\partial v_k}{\partial S_j} \tag{5.365a}$$

and

$$\tilde{R}_{ij}^J = \sum_{k=1}^{r} {}^k\tilde{R}_{ij}^J = \sum_{k=1}^{r} C_{ik}^J \varepsilon_{kj} = \sum_{k=1}^{r} C_{ik}^J \frac{\partial v_k}{\partial S_j}. \tag{5.365b}$$

The individual terms of the sums in Eq. (5.365) have a structure which is very similar to that of the terms which enter the response coefficients for parameter perturbations [Eqs. (5.28) and (5.29)]. Both are products of control coefficients and elasticity coefficients. However, in definitions (5.365a) and (5.365b) of the internal response coefficients, the parameter elasticities are replaced by elasticities with respect to concentrations of internal metabolites. Therefore, it seems appropriate to assume that $\tilde{R}_{ij}^S$ and $\tilde{R}_{ij}^J$ are related to the effect of a perturbation of a concentration S_j on a metabolite concentration S_i and a flux J_i, respectively. Moreover, the individual terms ${}^k\tilde{R}_{ij}^S$ and ${}^k\tilde{R}_{ij}^J$ may be considered as *partial internal response coefficients* (Kholodenko, 1990) quantifying the contributions of different regulatory routes to the total response. Previously, these terms have been called *regulatory strengths* (Kahn and Westerhoff, 1993a, 1993b; Hofmeyr and Cornish-Bowden, 1993).

In analogy to Eqs. (5.31) and (5.32), which are valid for parameter perturbations, one obtains with Eqs. (5.365) and (5.357b)

$$\delta \boldsymbol{S}^{\mathrm{resp}} = \tilde{\mathbf{R}}^S \delta \boldsymbol{S}^0 = \mathbf{C}^S \boldsymbol{\varepsilon} \delta \boldsymbol{S}^0 = -\delta \boldsymbol{S}^0 \tag{5.366}$$

for the response after perturbations of metabolite concentrations, and with Eq. (5.359b),

$$\delta \boldsymbol{J}^{\text{resp}} = \tilde{\mathbf{R}}^{J}\delta \boldsymbol{S}^{0} = \mathbf{C}^{J}\varepsilon\delta \boldsymbol{S}^{0} = \boldsymbol{0}, \tag{5.367}$$

where the matrices $\tilde{\mathbf{R}}^{S}$ and $\tilde{\mathbf{R}}^{J}$ contain the internal response coefficients defined in Eq. (5.365) as elements. Applying Eqs. (5.366) and (5.367) to perturbations $\delta S_i^0 \neq 0$, $\delta S_j^0 = 0$ $(i \neq j)$ leads to the relations

$$\tilde{R}_{ij}^{S} = \sum_{k=1}^{r} {}^{k}\tilde{R}_{ij}^{S} = \sum_{k=1}^{r} C_{ik}^{S}\varepsilon_{kj} = -\delta_{ij} \tag{5.368}$$

and

$$\tilde{R}_{ij}^{J} = \sum_{k=1}^{r} {}^{k}\tilde{R}_{ij}^{J} = \sum_{k=1}^{r} C_{ik}^{J}\varepsilon_{kj} = 0, \tag{5.369}$$

which are identical to the connectivity theorems given in Eqs. (5.51a) and (5.51b) for systems without conserved quantities ($\mathbf{L} = \mathbf{I}$). One may conclude, therefore, that the connectivity theorems may be physically interpreted by consideration of the response of a metabolic system toward perturbations of internal variables (see Westerhoff and Chen, 1984). In the case that the system involves conserved quantities, a similar reasoning applies, provided that the perturbations do not violate the conservation relationships.

For a further discussion of Eq. (5.368) we consider first the case $i \neq j$. A vanishing response of a metabolite S_i after a perturbation δS_j^0 means that within the sum (5.368) the individual terms ${}^{k}\tilde{R}_{ij}^{S}$ which characterize the response *via* different reactions k cancel each other.

It is worth distinguishing a situation where the sum of all positive partial response coefficients ${}^{k}\tilde{R}_{ij}$ is high from that where this sum is low. Grouping together positive and negative partial response coefficients, respectively, Eq. (5.368) may be rewritten as follows:

$$P_{ij}^{S} = \left(\sum_{k} {}^{k}\tilde{R}_{ij}^{S}\right)_{\text{positive}} = -\left(\sum_{k} {}^{k}\tilde{R}_{ij}^{S}\right)_{\text{negative}}. \tag{5.370}$$

For P_{ij}^{S} the name *regulatory potential* of the concentration response has been proposed (Hofmeyr and Cornish-Bowden, 1993). An analogous equation is obtained for the regulatory potential of the flux response if in Eq. (5.370) the superscript S is replaced by J.

For $i = j$, one obtains from Eq. (5.368)

$$\sum_{k=1}^{r} {}^{k}\tilde{R}_{ii}^{S} \delta S_i^0 = -\delta S_i^0. \tag{5.371}$$

The nonvanishing response of S_i after perturbations δS_i^0 may be considered as a result of the homeostatic effect of regulatory loops in such a way that S_i eventually reaches the same concentration as before the perturbation. From Eq. (5.371) follows

$$-\sum_{k=1}^{r} {}^{k}\tilde{R}_{ii}^{S} \equiv \sum_{k=1}^{r} {}^{k}H_{ii} = 1. \tag{5.372}$$

The term ${}^{k}H_{ii}$ may serve as a quantitative measure of the extent to which a certain reaction k counteracts the initial perturbation δS_i^0. Correspondingly, ${}^{k}H_{ii}$ has been called *homeostatic strength* (Kahn and Westerhoff, 1993a).

5.10.4. Rephrasing the Basic Equations of Metabolic Control Analysis in Terms of Coresponse Coefficients and Internal Response Coefficients

It has been shown by Hofmeyr *et al.* (1993) that the basic equations of metabolic control may be rewritten in terms of coresponse coefficients and internal response coefficients. As outlined in Section 5.3.3, the summation and connectivity relationships (5.44) and (5.51) may be used to calculate all the control coefficients in terms of elasticities [cf. Eq. (5.54)].

Let us define an $(r \times r)$ diagonal matrix $\mathbf{\Delta}_i$ whose elements are nonzero control coefficients. Possible representations for $\mathbf{\Delta}_i$ are

$$\mathbf{\Delta}_i = \mathrm{dg}(\boldsymbol{C}_i^J), \tag{5.373a}$$

$$\mathbf{\Delta}_i = \mathrm{dg}(\boldsymbol{C}_i^S), \tag{5.373b}$$

where the elements of the vectors $\boldsymbol{C}_i^J$ and $\boldsymbol{C}_i^S$ are the r flux control coefficients of J_i and the r concentration control coefficients of S_i, respectively. Another possibility would be that $\mathbf{\Delta}_i$ consists of a mixture of j flux control coefficients and $r - j$ concentration control coefficients.

Because the number of columns of the first matrix and the number of rows of the second matrix on the left-hand side of Eq. (5.54) equals the number r of reactions of the given metabolic system, this equation may be rewritten in the following way

$$\begin{pmatrix} \mathbf{C}^J \mathbf{\Delta}_i^{-1} \\ \mathbf{C}^S \mathbf{\Delta}_i^{-1} \end{pmatrix} (\mathbf{\Delta}_i \mathbf{K} \quad \mathbf{\Delta}_i \boldsymbol{\varepsilon}) = \begin{pmatrix} \mathbf{K} & \mathbf{0} \\ \mathbf{0} & -\mathbf{I} \end{pmatrix} \tag{5.374}$$

under the simplification that the system has no conserved quantities. It is easy to see that the elements of the matrices $\mathbf{C}^J\boldsymbol{\Delta}_i^{-1}$ and $\mathbf{C}^S\boldsymbol{\Delta}_i^{-1}$ are coresponse coefficients, whereas the matrix $\boldsymbol{\Delta}_i\boldsymbol{\varepsilon}$ contains the internal response coefficients.

Example. For the reaction system depicted in Scheme 10 (Section 5.3.4) one obtains, with Δ_i consisting of flux control coefficients,

$$\mathbf{C}^J\boldsymbol{\Delta}_i^{-1} = \mathbf{O}^{J,J} = \begin{pmatrix} \dfrac{C_{11}^J}{C_{i1}^J} & \dfrac{C_{12}^J}{C_{i2}^J} \\ \\ \dfrac{C_{21}^J}{C_{i1}^J} & \dfrac{C_{22}^J}{C_{i2}^J} \end{pmatrix} = \begin{pmatrix} 1 & 1 \\ 1 & 1 \end{pmatrix}, \tag{5.375a}$$

where it has been taken into account that $C_{1j}^J = C_{2j}^J$ for the unbranched reaction chain depicted in Scheme 10. Furthermore,

$$\mathbf{C}^S\boldsymbol{\Delta}_i^{-1} = \mathbf{O}^{S,J} = \begin{pmatrix} \dfrac{C_{11}^S}{C_{i1}^J}, & \dfrac{C_{12}^J}{C_{i2}^J} \end{pmatrix} \tag{5.375b}$$

and

$$\boldsymbol{\Delta}_i\boldsymbol{\varepsilon} = \begin{pmatrix} C_{i1}^J\varepsilon_{11} \\ \\ C_{i2}^J\varepsilon_{21} \end{pmatrix} = \begin{pmatrix} {}^1\tilde{R}_{i1}^J \\ \\ {}^2\tilde{R}_{i2}^J \end{pmatrix}. \tag{5.375c}$$

Likewise, if the diagonal matrix Δ_i contains concentration control coefficients

$$\mathbf{C}^J\boldsymbol{\Delta}_i^{-1} = \mathbf{O}^{J,S} = \begin{pmatrix} \dfrac{C_{11}^J}{C_{11}^S} & \dfrac{C_{12}^J}{C_{12}^S} \\ \\ \dfrac{C_{21}^J}{C_{11}^S} & \dfrac{C_{22}^J}{C_{12}^J} \end{pmatrix}, \tag{5.376a}$$

$$\mathbf{C}^S\boldsymbol{\Delta}_i^{-1} = (1 \quad 1), \tag{5.376b}$$

and

$$\boldsymbol{\Delta}_i\boldsymbol{\varepsilon} = \begin{pmatrix} C_{11}^S\varepsilon_{11} \\ \\ C_{12}^S\varepsilon_{21} \end{pmatrix} = \begin{pmatrix} {}^1\tilde{R}_{11}^S \\ \\ {}^2\tilde{R}_{11}^S \end{pmatrix}. \tag{5.376c}$$

At first glance, introduction of coresponse coefficients as new variables may seem to be questionable because they are simply combined of existing quantities and represent a large number of additional variables. However, it has been observed that the ratio of control coefficients in response to a modulation of a specific reaction is often independent of the choice of this reaction (Cornish-

Bowden and Hofmeyr, 1994). Therefore, the set of coresponse coefficients different from each other can considerably be reduced.

Traditionally, control coefficients have been determined in two different ways. One possibility is based on Eqs. (5.25b) and (5.26b) or on the summation and connectivity theorems and requires knowledge of the elasticities. The second method starts from the definitions of control coefficients and requires experimental determination of the dependence of steady-state variables and isolated reaction rates on a chosen parameter. In experimental practice, it is sometimes difficult to measure these dependences or the elasticities. Coresponse analysis paves the way to a third possibility of determining control coefficients, based on Eq. (5.374), which is equivalent to

$$(\boldsymbol{\Delta}_i\mathbf{K} \quad \boldsymbol{\Delta}_i\boldsymbol{\varepsilon}) = \begin{pmatrix} \mathbf{C}^J\boldsymbol{\Delta}_i^{-1} \\ \mathbf{C}^S\boldsymbol{\Delta}_i^{-1} \end{pmatrix}^{-1} \begin{pmatrix} \mathbf{K} & 0 \\ 0 & -\mathbf{I} \end{pmatrix}. \tag{5.377}$$

When **K** is chosen as indicated in Eq. (3.47), it contains the identity matrix. Therefore, a submatrix of the left-hand side of Eq. (5.377) consists of explicit control coefficients. They can be computed if the right-hand side of Eq. (5.377) is known. This requires determination of coresponse coefficients, which is feasible in experiment without measuring the fractional change in the perturbation parameter (e.g., enzyme activity). All that is required is to be able to modulate each enzyme activity around its normal value and measure the steady-state flux and concentration changes; knowledge of actual enzyme activities is unnecessary. With inhibitor studies, it is only necessary to know that the inhibitor acts on one particular enzyme, rather than to know the type of inhibitor or its concentration. Coresponse coefficients are obtained by plotting appropriate combinations of fluxes and concentrations against one another and, from that, control coefficients and response coefficients (and hence, also elasticities) can be calculated by use of Eq. (5.377). This procedure is presented in more detail in Cornish-Bowden and Hofmeyr (1994).

5.11. CONTROL WITHIN AND BETWEEN SUBSYSTEMS

It is frequently appropriate to group the body of enzyme data into classes corresponding to subsystems of the biochemical network. This is particularly useful when the network consists of several parts that interact in a restricted way, in that many elasticities are zero or that there is no mass flow between these parts. Examples are provided by cascades involving hierarchies of regulatory proteins modifying each other [e.g., the glutamine synthetase cascade (cf. Chock *et al.*, 1980)] and by the well-known hierarchic organization of genetic and metabolic processes in the living cell (cf. Scheme 15).

DNA -------► mRNA -------► enzymes -------► metabolites Scheme 15

Kahn and Westerhoff (1991) presented an approach to cope with the control of regulatory cascades. The basic idea is to calculate control coefficients for the particular levels of the hierarchy in terms of the elasticities within the levels and then to determine the control coefficients of the whole system from the intrinsic control coefficients of the subsystems and the elasticities describing the regulatory interactions between these.

We start from a decomposition of the reaction system into subsystems, which is represented by a partitioning of the stoichiometry matrix into blocks,

$$\mathbf{N} = \begin{pmatrix} \mathbf{N}_{11} & \mathbf{N}_{12} & \cdots & \mathbf{N}_{1p} \\ \mathbf{N}_{21} & \mathbf{N}_{22} & \cdots & \mathbf{N}_{2p} \\ \vdots & \vdots & \ddots & \vdots \\ \mathbf{N}_{p1} & \mathbf{N}_{p2} & \cdots & \mathbf{N}_{pp} \end{pmatrix} \tag{5.378}$$

with $\mathbf{N}_{ii}$ being the stoichiometry matrix of the ith subsystem and $\mathbf{N}_{ij}$ $(i \neq j)$ reflecting the involvement of the reactions belonging to subsystem j in the production or degradation of the species in subsystem i. p denotes the number of subsystems. Note that the number of rows in the arrangement of submatrices in Eq. (5.378) equals the number of columns because, for any given decomposition, each substance and each reaction uniquely belong to one subsystem.

Consider, for example, the reaction system shown in Figure 5.12A. It represents two pathways interconnected by a cycle involving the substances S_1 and S_3. For example, S_1 might be an enzyme. In that case, it would be sensible to simplify the scheme as shown in Figure 5.12B, where the broken arrow signifies the catalytic effect on the production of S_2. Identifying the two pathways with two subsystems, we can partition the complete stoichiometry matrix as

$$\mathbf{N} = \left(\begin{array}{cc:ccc} 1 & -1 & -1 & 1 & 0 \\ \cdots & \cdots & \cdots & \cdots & \cdots \\ 0 & 0 & 0 & 1 & -1 \\ 0 & 0 & 1 & -1 & 0 \end{array}\right). \tag{5.379}$$

A necessary condition for the present approach is that there be no net mass flow between the subsystems in steady state. This is, for example, the case for the levels of mRNA and enzymes in Scheme 15. This condition can be written as a block-diagonalization of the null-space matrix **K**, as expressed by Eq. (3.48). For example, for the scheme shown in Figure 5.12 A, the corresponding null-space matrix can be partitioned as

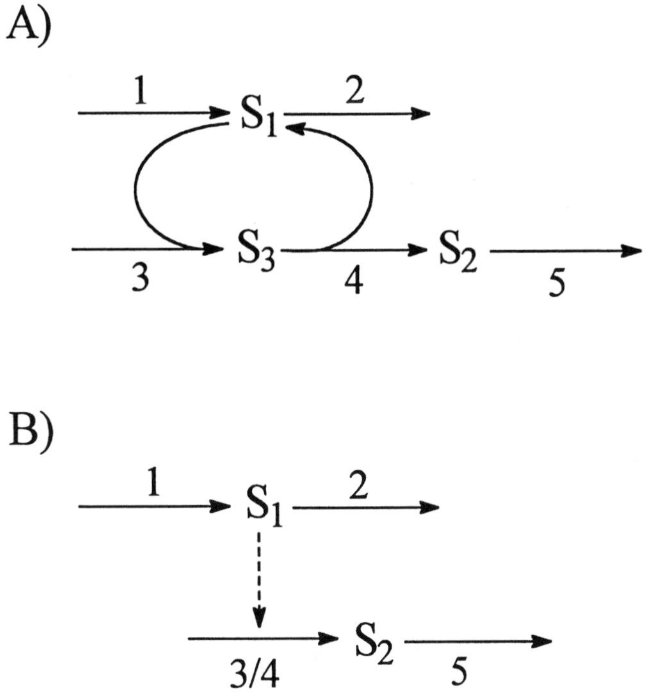

Figure 5.12 Simple example of a two-level system. (A) detailed scheme, (B) simplified representation.

$$\mathbf{K} = \begin{pmatrix} \mathbf{K}_1 & \mathbf{0} \\ \mathbf{0} & \mathbf{K}_2 \end{pmatrix} = \begin{pmatrix} 1 & \vdots & 0 \\ 1 & \vdots & 0 \\ \cdots & \vdots & \cdots \\ 0 & \vdots & 1 \\ 0 & \vdots & 1 \\ 0 & \vdots & 1 \end{pmatrix}. \tag{5.380}$$

A second condition for the present approach is that the link matrix **L** be decomposable into diagonal blocks, possibly after rearrangement of its rows,

$$\mathbf{L} = \begin{pmatrix} \mathbf{L}_1 & \mathbf{0} & \cdots & \mathbf{0} \\ \mathbf{0} & \mathbf{L}_2 & \cdots & \mathbf{0} \\ \vdots & \vdots & \ddots & \vdots \\ \mathbf{0} & \mathbf{0} & \cdots & \mathbf{L}_\mu \end{pmatrix}. \tag{5.381}$$

This means that each subsystem has separate conservation relations; that is, there are no conservation relations linking different subsystems. With this structure of $\mathbf{L}$, the equation $\mathbf{N} = \mathbf{L}\mathbf{N}^0$ [Eq. (3.7)] implies the equations

$$\mathbf{N}_{ii} = \mathbf{L}_i\mathbf{N}_{ii}^0, \quad i = 1, \ldots, \rho. \tag{5.382}$$

Note that for a given reaction system, there may be different ways of decomposition into subsystems that fulfill the two conditions (3.48) and (5.381), in particular if there are no conservation relations (i.e., $\mathbf{L} = \mathbf{I}$). One should preferably choose such a decomposition that the subsystems can be investigated separately in experiment. For the exemplifying scheme in Figure 5.12A, condition (5.381) is trivially fulfilled because it does not involve any conservation relations. Another decomposition is obtained by including S_3 into subsystem 1 instead of subsystem 2.

The elasticity matrix $\boldsymbol{\varepsilon}$ can be partitioned into blocks according to the decomposition of the system:

$$\boldsymbol{\varepsilon} = \begin{pmatrix} \boldsymbol{\varepsilon}_{11} & \boldsymbol{\varepsilon}_{12} & \cdots & \boldsymbol{\varepsilon}_{1\mu} \\ \vdots & \vdots & \ddots & \vdots \\ \boldsymbol{\varepsilon}_{\mu 1} & \boldsymbol{\varepsilon}_{\mu 2} & \cdots & \boldsymbol{\varepsilon}_{\mu\mu} \end{pmatrix}. \tag{5.383}$$

One assumes that each subsystem reaches an asymptotically stable steady state when all concentrations in the other subsystems are clamped. For such steady states, one can define intrinsic control coefficients of subsystem i and can assemble them into matrices $\mathbf{C}_i^{S(\mathrm{intr})}$ and $\mathbf{C}_i^{J(\mathrm{intr})}$. Their elements reflect the control behavior of subsystem i when the state of the other systems is kept constant. The corresponding stoichiometry matrix is $\mathbf{N}_{ii}$.

The decompositions (3.48) and (5.381) imply Eq. (5.382) and

$$\mathbf{N}_{ii}\mathbf{K}_i = \mathbf{0}. \tag{5.384}$$

Therefore, the summation and connectivity theorems for the intrinsic coefficients can, in analogy to Eq. (5.54), be written as

$$\begin{pmatrix} \mathbf{C}_i^{J(\mathrm{intr})} \\ \mathbf{C}_i^{S(\mathrm{intr})} \end{pmatrix} (\mathbf{K}_i \quad \boldsymbol{\varepsilon}_{ii}\mathbf{L}_i) = \begin{pmatrix} \mathbf{K}_i & \mathbf{0} \\ \mathbf{0} & -\mathbf{L}_i \end{pmatrix}. \tag{5.385}$$

The $\mathbf{K}_i$ and $\mathbf{L}_i$ are the null-space matrices and link matrices, respectively, of the subsystems, which are identical to the diagonal blocks in Eqs. (3.48) and (5.381). This leads to

$$\begin{pmatrix} \mathbf{C}_i^{J(\text{intr})} \\ \mathbf{C}_i^{S(\text{intr})} \end{pmatrix} = \begin{pmatrix} \mathbf{K}_i & \mathbf{0} \\ \mathbf{0} & -\mathbf{L}_i \end{pmatrix} (\mathbf{K}_i \quad \varepsilon_{ii}\mathbf{L}_i)^{-1}. \tag{5.386}$$

From Eqs. (5.382) and (5.384), it follows that $\mathbf{N}_{ii}^0\mathbf{K}_i = \mathbf{0}$. Therefore, one can show by a similar reasoning as used in Section 5.3.3 that

$$(\mathbf{K}_i \quad \varepsilon_{ii}\mathbf{L}_i)^{-1} = \begin{pmatrix} (\mathbf{K}_i^{\mathrm{T}}\mathbf{K})^{-1}\mathbf{K}_i^{\mathrm{T}}(\mathbf{I} - \varepsilon_{ii}\mathbf{L}_i(\mathbf{N}_{ii}^0\varepsilon_{ii}\mathbf{L}_i)^{-1}\mathbf{N}_{ii}^0) \\ (\mathbf{N}_{ii}^0\varepsilon_{ii}\mathbf{L}_i)^{-1}\mathbf{N}_{ii}^0 \end{pmatrix}. \tag{5.387}$$

Substitution of Eq. (5.387) into Eq. (5.386) yields

$$\mathbf{C}_i^{S(\text{intr})} = -\mathbf{L}_i(\mathbf{N}_{ii}^0\varepsilon_{ii}\mathbf{L}_i)^{-1}\mathbf{N}_{ii}^0, \tag{5.388a}$$

$$\mathbf{C}_i^{J(\text{intr})} = \mathbf{I} + \varepsilon_{ii}\mathbf{C}_i^S. \tag{5.388b}$$

Now we consider the situation that no subsystems are clamped; that is, all concentrations in the network are allowed to attain a new steady state after parameter perturbation. Instead of the intrinsic control coefficients, we should now use the control matrices of the whole system, which can, according to the decomposition of the system, be partitioned as

$$\mathbf{C}^S = \begin{pmatrix} \mathbf{C}_{11}^S & \cdots & \mathbf{C}_{1p}^S \\ \vdots & \ddots & \vdots \\ \mathbf{C}_{p1}^S & \cdots & \mathbf{C}_{pp}^S \end{pmatrix}, \tag{5.389a}$$

$$\mathbf{C}^J = \begin{pmatrix} \mathbf{C}_{11}^J & \cdots & \mathbf{C}_{1p}^J \\ \vdots & \ddots & \vdots \\ \mathbf{C}_{p1}^J & \cdots & \mathbf{C}_{pp}^J \end{pmatrix}. \tag{5.389b}$$

Due to the decompositions (3.48) and (5.381), the summation and connectivity relationships (5.44) and (5.51) imply the *block summation theorems*

$$\mathbf{C}_{ij}^S\mathbf{K}_j = \mathbf{0}, \tag{5.390a}$$

$$\mathbf{C}_{ij}^J\mathbf{K}_j = \begin{cases} \mathbf{0} & \text{if } i \neq j \\ \mathbf{K}_j & \text{if } i = j \end{cases} \tag{5.390b}$$

and the *block connectivity theorems*

$$\sum_k \mathbf{C}_{ik}^S\varepsilon_{kj}\mathbf{L}_j = \begin{cases} \mathbf{0} & \text{if } i \neq j \\ -\mathbf{L}_j & \text{if } i = j, \end{cases} \tag{5.391a}$$

$$\sum_{\mathbf{k}} \mathbf{C}^J_{ik}\boldsymbol{\varepsilon}_{kj}\mathbf{L}_j = \mathbf{0}, \tag{5.391b}$$

with $i, j = 1, \ldots, p$.

It is of interest to inquire what information can be derived from the stoichiometry matrices of the subsystems, $\mathbf{N}_{ii}$, only (i.e., by neglecting the stoichiometric interactions between the subsystems). This question arises because it has been invoked that there be no mass flow between different blocks at steady state. Let us define a stoichiometry matrix, $\hat{\mathbf{N}}$, of a stoichiometrically disconnected system by

$$\hat{\mathbf{N}} = \begin{pmatrix} \mathbf{N}_{11} & \mathbf{0} & \cdots & \mathbf{0} \\ \mathbf{0} & \mathbf{N}_{22} & \cdots & \mathbf{0} \\ \vdots & \vdots & \ddots & \vdots \\ \mathbf{0} & \mathbf{0} & \cdots & \mathbf{N}_{pp} \end{pmatrix} \tag{5.392}$$

(i.e., $\hat{\mathbf{N}}_{ii} = \mathbf{N}_{ii}$ and $\hat{\mathbf{N}}_{ij} = \mathbf{0}$ for any $i \neq j$). Similar relations then hold among the reduced matrices $\hat{\mathbf{N}}^0$ and $\mathbf{N}^0$. Note that the elasticity matrix is to remain the same upon replacement of $\mathbf{N}$ by $\hat{\mathbf{N}}$. Applying Eqs. (5.25b) and (5.26b) gives the control coefficients of the stoichiometrically disconnected reaction system

$$\hat{\mathbf{C}}^S = -\mathbf{L}(\hat{\mathbf{N}}^0\boldsymbol{\varepsilon}\mathbf{L})^{-1}\hat{\mathbf{N}}^0, \tag{5.393a}$$

$$\hat{\mathbf{C}}^J = \mathbf{I} - \boldsymbol{\varepsilon}\mathbf{L}(\hat{\mathbf{N}}^0\boldsymbol{\varepsilon}\mathbf{L})^{-1}\hat{\mathbf{N}}^0. \tag{5.393b}$$

In these equations, the link matrix $\mathbf{L}$ of the original system can be used, because the new matrix $\hat{\mathbf{N}}$ has the same link matrix $\mathbf{L}$ due to the decomposability condition (5.381). Because the same null-space matrix $\mathbf{K}$ as for the original system can be chosen [due to condition (3.48)], Eqs. (5.393a) and (5.393b) lead, by postmultiplication by $\mathbf{K}$ and $\boldsymbol{\varepsilon}\mathbf{L}$ to the same block summation and connectivity theorems as belonging to the original system [Eqs. (5.390) and (5.391)]. As was shown in Section 5.3.3, the control coefficients are uniquely determined by the theorems. Consequently, we have

$$\hat{\mathbf{C}}^S = \mathbf{C}^S, \quad \hat{\mathbf{C}}^J = \mathbf{C}^J. \tag{5.394}$$

Thus, we have arrived at the interesting result that under the decomposability conditions (3.48) and (5.381), the control coefficients of the original system and the stoichiometrically disconnected system are the same.

Note that Eqs. (5.388a), (5.388b) and (5.393a), (5.393b) differ in that the former only contain a subset of the elasticities, namely those expressing the effects within the subsystems ($\boldsymbol{\varepsilon}_{ii}$), whereas the latter also contain the elasticities expressing the cross-effects between subsystems ($\boldsymbol{\varepsilon}_{ij}$ for $i \neq j$).

Kahn and Westerhoff (1991) treated the control analysis of subsystems by considering, from the very beginning, stoichiometrically disconnected systems, that is, stoichiometry matrices of the form (5.392). Because of Eq. (5.394), all results derived by Kahn and Westerhoff (1991) for systems satisfying condition (5.392) are thus valid also for systems fulfilling conditions (3.48) and (5.381). They are weaker than Eq. (5.392) because when **K** and **L** are block-diagonalizable, **N** need not be.

One of their results is the "Block Composition Theorem," consisting of the following relationships:

$$\mathbf{C}^S_{ij} = \mathbf{C}_i^{S(\text{intr})}\left(\mathbf{I}_{ij} + \sum_{k\neq i} \boldsymbol{\varepsilon}_{ik}\mathbf{C}^S_{kj}\right), \tag{5.395a}$$

$$\mathbf{C}^S_{ij} = \left(\mathbf{I}_{ij} + \sum_{k\neq j} \mathbf{C}^S_{ik}\boldsymbol{\varepsilon}_{kj}\right)\mathbf{C}_j^{S(\text{intr})}, \tag{5.395b}$$

$$\mathbf{C}^J_{ij} = \mathbf{C}_i^{J(\text{intr})}\left(\mathbf{I}_{ij} + \sum_{k\neq i} \boldsymbol{\varepsilon}_{ik}\mathbf{C}^S_{kj}\right), \tag{5.396a}$$

$$\mathbf{C}^J_{ij} = \mathbf{I}_{ij}\mathbf{C}^J_j + \sum_{k\neq j} \mathbf{C}^J_{ik}\boldsymbol{\varepsilon}_{kj}\mathbf{C}_j^{S(\text{intr})}. \tag{5.396b}$$

These equations relate the control coefficients of the whole system to the intrinsic control coefficients of the subsystems. In the case $i = j$, Eqs. (5.395) and (5.396) express the fact that the control exerted by the reactions of a subsystem on this subsystem itself is composed in an additive way of the intrinsic control within this subsystem and the indirect effects via all other subsystems. If $i \neq j$, Eqs. (5.395) and (5.396) state that the control by some subsystem on another is again the sum of all the effects via all subsystems.

Note the difference between Eqs. (5.395a) and (5.395b) and likewise between Eqs. (5.396a) and (5.396b). It appears that the multiplication of control coefficients and intrinsic control coefficients shows certain commutativity properties, which are probably linked with the property of control matrices to be projection matrices (cf. Section 5.3.4).

As mentioned earlier, the goal of the approach dealt with in this section is to determine the control properties of a metabolic system from the intrinsic control coefficients and the elasticities describing the regulatory interactions between subsystems. This has been achieved until now only for systems with certain architectures in terms of their subsystems, whereas Eqs. (5.395) and (5.396) hold for systems of any structure. For example, generalized cascades (convergent, divergent, or nested) can be treated. These cascades have the property that their subsystems can be numbered in such a way that the concentrations of subsystem j only affect reactions in subsystems i with $i \geq j$ [i.e., $\boldsymbol{\varepsilon}_{kj} = \mathbf{0}$ for any $k < j$ (no feedback)]. For systems of this property, the control matrices can be computed as

$$\mathbf{C}_{ii}^{Y} = \mathbf{C}_{i}^{Y(\text{intr})}, \tag{5.397}$$

$$\mathbf{C}_{ij}^{Y} = \mathbf{0} \quad \text{for } i < j, \tag{5.398}$$

$$\mathbf{C}_{ij}^{Y} = \mathbf{C}_{i}^{Y(\text{intr})}\left(\sum_{L(i,j)} \boldsymbol{\varepsilon}_{ik}\, \mathbf{C}_{k}^{S(\text{intr})}\boldsymbol{\varepsilon}_{kl} \cdots \mathbf{C}_{m}^{S(\text{intr})}\boldsymbol{\varepsilon}_{mj}\right)\mathbf{C}_{j}^{S(\text{intr})} \quad \text{for } i > j, \tag{5.399}$$

where the sum runs over all regulatory loops, $L(i,j)$, connecting module j to module i, and Y stands for either S or J. The proof of this *Generalized Cascade Control Theorem* was given by Kahn and Westerhoff (1991).

Equation (5.397) means that in the absence of feedback, the internal control behavior of each subsystem is unaffected by external regulatory interactions. Equation (5.398) expresses the fact that no reaction is able to control any concentration or flux in a subsystem upstream in the hierarchy. Control in the downstream direction proceeds via all routes of regulatory effects [Eq. (5.399)].

Moreover, Kahn and Westerhoff (1991) derived formulas analogous to Eqs. (5.397)–(5.399) for linear cascades in which one subsystem may regulate a subsystem higher in the hierarchy by feedback.

5.12. MODULAR APPROACH

5.12.1. Overall Elasticities

One usually discerns functional units in cell metabolism, such as amino acid synthesis, protein synthesis, and protein degradation, or cytosol and mitochondrion. Accordingly, it is desirable to carry out metabolic control analysis in terms of control features of these functional subunits (i.e., at a higher level of organization), rather than to discuss control only in terms of kinetic properties of the individual enzymes. For example, one could try to explain the control of the intracellular glucose concentration as being the result of the elasticities of glucose uptake and glycolysis (and possibly gluconeogenesis) versus glucose, instead of discussing such control in terms of all the contributions of all enzymes involved. Moreover, it should be acknowledged that for large biochemical networks, the structural and kinetic data characterizing the interior of the functional units represent a huge amount of information, which often is not readily measurable or even if so, is difficult to handle.

In this section, an approach is outlined in which metabolic systems are decomposed into subsystems, some of which are incompletely observable in the sense that the stoichiometric structure within these subsystems is not fully known and/or not all of the elasticities are measurable. This approach differs from the decomposition method set out in Section 5.11, where all subsystems are assumed to be completely observable.

A situation where a decomposition into functional subunits is sensible is mitochondrial oxidative phosphorylation. For this system, a solution was devised which groups all the reactions involved into three parts: those connected with respiration and generating proton-motive force ($\Delta\tilde{\mu}_{H^+}$), those connected with synthesis of extramitochondrial ATP using the force $\Delta\tilde{\mu}_{H^+}$, and the proton leak (Westerhoff et al., 1983). The control of mitochondrial respiration was described as divided over these three units. In this way, control of oxidative phosphorylation could be understood in terms of regulatory interactions between three *modules,* the internal regulations of which were not completely known. Note that also in Section 5.4.5, we used a similar approach by grouping all the enzymes of the respiratory chain into one unit.

Control coefficients of enzyme sequences had been defined already by Heinrich and Rapoport (1973, 1974a) and later by Fell and Snell (1988); Kacser (1983) elaborated the idea for branched metabolic pathways. More recently, the approach has been extended and renamed the *top-down approach* (Brown *et al.,* 1990; Hafner *et al.,* 1990; Quant, 1993). All of these approaches are limited to cases in which any two subunits into which one divides metabolism are linked by only one flux. This drawback was eliminated in a recent further development of the modular approach (S. Schuster *et al.,* 1993a), which is outlined in the following.

In view of the above-mentioned fact that the elasticities of many enzymes are not available, the first step in the modular approach is the decomposition of the metabolic network under study into modules of two types. Modules of type I (*black boxes*) are subsystems for which we are only able to observe the reactions that link those subsystems with their surroundings but not internal reactions and metabolites. Type-II modules are subsystems subject to explicit observation. The modular partitioning may or may not correspond to a spatial decomposition into compartments. The question of under what conditions we are able to determine the control properties of modules I and II arises.

To begin with, we consider a decomposition of a network into one module of type I and one module of type II (Figure 5.13).

The reactions can be classified into three types: reactions proceeding inside of module I, reactions bridging the two modules, and reactions internal to subsystem II. Reactions connecting module I with the surroundings of the whole system can formally be included in the set of bridging reactions (cf. Figure 5.13). According to this decomposition, the stoichiometry matrix can be partitioned as

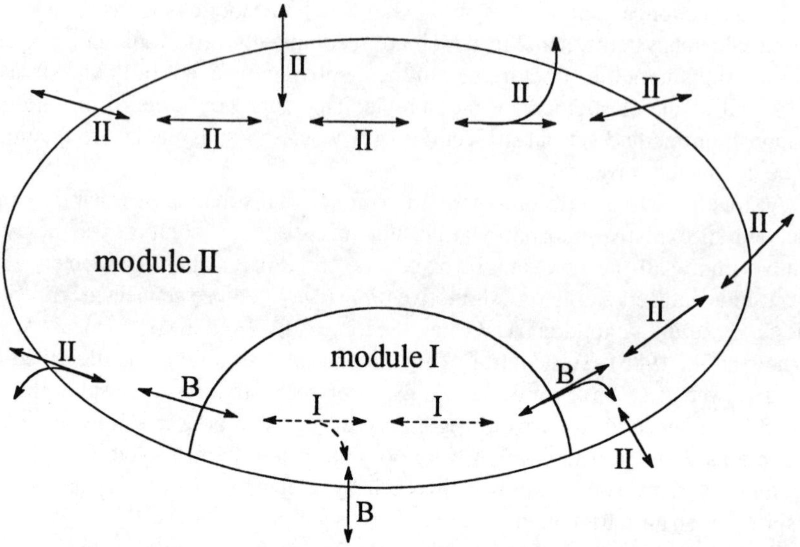

Figure 5.13 Schematic representation of a modular decomposition of a metabolic network. Reactions within module I (dotted arrows) are not necessarily observable, whereas reactions in module II and the reactions bridging the two modules (index B) are.

$$\mathbf{N} = \begin{pmatrix} \mathbf{N}_{\mathrm{I,I}} & \mathbf{N}_{\mathrm{I,B}} & \mathbf{0} \\ \mathbf{0} & \mathbf{N}_{\mathrm{II,B}} & \mathbf{N}_{\mathrm{II,II}} \end{pmatrix}, \tag{5.400}$$

and, accordingly, the concentration vector $\boldsymbol{S} = (\boldsymbol{S}_{\mathrm{I}},\boldsymbol{S}_{\mathrm{II}})^{\mathrm{T}}$ and rate vector $\boldsymbol{v} = (\boldsymbol{v}_{\mathrm{I}},\boldsymbol{v}_{\mathrm{B}},\boldsymbol{v}_{\mathrm{II}})^{\mathrm{T}}$. The matrix of non-normalized elasticities can be decomposed as

$$\boldsymbol{\varepsilon} = \begin{pmatrix} \boldsymbol{\varepsilon}_{\mathrm{I,I}} & \boldsymbol{\varepsilon}_{\mathrm{I,II}} \\ \boldsymbol{\varepsilon}_{\mathrm{B,I}} & \boldsymbol{\varepsilon}_{\mathrm{B,II}} \\ \boldsymbol{\varepsilon}_{\mathrm{II,I}} & \boldsymbol{\varepsilon}_{\mathrm{II,II}} \end{pmatrix}. \tag{5.401}$$

The following calculations will show that one can determine certain control properties without knowledge of the internal details of module I, that is, the elasticity submatrices $\boldsymbol{\varepsilon}_{\mathrm{I,I}}$, $\boldsymbol{\varepsilon}_{\mathrm{I,II}}$, $\boldsymbol{\varepsilon}_{\mathrm{B,I}}$, and $\boldsymbol{\varepsilon}_{\mathrm{II,I}}$ and the stoichiometry matrices $\mathbf{N}_{\mathrm{I,I}}$ and $\mathbf{N}_{\mathrm{I,B}}$. These quantities will not enter the final results concerning the overall control coefficients. In contrast, the bridging reactions are assumed to be observable, in the sense that their response to changes in $\boldsymbol{S}_{\mathrm{II}}$ can be measured. This flux response is meant to imply that the black-box module can attain a new steady state while the concentrations in the observable module are clamped. The respective elasticities are to be called overall elasticities and to be gathered in a matrix

$$^{*}\varepsilon_{B,II} = \left(\frac{\partial^{*}\boldsymbol{v}_B}{\partial \boldsymbol{S}_{II}}\right). \tag{5.402}$$

The asterisk superscript refers to the situation in which the black-box module is allowed to attain a steady state on its own. It is important to distinguish the definition of overall elasticities from that of intrinsic control coefficients (see Section 5.11). Indeed, in both cases the surroundings of the subsystem are clamped, but in the former case, an activity of an internal reaction is changed, whereas in the latter case, a concentration outside of the subsystem is altered.

Consider, by way of example, the mitochondria and cytosol of a cell as modules I and II, respectively. The response coefficients $^{*}\varepsilon_{B,II}$ can then be measured experimentally by resuspending the mitochondria in a sufficiently large incubation medium, where the substances of interest thus have concentrations independent of the reactions within the black-box module. By changing experimentally these concentrations and measuring concomitantly the fluxes linking the mitochondria with their surroundings (e.g. the rate of oxygen consumption) gives the above-mentioned overall elasticities.

Another relevant situation is when the black-box module is a fast subsystem, that is, if it gains steady state much faster than the entire system. In this situation, $^{*}\varepsilon_{B,II}$ expresses the response of the black-box module toward changes in $\boldsymbol{S}_{II}$, with the response measured in a time scale long enough to allow the black-box module to reach a new stationary state but short enough so that $\boldsymbol{S}_{II}$ has not yet relaxed to the original values.

Because at the steady state of the black-box module, input and output fluxes of this subsystem must balance each other, the bridging fluxes are usually linearly interdependent. This dependence can be expressed by a matrix $\mathbf{Q}$, such that

$$^{*}\boldsymbol{v}_B = \mathbf{Q}\,^{*}\boldsymbol{v}_R, \tag{5.403}$$

where $^{*}\boldsymbol{v}_R$ is a vector of linearly independent bridging fluxes. The matrix $\mathbf{Q}$ can, in principle, be obtained by a null-space analysis of the submatrix $(\mathbf{N}_{I,I}\ \mathbf{N}_{I,B}\ \mathbf{0})$ corresponding to module I. As this information may not, however, be available, it is assumed that $\mathbf{Q}$ can be determined by observation from the outside (e.g., by determining the balance of atom groups entering and leaving module I).

Consider, for example, a branched reaction scheme as shown in Figure 5.14A. Let S_1 be identified as the black-box module, and the three adjacent reactions as bridging reactions. Because at the steady state of the black-box module, the flux J_1 is the sum of J_2 and J_3, only two of these fluxes are linearly independent. So it makes sense to redraw the scheme as depicted in Figure 5.14B.

We may take $^{*}v_2$ and $^{*}v_3$ as independent fluxes, so that Eq. (5.403) reads

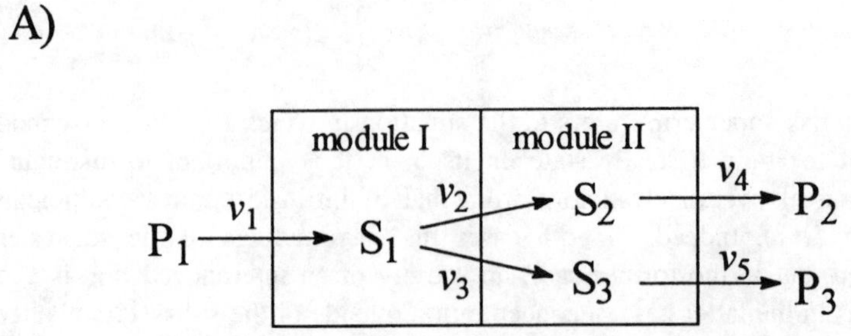

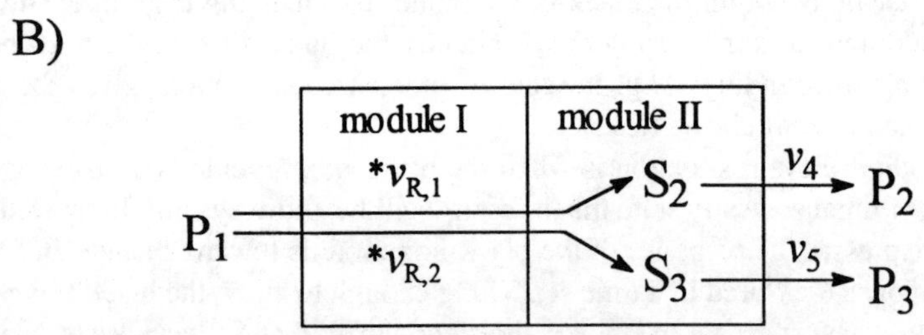

Figure 5.14 Lumping of bridging fluxes in a simple branched reaction chain. (A) Complete scheme; (B) simplified scheme. The arrows crossing module I in scheme B correspond to the null-space vectors $(1\ 1\ 0)^T$ and $(1\ 0\ 1)^T$ [cf. Eqs. (5.119) and (5.404)] and signify degrees of freedom in flux of module I.

$$*\boldsymbol{v}_B = \begin{pmatrix} *v_1 \\ *v_2 \\ *v_3 \end{pmatrix} = \mathbf{Q}\, *\boldsymbol{v}_R = \begin{pmatrix} 1 & 1 \\ 1 & 0 \\ 0 & 1 \end{pmatrix} \begin{pmatrix} *v_2 \\ *v_3 \end{pmatrix} . \tag{5.404}$$

The lumped bridging fluxes as well as the fluxes in the observable module are functions of $\boldsymbol{p}$, $\boldsymbol{S}_I$ and $\boldsymbol{S}_{II}$. The assumption that the black-box module subsists in a steady state on its own implies that $\boldsymbol{S}_I$ is a function of $\boldsymbol{S}_{II}$, so that the observable fluxes can be written as

$$*\boldsymbol{v}_R = *\boldsymbol{v}_R(\boldsymbol{p}, \boldsymbol{S}_{II}) = \boldsymbol{v}_R(\boldsymbol{p}, \boldsymbol{S}_I(\boldsymbol{S}_{II}, \boldsymbol{p}), \boldsymbol{S}_{II}) , \tag{5.405}$$

$$*\boldsymbol{v}_{II} = *\boldsymbol{v}_{II}(\boldsymbol{p}, \boldsymbol{S}_{II}) = \boldsymbol{v}_{II}(\boldsymbol{p}, \boldsymbol{S}_I(\boldsymbol{S}_{II}, \boldsymbol{p}), \boldsymbol{S}_{II}) . \tag{5.406}$$

The functions $*\boldsymbol{v}_R = *\boldsymbol{v}_R(\boldsymbol{S}_{II},\boldsymbol{p})$ represent *overall rate laws* of module I. An example is provided by the overall rate law of the hexokinase-phosphofructokinase system derived by Heinrich *et al.* (1977). The reaction rates $^*\boldsymbol{v}_R$ can be regarded from two different viewpoints. They play the role of "isolated" reaction rates, in the sense that the black-box module may be looked upon as a *superreaction* embedded in a larger metabolic system. If, on the other hand, the black-box module is regarded as a system on its own (with $\boldsymbol{S}_{II}$ clamped), $^*\boldsymbol{v}_R$ plays the role of a steady-state flux vector.

Differentiation of Eq. (5.406) with respect to $\boldsymbol{S}_{II}$ yields

$$^*\boldsymbol{\varepsilon}_{II,II} = \left(\frac{\partial^* \boldsymbol{v}_{II}}{\partial \boldsymbol{S}_{II}}\right) = \boldsymbol{\varepsilon}_{II,I} \frac{\partial \boldsymbol{S}_I}{\partial \boldsymbol{S}_{II}} + \boldsymbol{\varepsilon}_{II,II}. \tag{5.407}$$

As substances in the black-box module are not converted by reactions in the observable module ($\mathbf{N}_{I,II} = \mathbf{0}$), the elasticities $\boldsymbol{\varepsilon}_{II,I}$ are made up of effector influences of $\boldsymbol{S}_I$ on $\boldsymbol{v}_{II}$.

Because of Eq. (5.403), the matrix of overall elasticities, $^*\boldsymbol{\varepsilon}_{R,II}$, pertaining to the reduced bridging fluxes, is linked with $^*\boldsymbol{\varepsilon}_{B,II}$ by

$$^*\boldsymbol{\varepsilon}_{B,II} = \mathbf{Q}\, ^*\boldsymbol{\varepsilon}_{R,II}. \tag{5.408}$$

As the matrix $\mathbf{Q}$ and the elasticities $^*\boldsymbol{\varepsilon}_{B,II}$ are assumed to be known, the same holds for $^*\boldsymbol{\varepsilon}_{R,II}$.

5.12.2. Overall Control Coefficients

Let $\boldsymbol{p}_{II}$ be a vector of parameters for which the matrix $\partial \boldsymbol{v}_{II}/\partial \boldsymbol{p}_{II}$ is a nonsingular square matrix. The matrices of non-normalized control coefficients expressing the control exerted by the processes in the observable module can be defined as

$$\mathbf{C}^S_{II,II} = \left(\frac{\partial \boldsymbol{S}_{II}}{\partial \boldsymbol{p}_{II}}\right)\left(\frac{\partial \boldsymbol{v}_{II}}{\partial \boldsymbol{p}_{II}}\right)^{-1}, \tag{5.409a}$$

$$\mathbf{C}^J_{R,II} = \left(\frac{\partial \boldsymbol{J}_R}{\partial \boldsymbol{p}_{II}}\right)\left(\frac{\partial \boldsymbol{v}_{II}}{\partial \boldsymbol{p}_{II}}\right)^{-1}, \tag{5.409b}$$

$$\boldsymbol{C}^J_{II,II} = \left(\frac{\partial \boldsymbol{J}_{II}}{\partial \boldsymbol{p}_{II}}\right)\left(\frac{\partial \boldsymbol{v}_{II}}{\partial \boldsymbol{p}_{II}}\right)^{-1}. \tag{5.409c}$$

In a straightforward analogy, coefficients expressing the control exerted by the lumped bridging reactions can be defined as,

$$^*\mathbf{C}^S_{II,R} = \left(\frac{\partial \boldsymbol{S}_{II}}{\partial \boldsymbol{p}_R}\right)\left(\frac{\partial\, ^*\boldsymbol{v}_R}{\partial \boldsymbol{p}_R}\right)^{-1}, \tag{5.410a}$$

$$^{*}\mathbf{C}^{J}_{\mathrm{R,R}} = \left(\frac{\partial \boldsymbol{J}_{\mathrm{R}}}{\partial \boldsymbol{p}_{\mathrm{R}}}\right)\left(\frac{\partial\, ^{*}\boldsymbol{v}_{\mathrm{R}}}{\partial \boldsymbol{p}_{\mathrm{R}}}\right)^{-1}, \tag{5.410b}$$

$$^{*}\mathbf{C}^{J}_{\mathrm{II,R}} = \left(\frac{\partial \boldsymbol{J}_{\mathrm{II}}}{\partial \boldsymbol{p}_{\mathrm{R}}}\right)\left(\frac{\partial\, ^{*}\boldsymbol{v}_{\mathrm{R}}}{\partial \boldsymbol{p}_{\mathrm{R}}}\right)^{-1}. \tag{5.410c}$$

The asterisk superscript in Eqs. (5.410a)–(5.410c) refers to the fact that we deal with *overall* control coefficients. Here, the components of $^{*}\boldsymbol{v}_{\mathrm{R}}$ play the role of "isolated" reaction rates corresponding to the different degrees of freedom of module I. $\boldsymbol{p}_{\mathrm{R}}$ denotes a vector of parameters affecting the reduced bridging fluxes only. Importantly, when there are effector influences from within the black-box module acting on the observable module, changes in the parameters pertaining to module I generally affect, via $\boldsymbol{S}_{\mathrm{I}}$, the rates $^{*}\boldsymbol{v}_{\mathrm{II}}$, even if $\boldsymbol{S}_{\mathrm{II}}$ is clamped. Thus, it may occur that no parameter vector $\boldsymbol{p}_{\mathrm{II}}$ can be found that affects $^{*}\boldsymbol{v}_{\mathrm{R}}$ but not $^{*}\boldsymbol{v}_{\mathrm{II}}$. This would make it difficult to employ definition (5.410). One way of coping with this problem is by choosing the parameter vector $\boldsymbol{p}_{\mathrm{II}}$ in such a way that any changes in this vector do not alter $\boldsymbol{S}_{\mathrm{I}}$. In this way, however, the favorable property of control coefficients to be independent of the special choice of the perturbation parameter could not be guaranteed (see Sections 5.2 and 5.6). Therefore, we prefer to impose the condition that the black-box module does not exert effector influences on module II,

$$\boldsymbol{\varepsilon}_{\mathrm{II,I}} = \mathbf{0}. \tag{5.411}$$

If, in a given scheme, such influences occur, one can often draw a more explicit reaction scheme in which the effector influences are represented as bridging reactions, so that condition (5.411) is fulfilled.

It is worth noting that at variance with the widely used definitions of control coefficients (5.3) and (5.5), the matrices in the denominators in Eq. (5.410) are not normally diagonal, because for a given parametrization it is not, in general, possible to find parameters influencing the components of $^{*}\boldsymbol{v}_{\mathrm{R}}$ specifically. This fact does not, however, restrict the applicability of these definitions, as long as the matrices in question are nonsingular (see Reder, 1988).

The overall control properties of a metabolic system can be calculated in terms of the overall elasticity properties of its modules. Due to Eq. (5.403), the steady-state condition for the observable module reads

$$\mathbf{N}_{\mathrm{II,B}}\mathbf{Q}\, ^{*}\boldsymbol{v}_{\mathrm{R}} + \mathbf{N}_{\mathrm{II,II}}\, ^{*}\boldsymbol{v}_{\mathrm{II}} = \boldsymbol{0}. \tag{5.412}$$

Differentiation of this equation with respect to any parameter vector $\boldsymbol{p}$ yields, due to Eqs. (5.407) and (5.411),

$$\mathbf{N}_{\mathrm{II,B}}\mathbf{Q}\left(\frac{\partial\,^*\boldsymbol{v}_{\mathrm{R}}}{\partial \boldsymbol{p}} + {}^*\boldsymbol{\varepsilon}_{\mathrm{R,II}}\frac{\partial \boldsymbol{S}_{\mathrm{II}}}{\partial \boldsymbol{p}}\right) + \mathbf{N}_{\mathrm{II,II}}\left(\frac{\partial\,^*\boldsymbol{v}_{\mathrm{II}}}{\partial \boldsymbol{p}} + {}^*\boldsymbol{\varepsilon}_{\mathrm{II,II}}\frac{\partial \boldsymbol{S}_{\mathrm{II}}}{\partial \boldsymbol{p}}\right) = \mathbf{0}. \tag{5.413}$$

The asterisk for the term $^*\boldsymbol{\varepsilon}_{\mathrm{II,II}}$ can be dropped because of Eq. (5.407) and condition (5.411).

In order to solve Eq. (5.413) for $\partial \boldsymbol{S}_{\mathrm{II}}/\partial \boldsymbol{p}$, one has to reduce the matrix $\mathbf{N}_{\mathrm{II,B}}\mathbf{Q}^*\boldsymbol{\varepsilon}_{\mathrm{R,II}} + \mathbf{N}_{\mathrm{II,II}}{}^*\boldsymbol{\varepsilon}_{\mathrm{II,II}} = \mathbf{0}$ to a nonsingular square matrix. This is achieved by considering the conservation relations imposed on the concentrations $\boldsymbol{S}_{\mathrm{II}}$. In analogy to Eq. (3.7), we reduce the stoichiometry matrix of module II [cf. Eq. (5.400)] to its linearly independent rows,

$$(\mathbf{0}\ \ \mathbf{N}_{\mathrm{II,B}}\ \ \mathbf{N}_{\mathrm{II,II}}) = \begin{pmatrix}\mathbf{I}\\ \mathbf{L}'_{\mathrm{II,II}}\end{pmatrix}(\mathbf{0}\ \ \mathbf{N}^0_{\mathrm{II,B}}\ \ \mathbf{N}^0_{\mathrm{II,II}})\,. \tag{5.414}$$

In what follows, we show that there must be no conservation relations linking concentrations inside the black-box module with concentrations inside the observable module in order that "parameter-independent" overall control coefficients can be defined.

Let ζ_{I} and ζ_{II} denote respectively the ranks of the submatrices $(\mathbf{N}_{\mathrm{I,I}}\ \mathbf{N}_{\mathrm{I,B}}\ \mathbf{0})$ and $(\mathbf{0}\ \mathbf{N}_{\mathrm{II,B}}\ \mathbf{N}_{\mathrm{II,II}})$ [cf. Eq. (5.400)]. After a straightforward decomposition of the concentration vectors into independent variables, the conservation relations of the black-box module I and module II taken separately can be written as

$$(-\mathbf{L}'_{\mathrm{I,I}}\ \ \mathbf{I})\begin{pmatrix}\boldsymbol{S}_{\mathrm{I,a}}\\ \boldsymbol{S}_{\mathrm{I,b}}\end{pmatrix} = \text{const.}, \tag{5.415a}$$

$$(-\mathbf{L}'_{\mathrm{II,II}}\ \ \mathbf{I})\begin{pmatrix}\boldsymbol{S}_{\mathrm{II,a}}\\ \boldsymbol{S}_{\mathrm{II,b}}\end{pmatrix} = \text{const.} \tag{5.415b}$$

The steady-state equation for the black-box module reads

$$(\mathbf{N}^0_{\mathrm{I,I}}\ \ \mathbf{N}^0_{\mathrm{I,B}})\begin{pmatrix}\boldsymbol{v}_{\mathrm{I}}(\boldsymbol{S}_{\mathrm{I}}, \boldsymbol{S}_{\mathrm{II}}, \boldsymbol{p})\\ \boldsymbol{v}_{\mathrm{B}}(\boldsymbol{S}_{\mathrm{I}}, \boldsymbol{S}_{\mathrm{II}}, \boldsymbol{p})\end{pmatrix} = \boldsymbol{0}. \tag{5.416}$$

The assumption that the black-box module can attain a stable steady state on its own implies not only that Eq. (5.416) has a solution for $\boldsymbol{S}_{\mathrm{I}}$, but also that the real parts of all eigenvalues of the Jacobian matrix $(\mathbf{N}^0_{\mathrm{I,I}}\ \mathbf{N}^0_{\mathrm{I,B}}\ \mathbf{0})(\partial \boldsymbol{v}/\partial \boldsymbol{S}_{\mathrm{I}})\mathbf{L}_{\mathrm{I,I}}$ be negative. So this matrix must have full rank, ζ_{I}. This ensures that Eq. (5.416) (which encompasses ζ_{I} independent equations) and Eq. (5.415a) are, in general, sufficient to express the steady-state concentrations $\boldsymbol{S}_{\mathrm{I}}$ and, hence, the fluxes $^*\boldsymbol{v}_{\mathrm{R}}$ as functions of $\boldsymbol{S}_{\mathrm{II}}$ and $\boldsymbol{p}$.

The number of independent equations contained in Eq. (5.412) equals ζ_{II}.

Together with the $\dim(\boldsymbol{S}_{\mathrm{II}}) - \zeta_{\mathrm{II}}$ independent conservation relations contained in Eq. (5.415b), Eq. (5.412) determines the stationary concentrations $\boldsymbol{S}_{\mathrm{II}}$.

Connecting the two modules does not change the conservation relations within the modules but may add conservation relations involving both modules. In algebraic terms, the linear dependencies between the rows of submatrices remain valid if the submatrices are combined. Linear dependencies between the rows of the whole matrix may arise in addition. Hence,

$$\zeta_{\mathrm{I}} + \zeta_{\mathrm{II}} \geq \mathrm{rank}(\mathbf{N}). \tag{5.417}$$

In the case that relation (5.417) is fulfilled as a strict inequality, the equation system (5.412), (5.415), and (5.416) is overdetermined and has, in general, no solution for $\boldsymbol{S}_{\mathrm{I}}$ and $\boldsymbol{S}_{\mathrm{II}}$. From this reasoning, we conclude that we should impose the condition

$$\mathbf{L} = \begin{pmatrix} \mathbf{L}_{\mathrm{I,I}} & \mathbf{0} \\ \mathbf{0} & \mathbf{L}_{\mathrm{II,II}} \end{pmatrix}; \tag{5.418}$$

that is, there should be no conservation relations linking the two modules. Equations (5.413), (5.414), and (5.418) can be combined to obtain

$$\mathbf{N}^0_{\mathrm{II,B}}\mathbf{Q}\frac{\partial\, {}^*\boldsymbol{v}_{\mathrm{R}}}{\partial \boldsymbol{p}} + \mathbf{N}^0_{\mathrm{II,II}}\frac{\partial\, {}^*\boldsymbol{v}_{\mathrm{II}}}{\partial \boldsymbol{p}} + {}^*\mathbf{M}\frac{\partial \boldsymbol{S}_{\mathrm{II,a}}}{\partial \boldsymbol{p}} = \mathbf{0}, \tag{5.419}$$

with

$${}^*\mathbf{M} = (\mathbf{N}^0_{\mathrm{II,B}}\mathbf{Q}\,{}^*\boldsymbol{\varepsilon}_{\mathrm{R,II}} + \mathbf{N}^0_{\mathrm{II,II}}\boldsymbol{\varepsilon}_{\mathrm{II,II}})\mathbf{L}_{\mathrm{II,II}}. \tag{5.420}$$

The matrix $^*\mathbf{M}$ is the Jacobian matrix of the observable module taking into account that the black-box module attains a new steady state after a change of a concentration in the former module.

Specifying $\boldsymbol{p}$, consecutively, to be a vector of parameters only affecting the bridging reactions and a vector of parameters only acting on reactions in the observable module, we can derive the concentration control coefficients defined in eqs (5.409a) and (5.410a) from Eq. (5.419),

$$\mathbf{C}^S_{\mathrm{II,II}} = -\mathbf{L}_{\mathrm{II,II}}({}^*\mathbf{M})^{-1}\mathbf{N}^0_{\mathrm{II,II}}, \tag{5.421a}$$

$${}^*\mathbf{C}^S_{\mathrm{II,R}} = -\mathbf{L}_{\mathrm{II,II}}({}^*\mathbf{M})^{-1}\mathbf{N}^0_{\mathrm{II,B}}\mathbf{Q}. \tag{5.421b}$$

To obtain the flux control coefficients, we differentiate the equations $\boldsymbol{J}_{\mathrm{R}}(\boldsymbol{p}) = {}^*\boldsymbol{v}_{\mathrm{R}}(\boldsymbol{p},\boldsymbol{S}_{\mathrm{II}})$ and $\boldsymbol{J}_{\mathrm{II}}(\boldsymbol{p}) = {}^*\boldsymbol{v}_{\mathrm{II}}(\boldsymbol{p},\boldsymbol{S}_{\mathrm{II}})$ with respect to $\boldsymbol{p}$. By Eq. (5.419), one obtains

$$\mathbf{C}^J_{\mathrm{R,II}} = {}^*\boldsymbol{\varepsilon}_{\mathrm{R,II}}\mathbf{C}^S_{\mathrm{II,II}} \tag{5.422a}$$

$$ {}^*\mathbf{C}^J_{\mathrm{R,R}} = \mathbf{I} + {}^*\boldsymbol{\varepsilon}_{\mathrm{R,II}}\mathbf{C}^S_{\mathrm{II,R}}, \tag{5.422b}$$

$$\mathbf{C}^J_{\mathrm{II,II}} = \mathbf{I} + \boldsymbol{\varepsilon}_{\mathrm{II,II}}\mathbf{C}^S_{\mathrm{II,II}}, \tag{5.423a}$$

$$ {}^*\mathbf{C}^J_{\mathrm{II,R}} = \boldsymbol{\varepsilon}_{\mathrm{II,II}}\,{}^*\mathbf{C}^S_{\mathrm{II,R}}. \tag{5.423b}$$

The normalized overall control coefficients are obtained in a straightforward way,

$$(\mathrm{dg}\,\boldsymbol{J}_{\mathrm{R}})^{-1}\,{}^*\mathbf{C}^J_{\mathrm{R,II}}(\mathrm{dg}\,\boldsymbol{J}_{\mathrm{II}}) \rightarrow {}^*\mathbf{C}^J_{\mathrm{R,II}}, \tag{5.424}$$

and similarly for the other overall coefficients.

Now we have derived expressions for control coefficients in terms of quantities assumed to be known. These results show that one is able to determine a considerable number of control properties even without knowing the internal details of the black-box module. These control properties include the control coefficients related to the observable module and the overall coefficients expressing the control exerted by the degrees of freedom of the black-box module on the fluxes of the bridging reactions and on the variables of the observable module. The information of the inside of the black-box module that is relevant for determining these control properties is fully represented by the overall elasticity coefficients $^*\boldsymbol{\varepsilon}_{\mathrm{R,II}}$ and matrix $\mathbf{Q}$, which expresses the linear dependencies between the bridging fluxes.

It is worth noting that, because formulas (5.421b), (5.422b), and (5.423b) do not contain the parameters $\boldsymbol{p}_{\mathrm{R}}$ used in the definitions of overall control coefficients, these coefficients have the favorable property of not depending on the choice of the perturbation parameter. This is, however, only true under conditions (5.411) and (5.418).

Overall control coefficients fulfill summation and connectivity theorems similar to the theorems presented in Section 5.3. For example, Eqs. (5.421a) and (5.421b) imply the summation theorem

$$ {}^*\mathbf{C}^S_{\mathrm{II,R}}\boldsymbol{J}_{\mathrm{R}} + \mathbf{C}^S_{\mathrm{II,II}}\,\boldsymbol{J}_{\mathrm{II}} = \boldsymbol{0}. \tag{5.425}$$

A connectivity relationship reads

$$ {}^*\mathbf{C}^S_{\mathrm{II,R}}\,{}^*\boldsymbol{\varepsilon}_{\mathrm{R,II}}\mathbf{L}_{\mathrm{II,II}} + \mathbf{C}^S_{\mathrm{II,II}}\,{}^*\boldsymbol{\varepsilon}_{\mathrm{II,II}}\mathbf{L}_{\mathrm{II,II}} = -\mathbf{L}_{\mathrm{II,II}}. \tag{5.426}$$

A comparison of these theorems with those of traditional metabolic control analysis [Eqs. (5.42), (5.43) and (5.51)] shows that there exist relations among overall control coefficients and the usual control coefficients pertaining to module I. When module I is, for example, an unbranched pathway, its overall control coefficient with respect to any concentration or flux is simply the sum of the respective particular control coefficients of the reactions inside module I (cf. Brown *et al.*, 1990).

A frequently occurring situation is when the fluxes through all lumped bridging reactions are changed by the same fractional amount, a (e.g., by changing the number of mitochondria in a suspension):

$$\delta \ln {}^{*}v_{\mathrm{R},k} = a \quad \text{for any } k. \tag{5.427}$$

In that situation, the control exerted by the black-box module as a whole can, due to Eq. (5.424), be expressed as the sum of all normalized overall control coefficients belonging to this module:

$$\delta \ln J_{\mathrm{R},i} = a \sum_{k} {}^{*}C^{J}_{ik}, \tag{5.428a}$$

$$\delta \ln J_{\mathrm{II},j} = a \sum_{k} {}^{*}C^{J}_{jk}, \tag{5.428b}$$

where the sum runs over all the degrees of freedom of module I.

In the above treatment, we assumed that the system under study only involves one black-box module. The approach can readily be generalized to cases with several black-box modules. Then the problem arises that the modular approach requires that on determining the overall elasticities of some black-box module, the metabolites in all other black-box units have to attain steady state as well. This might be difficult to achieve in experiment. This problem can be circumvented by confining the black-box modules so that there are no effector interactions between them. A more detailed analysis is given in S. Schuster *et al.* (1993a).

Examples of overall control coefficients have, in fact, been given in some of the previous sections, in particular the control coefficients pertaining to the HK-PFK system (Section 5.4.4) and to the subsystem consisting of 3-phosphoglycerate dehydrogenase and phosphoserine transaminase (Section 5.4.6).

Enzymic reactions as composed of elementary steps can often be treated as steady-state modules in the sense defined above, as was done in Section 5.6. Therefore, the modular approach may be considered as a generalization of the control analysis set out in Sections 5.1–5.3. Conventional control coefficients are then identical to the overall control coefficients pertaining to the catalytic cycle of an enzyme.

Application of the modular approach to single enzymes is particularly useful when applying it to slipping enzymes (i.e., enzymes that catalyse distinct processes which are incompletely coupled). Examples are provided by the various H^{+}-ATPases and Na/K-ATPases and the enzymes involved in the respiratory chain. When treating a slipping enzyme as a black box, the coupled portion (such as ATP synthesis) and uncoupled portion (slip) can be taken as reduced bridging fluxes, and appropriate overall control coefficients can be calculated.

5.13. FLUX CONTROL INSUSCEPTIBILITY

As pointed out in the sections on stoichiometric analysis (Chapter 3), an important step in metabolic modeling is to determine invariant properties by only using those parameters that are relatively constant and are known to a satisfactory accuracy. This information may concern stoichiometry, separation of time constants, thermodynamic properties, and patterns of nonstoichiometric effector interactions, and is, in most cases, easier to obtain than the exact values of elasticities. As for control analysis, this implies that the values of control coefficients sometimes cannot be calculated. One may, however, attempt to make qualitative statements about the control structure on the basis of incomplete knowledge about the elasticities. For example, one may analyze which fluxes are insusceptible to control by which reactions, that is, which flux control coefficients are always zero, irrespective of the special values of kinetic parameters.

Knowledge of the effector interactions together with information about what substrates and products of reactions enter the kinetic equations can be compiled in a *qualitative elasticity matrix,* the elements of which are defined by

$$\varepsilon_{ji}^{\text{qual}} = \begin{cases} 0 & \text{if } \partial v_j/\partial S_i = 0 \text{ for any admissible steady-state vector } \mathbf{S} \\ \varepsilon & \text{otherwise.} \end{cases} \tag{5.429}$$

ε is used as a mathematical symbol which stands for a variable that can adopt different values rather than to be always equal to zero. Equation (5.429) means that the qualitative elasticity $\varepsilon_{ji}^{\text{qual}}$ is zero if, and only if, S_i does not enter the rate law $v_j(\mathbf{S})$. This condition can be fulfilled in one of the following cases:

(a) The metabolite S_i does not participate in reaction R_j (i.e., $n_{ij} = 0$)

(b) The respective enzyme is saturated with S_i

(c) The reaction is irreversible in the direction of formation of S_i.

In all three cases, S_i must not influence reaction R_j as a catalyst or effector, that is, in a way that is not reflected in the stoichiometry matrix.

The matrix $\boldsymbol{\varepsilon}^{\text{qual}}$ is an example of what is called in mathematics a *structured matrix,* which has fixed zeros in certain locations and arbitrary elements in the remaining locations (see Wonham, 1974; Shields and Pearson, 1976). In the analysis of structured matrices, the concept of rank has to be generalized. One defines the *generic rank* of $\boldsymbol{\varepsilon}^{\text{qual}}$ as the maximum rank which can be achieved as a function of the variable elements in an (ordinary) matrix $\boldsymbol{\varepsilon}$ that has zeros at the same locations as $\boldsymbol{\varepsilon}^{\text{qual}}$. The generic rank of any structured matrix can be determined by an algorithm given by Shields and Pearson (1976).

The incomplete knowledge of kinetic parameters can be taken into account by considering classes of reaction systems, $\Xi(\mathbf{N},\boldsymbol{\varepsilon}^{\text{qual}})$, having the same stoichiometry matrix **N** and the same qualitative elasticity matrix $\boldsymbol{\varepsilon}^{\text{qual}}$. Let $\boldsymbol{\Gamma}$ denote the set of all reactions, R_j ($j = 1, \ldots, r$) of any system belonging to Ξ.

Recent results concerning zero flux control can be phrased in various theorems. The proofs were given in S. Schuster and R. Schuster (1992) on the basis of the generalized mass-action kinetics (2.15).

Theorem 5A. *If one flux control coefficient is zero for all reaction systems belonging to a given class* $\Xi(\mathbf{N},\boldsymbol{\varepsilon}^{\text{qual}})$, *then each reaction network belonging to* Ξ *can be subdivided into two subsystems* $\boldsymbol{\Gamma}_1$ and $\boldsymbol{\Gamma}_2$, *in such a way that the subsystem* $\boldsymbol{\Gamma}_1$ *of reactions is not controlled by the reactions belonging to the subsystem* $\boldsymbol{\Gamma}_2$,

$$C^J_{ij} = 0 \quad \text{for all } R_i \in \boldsymbol{\Gamma}_1,\ R_j \in \boldsymbol{\Gamma}_2, \tag{5.430a}$$

$$\boldsymbol{\Gamma}_1 \cup \boldsymbol{\Gamma}_2 = \boldsymbol{\Gamma}, \tag{5.430b}$$

$$\boldsymbol{\Gamma}_1 \cap \boldsymbol{\Gamma}_2 = \varnothing. \tag{5.430c}$$

For systems of more than two reactions, this implies that when one reaction R_i is insusceptible to flux control by another reaction R_j, which is to say that the control coefficient C^J_{ij} is zero irrespective of the values of kinetic parameters, then more flux control coefficients than just C^J_{ij} are zero.

Now we decompose $\boldsymbol{v}$, **N** and $\mathbf{N}^0$ according to a given partition of $\boldsymbol{\Gamma}$,

$$\boldsymbol{v} = \begin{pmatrix} \boldsymbol{v}_1 \\ \boldsymbol{v}_2 \end{pmatrix}, \tag{5.431a}$$

$$\mathbf{N} = (\mathbf{N}_1\ \ \mathbf{N}_2), \tag{5.431b}$$

$$\mathbf{N}^0 = (\mathbf{N}^0_1\ \ \mathbf{N}^0_2). \tag{5.431c}$$

Let

$$\zeta_i = \text{rank}(\mathbf{N}_i), \quad i = 1, 2. \tag{5.432}$$

Note that the decomposition into subsystems differs from that used in Sections 5.11 and 5.12 in that here the set of reactions is subdivided into two classes, whereas no grouping of the substances is made.

Theorem 5B. *For a given class* $\Xi(\mathbf{N},\boldsymbol{\varepsilon}^{\text{qual}})$, *a necessary and sufficient condition for a subsystem* $\boldsymbol{\Gamma}_1$ *not to be susceptible to flux control by the remaining subsystem is*

$$\mathrm{rank}(\boldsymbol{\varepsilon}_1^{\mathrm{qual}}\ \mathbf{L}) = \mathrm{rank}(\mathbf{N}) - \zeta_2. \tag{5.433}$$

Note that upon multiplication of a structured matrix and an ordinary matrix, as carried out in Eq. (5.433), a structured matrix arises.

Theorem 5C. *For a given class $\Xi(\mathbf{N},\boldsymbol{\varepsilon}^{\mathrm{qual}})$ with all reactions being reversible, a necessary condition for a subsystem $\boldsymbol{\Gamma}_1$ not to be susceptible to flux control by the remaining subsystem is: The null-space matrix $\mathbf{K}$ can be chosen to be block-diagonal, with the diagonal blocks corresponding to the subsystems $\boldsymbol{\Gamma}_1$ and $\boldsymbol{\Gamma}_2$.*

As the fluxes in strictly detailed balanced subnetworks are always zero when the whole system is at steady state (see Section 3.3), it is clear that these subnetworks are not susceptible to flux control. This assertion can be proved by Theorem 5B.

Now we assume that all strictly detailed balanced reactions have been detected and excluded from the further analysis.

Theorem 5D. *In the absence of strictly detailed balanced reactions, some subsystem $\boldsymbol{\Gamma}_1$ is insusceptible to flux control by the reactions of the subsystem $\boldsymbol{\Gamma}_2$ with $\boldsymbol{\Gamma}_1$ and $\boldsymbol{\Gamma}_2$ fulfilling Eqs. (5.430a)–(5.430c) if, and only if, the following conditions are satisfied:*

(D1) *The same condition as in Theorem 5C.*

(D2) *The link matrix $\mathbf{L}$ can be rearranged to give*

$$\mathbf{L} = \begin{pmatrix} \mathbf{L}_1 & \mathbf{0} \\ \mathbf{0} & \mathbf{L}_2 \end{pmatrix} \tag{5.434}$$

with $\mathbf{L}_1$ being the matrix expressing the conservation relations of $\boldsymbol{\Gamma}_1$.

(D3) $$\varepsilon_{ji}^{\mathrm{qual}} = 0 \tag{5.435}$$

for any i,j with $R_j \in \Gamma_1$ and $i > a$, where a is the number of rows of $\mathbf{L}_1$.

Condition (D2) excludes any influence of subnetwork $\boldsymbol{\Gamma}_2$ on $\boldsymbol{\Gamma}_1$ *via* conservation relations which involve metabolites of both the subnetworks. Condition (D3) guarantees that changes in kinetic parameters of subsystem $\boldsymbol{\Gamma}_2$ do not influence the fluxes in $\boldsymbol{\Gamma}_1$ by effector influences.

The above results enable us to decide, for reaction systems of any complexity, what fluxes cannot be controlled by what reactions. After having constructed, by the algorithm given in S. Schuster and R. Schuster (1991), the representation of $\mathbf{K}$ with the maximum number of diagonal blocks, one cancels all rows of $\mathbf{K}$ that correspond to strictly detailed balanced reactions. Now all combinations of the remaining diagonal blocks into two submatrices have to be examined as for conditions (D2) and (D3).

Decomposability of the null-space matrix applies, in particular, to hierarchic reaction systems characteristic to living cells. An example is provided by the system shown in Scheme 15 (Section 5.11) representing the hierarchic organization of cellular processes. To demonstrate the applicability of the method to hierarchical systems, we again consider the simple two-level system shown in Figure 5.12B. S_1 and S_2 may stand, for instance, for an mRNA species and a protein, respectively. The dashed arrow signifies a catalytic influence. Conditions (D1) and (D2) are fulfilled with $\mathbf{\Gamma}_1$ and $\mathbf{\Gamma}_2$ corresponding to the upper and lower level in the scheme of Figure 5.12, respectively. If, and only if, S_2 does not influence the reactions involving S_1 nonstoichiometrically (e.g., if there is no feedback from the protein to gene expression), then condition (D3) is fulfilled and there is no upstream control in the hierarchy (see also Westerhoff *et al.*, 1990; Kahn and Westerhoff, 1991).

It is an intriguing question whether Theorem 5D still applies if hierarchic systems are studied at a more detailed level, by including some or all of the elementary reactions of enzyme action. By way of example, we consider the system shown in Figure 5.12A. The stoichiometry matrix of this more detailed scheme, which is given in Eq. (5.379), is not block-diagonal. However, both matrices $\mathbf{K}$ and $\mathbf{L}$ (the latter being the 3×3 identity matrix) can be chosen to be block-diagonal [cf. Eq. (5.380)]. Thus, conditions (D1) and (D2) are fulfilled irrespective of the way of description of such a type of hierarchic system.

Another class of hierarchic schemes are the interconvertible enzyme cascades (see Chock *et al.*, 1980; Goldbeter and Koshland Jr., 1984; Cárdenas and Cornish-Bowden, 1989). A bicyclic system is presented schematically in Figure 5.15A. S_1 and S_2 stand for two forms of a protein catalyzing the transformation (e.g., by

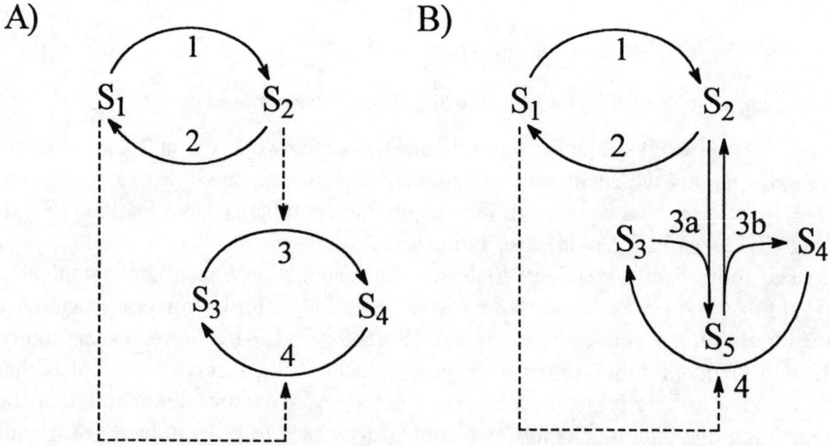

Figure 5.15 Example of a bicyclic system. (A) Simplified scheme; (B) detailed scheme.

phosphorylation and dephosphorylation) of another protein, with the forms S_3 and S_4.

This scheme may represent the glutamine synthetase cascade in *E. coli* (Chock *et al.*, 1980) or any two cycles within the glycogen phosphorylase-glycogen synthase system. In the latter instance, only one activatory loop is operative.

The stoichiometry matrix corresponding to Figure 5.15A is block-diagonal. According to whether or not S_1 is an effector of reaction R_4, the qualitative elasticity matrix reads

$$\boldsymbol{\varepsilon}^{\text{qual}} = \begin{pmatrix} \varepsilon & \varepsilon & 0 & 0 \\ \varepsilon & \varepsilon & 0 & 0 \\ 0 & \varepsilon & \varepsilon & \varepsilon \\ \varepsilon & 0 & \varepsilon & \varepsilon \end{pmatrix} \tag{5.436a}$$

or

$$\boldsymbol{\varepsilon}^{\text{qual}} = \begin{pmatrix} \varepsilon & \varepsilon & 0 & 0 \\ \varepsilon & \varepsilon & 0 & 0 \\ 0 & \varepsilon & \varepsilon & \varepsilon \\ 0 & 0 & \varepsilon & \varepsilon \end{pmatrix}, \tag{5.436b}$$

respectively. If the bottom cycle is chosen as subsystem $\boldsymbol{\Gamma}_2$, both of the matrices given in Eqs. (5.436) fulfill condition (D3). Therefore,

$$C^J_{13}, C^J_{14}, C^J_{23}, C^J_{24} = 0 \tag{5.437}$$

for any set of kinetic parameter values. This reaction scheme is an example of a cascade without feedback, as studied in Section 5.11. Therefore, Eq. (5.437) also follows from Eq. (5.398). On the other hand, Eq. (5.398) can be derived from Theorem 5D.

When reaction 3 in the enzyme cascade is split up into two elementary steps, as shown in Figure 5.15B, the null-space matrix can be chosen to be block-diagonal but the link matrix cannot. One may conclude that for hierarchic systems where the different levels are formed by moiety-conserved cycles (e.g. in the case of enzyme cascades), the values of control coefficients depend on the way of description (see also Fell and Sauro, 1990; Kholodenko *et al.*, 1993b).

Consider a simplified scheme of threonine synthesis in *E. coli* (see Gottschalk, 1986), as depicted in Figure 4.3. Aspartate, ATP, ADP, lysine, methionine, and the nicotinamide cofactors are treated as external metabolites. The reaction in which aspartate is phosphorylated is catalyzed by three enzymes: aspartokinase I, II, and III. This reaction and the homoserine kinase reaction are known to be practically irreversible. The aspartokinase I and homoserine dehydrogenase activities are carried in *E. coli* by a bifunctional protein. Both of these activities are in most strains inhibited by threonine (Patte *et al.*, 1966). Furthermore, threonine inhibits homoserine kinase (Thèze *et al.,* 1974). Neglecting further regulatory loops, we can write the qualitative elasticity matrix as

$$\boldsymbol{\varepsilon}^{\text{qual}} = \begin{pmatrix} 0 & 0 & 0 & 0 & \varepsilon \\ \varepsilon & \varepsilon & 0 & 0 & 0 \\ 0 & \varepsilon & \varepsilon & 0 & \varepsilon \\ 0 & 0 & \varepsilon & 0 & \varepsilon \\ 0 & 0 & 0 & \varepsilon & \varepsilon \\ 0 & 0 & 0 & 0 & \varepsilon \\ 0 & \varepsilon & 0 & 0 & 0 \\ 0 & 0 & \varepsilon & 0 & 0 \end{pmatrix}, \tag{5.438}$$

where reactions are numbered as indicated in the legend to Figure 4.3, and the internal metabolites are numbered as follows: AspP, 1; ASA, 2; HSer, 3; HSerP, 4; Thr, 5. We take the aspartate semialdehyde dehydrogenase reaction (R_2) as subsystem Γ_2. Because the whole network does not involve any conservation relations, the link matrix is an identity matrix. The generic rank of the matrix product $\boldsymbol{\varepsilon}_1^{\text{qual}}\,\mathbf{L}$ equals 4, whereas rank($\mathbf{N}$) $= 5$ and $\zeta_2 = 1$. Therefore, condition (5.433) is fulfilled. Similarly, this condition is satisfied for a decomposition with the threonine synthase reaction constituting subsystem Γ_2. Consequently, the flux of threonine synthesis is insusceptible to flux control by aspartate semialdehyde dehydrogenase and threonine synthase, $C^J_{k,2} = 0$ $(k \neq 2)$ and $C^J_{k,5} = 0$ $(k \neq 5)$, due to Theorem 5B. Moreover, it can be concluded, with the help of the equation $\mathbf{NC}^J = \mathbf{0}$, which follows from Eq. (5.26b), that these reactions do not control their own steady-state fluxes either ($C^J_{22} = 0$ and $C^J_{55} = 0$). This is in accord with simulations by Raïs *et al.* (1993).

Consider the hypothetical situation (which might occur in mutant strains) that there are no side pathways leading to lysine and methionine and no feedback exerted by threonine. By determining the matrix $\boldsymbol{\varepsilon}_1^{\text{qual}}\,\mathbf{L}$ for this case, one derives that all flux control coefficients with respect to reactions behind the aspartokinase reaction are zero. This is in line with the results for unbranched chains involving irreversible reactions (Section 5.4.3.1).

Suppose now that the feedback loops are operative. Then all reactions between aspartate 4-phosphate and threonine exhibit zero flux control coefficients. Including now the branches leading to lysine and methionine, we arrive at our above result, which can be generalized in that all reactions between an irreversible step and a metabolite acting as a feedback inhibitor, except those situated behind branching points, are not able to control any flux (cf. Eq. (5.112)).

An analysis of the structure of control insusceptibility of the reaction scheme of glycolysis including the phosphoglucomutase reaction and fructose-bisphosphate cycle can be found in the work of S. Schuster and R. Schuster (1992).

The above results on flux control insusceptibility are valid not only for infinitesimally small parameter perturbations but also for large changes. In fact, integration of Eq. (5.16b) gives zero if C^J is always zero irrespective of the parameter values.

In the situation that some enzyme E_i is saturated with its substrate [case (b) in the above classification], the reactions upstream this enzyme usually have very low flux control coefficients. When they are inhibited to a large extent, the point will eventually be reached where the substrate of E_i drops below the Michaelis constant of E_i, so that the control coefficients become nonzero. This gives rise to a threshold phenomenon in the effect of inhibition on a flux (Letellier *et al.*, 1994).

A frequent situation in biochemical systems is that some reactions are very fast, so that they subsist at quasi-equilibrium. As was pointed out for unbranched chains in Section 5.4.3.1, quasi-equilibrium reactions exert very weak flux control. This statement can be generalized for systems of any complexity, by again using the connectivity theorem (5.51b). Another substantiation might be based on Theorem 5B and a rescaling of elasticities. In the limit of infinitely fast reactions, such rescaling brings about that some elements of $\boldsymbol{\varepsilon}_1 \mathbf{L}$ tend to zero, which changes the generic rank of $\boldsymbol{\varepsilon}_1^{\text{qual}}\,\mathbf{L}$.

5.14. CONTROL EXERTED BY ELEMENTARY STEPS IN ENZYME CATALYSIS

In the modular approach, metabolic control analysis was generalized by considering functional units containing several enzymes. Another generalization may be achieved by going further into the details of the particular enzymic reactions. Elementary steps of enzyme catalysis (such as substrate–enzyme binding or isomerization of enzyme–substrate complexes) rather than overall enzymatic reactions are then the basic entities. Such an approach may answer the question of whether a particular step of an enzyme can be rate-limiting to the rate of that enzyme (Ray, 1983; Brown and Cooper, 1993; Kholodenko *et al.*, 1994).

The flux control coefficients pertaining to elementary steps in an enzyme scheme can be defined by

$$C_j^v = \frac{w_j}{v}\frac{\partial v}{\partial w_j} = \frac{w_j}{v}\frac{\partial v/\partial p_j}{\partial w_j/\partial p_j}, \tag{5.439}$$

where w_j stands for the rate of the jth elementary step. When the enzyme is embedded in a reaction network and the control over steady-state fluxes is considered, v has to be replaced in Eq. (5.439) by J_k. Analogously, concentration control coefficients are defined. They can be given not only for the free metabolites but also for enzyme intermediates. The parameter p_j, which enters Eq. (5.439) will usually be a rate constant of an elementary step, k_{+j} or k_{-j}.

An alternative definition of control coefficients pertaining to single enzymes refers to particular rate constants rather than to particular elementary steps (Brown and Cooper, 1993; Kholodenko *et al.*, 1994). In our terminology, these quantities are response coefficients,

$$R^v_{k_{\pm j}} = \frac{k_{\pm j}}{v}\frac{\partial v}{\partial k_{\pm j}}, \tag{5.440}$$

because they belong to the class defined by Eqs. (5.11) and (5.12).

As pointed out in Section 5.6.2, the control coefficients C^v_j have the property of being independent of the choice of the perturbation parameter. This will now be illustrated for the catalytic Scheme 1 (Section 2.2.2), with the substrate concentration $S_1 = S$ and product concentration $S_2 = P$.

For the net rates of the two steps, we have

$$w_1 = k_1 S \cdot E - k_{-1} ES, \tag{5.441a}$$

$$w_2 = k_2 \cdot ES - k_{-2} \cdot P \cdot E. \tag{5.441b}$$

For the quasi-steady-state of the enzyme ($w_1 = w_2$), with the conservation relation $E + ES = E_T$, one derives

$$E = \frac{E_T(k_{-1} + k_2)}{k_1 S + k_{-2}P + k_{-1} + k_2}, \tag{5.442a}$$

$$ES = \frac{E_T(k_1 S + k_{-2}P)}{k_1 S + k_{-2}P + k_{-1} + k_2}. \tag{5.442b}$$

For the enzyme rate, one obtains the reversible Michaelis–Menten kinetics in terms of elementary rate constants,

$$v = \frac{E_T}{D}(k_1 k_2 S - k_{-1}k_{-2}P), \tag{5.443a}$$

$$D = k_1 S + k_{-2}P + k_{-1} + k_2 \tag{5.443b}$$

[cf. Eq. (2.20)]. The unscaled flux control coefficient of the first step, for example, can be calculated as

$$C^v_1 = \left(\frac{\partial v/\partial k_1}{\partial w_1/\partial k_1}\right)_{k_{-1} = \text{const.}} \tag{5.444a}$$

or

$$C_1^v = \left(\frac{\partial v / \partial k_{-1}}{\partial w_1 / \partial k_{-1}} \right)_{k_1 = \text{const.}} . \tag{5.444b}$$

Straightforward differentiation of Eqs. (5.441) and (5.443) yields

$$\left. \frac{\partial w_1}{\partial k_1} \right|_{k_{-1}} = S \cdot E = \frac{S \cdot E_T}{D} (k_{-1} + k_2), \tag{5.445}$$

$$\left. \frac{\partial v}{\partial k_1} \right|_{k_{-1}} = \frac{E_T \cdot S}{D^2} (k_{-1} + k_2)(k_{-2}P + k_2) . \tag{5.446}$$

This leads to

$$C_1^v = \frac{k_{-2}P + k_2}{D} . \tag{5.447}$$

Using k_{-1} as a perturbation parameter, we have

$$\left. \frac{\partial w_1}{\partial k_{-1}} \right|_{k_1} = -ES = -\frac{E_T}{D} (k_1 S + k_{-2}P) , \tag{5.448}$$

$$\left. \frac{\partial v}{\partial k_{-1}} \right|_{k_1} = -\frac{E_T}{D^2} (k_{-2}P + k_2)(k_1 S + k_{-2}P) . \tag{5.449}$$

So we obtain the same expression for C_1^v as in Eq. (5.447).

In a similar way, by using either of the perturbation parameters k_2 and k_{-2}, one obtains for the control coefficient of the second step,

$$C_2^v = \frac{k_1 S + k_{-1}}{D} . \tag{5.450}$$

Note that the response coefficients defined in Eq. (5.440), which may easily be calculated for the considered example from Eqs. (5.446) and (5.449), do not have the property of being independent of the choice of perturbation parameter.

The elementary rate constants are linked with the equilibrium constant, q, of the overall reaction by the Haldane relation [cf. Eq. (2.26) and the relations between phenomenological and elementary parameters given in Eqs. (2.21) and (2.22)]. For the enzyme depicted in Scheme 1, this relation reads

$$\frac{k_1 k_2}{k_{-1} k_{-2}} = q. \tag{5.451}$$

Because the equilibrium constant is independent of the catalytic properties of the enzyme, one may argue that only those perturbations are allowed for which the equilibrium constant remains constant. This may be taken into account by the condition that changes in k_1 are always accompanied with opposite changes in k_{-1}. We then have

$$\left.\frac{\partial w_1}{\partial k_1}\right|_{q_1} = \frac{E_T}{D}\left(k_2 S + \frac{k_{-2}P}{q_1}\right), \tag{5.452}$$

$$\left.\frac{\partial v}{\partial k_1}\right|_{q_1} = \frac{E_T}{D^2}(k_{-2}P + k_2)\left(k_2 S - \frac{k_{-2}P}{q_1}\right). \tag{5.453}$$

Equations (5.453) and (5.452) yield the same expression for C_1^v as given in Eq. (5.447). Accordingly, the Haldane relation need not be considered when control coefficients are calculated. This point is of practical relevance because changes of the equilibrium constant can actually occur in practice, owing to changes in temperature, for example.

Response coefficients with respect to temperature can be written as

$$R_{\mathrm{T}}^{Y} = \sum_i C_i^Y \pi_{\mathrm{i,T}}, \tag{5.454}$$

where C_i^Y are the control coefficients of the elementary steps with respect to any steady-state variable Y and $\pi_{i,\mathrm{T}}$ are the scaled elasticities of the elementary steps with respect to temperature. Moreover, the rate constants $k_{\pm i}$ may incorporate concentrations of external metabolites or ions such as H^+. Changes in pH then change the (apparent) equilibrium constant.

Because the calculations in Section 5.3 also apply to the control coefficients pertaining to elementary steps, the flux control coefficients of these steps sum up to unity and the concentration control coefficients sum up to zero. For example, the sum of the flux control coefficients calculated in Eqs. (5.447) and (5.450) equals unity. As the velocities of elementary steps are usually linear functions of the rate constants, the response coefficients $R_{k_{\pm i}}^Y$ also satisfy summation theorems very similar to the theorems derived in Section 5.3,

$$\sum_i R_{j,k+i}^S + \sum_i R_{j,k-i}^S = 0, \tag{5.455a}$$

$$\sum_i R_{j,k+i}^J + \sum_i R_{j,k-i}^J = 1, \tag{5.455b}$$

as can be proven as follows. For any elementary step i, the reaction velocity can be written as $w_i = w_i^+ - w_i^-$, where the two terms are proportional to the corresponding elementary rate constants. Thus, we have

$$\pi_{i,k+i} + \pi_{i,k-i} = \frac{w_i^+}{w_i} - \frac{w_i^-}{w_i} = 1. \tag{5.456}$$

The response coefficients can be expressed as

$$R_{k+i}^Y = C_i^Y \pi_{i,k+i}, \tag{5.457a}$$

$$R_{k-i}^Y = C_i^Y \pi_{i,k-i}. \tag{5.457b}$$

Equations (5.456) and (5.457) and the summation theorems for control coefficients of elementary steps imply relations (5.455a) and (5.455b).

Response coefficients can also be defined to refer to particular elementary steps rather than to particular rate constants, by keeping constant the equilibrium constants of the elementary steps upon differentiation (Ray, 1983; Brown and Cooper, 1994; Kholodenko *et al.,* 1994). Because the reaction rates are then proportional to the perturbation parameters, the response coefficients so defined and the control coefficients of elementary steps are identical.

Brown and Cooper (1993) also defined coefficients expressing the effect of changes in elementary rate constants on the maximal activity, V_m, and Michaelis constant, K_m. However, these are no control coefficients in the sense of metabolic control analysis, because V_m and K_m are no steady-state variables.

Computation of control coefficients for elementary steps of triose phosphate isomerase (EC 5.3.1.1), carbamate kinase (EC 2.7.2.2) and lactate dehydrogenase (EC 1.1.1.27) from literature values of the rate constants shows that these enzymes do not have unique rate-limiting steps, but flux control is shared by several steps and varies with substrate, product, and effector concentrations (Brown and Cooper, 1993).

As the rates of elementary steps are linear functions of enzyme intermediate levels and rate constants, there are numerous simple relations between these quantities and the control coefficients with respect to rate constants or elementary steps (Brown and Cooper, 1994). These relations can be used to determine control coefficients from measurement of enzyme intermediates rather than of rate constants.

To illustrate an interrelation between the concentrations of enzyme forms and the response coefficients referring to rate constants, one may calculate the coefficients R_{k-1}^v and R_{k2}^v for Scheme 1 (Section 2.2.2), by differentiating Eq. (5.443). This gives, for the sum of these coefficients,

$$R_{k-1}^v + R_{k2}^v = \frac{k_1 S + k_{-2} P}{D}. \tag{5.458}$$

Comparison with Eq. (5.442) shows that

$$R^v_{k-1} + R^v_{k2} = \frac{ES}{E_T}. \tag{5.459}$$

This is a special case of a general relation found and proven by Kholodenko *et al.* (1994), which states that the sum of the control coefficients referring to the rate constants of the processes leading away from (consuming) any enzyme subform is equal to the concentration of that subform divided by the total enzyme concentration,

$$\sum_i R^v_{k\pm i} = \frac{E_k}{E_T}, \tag{5.460}$$

where the sum runs over all elementary rates flowing away from E_k. The basic idea of the proof is to consider a hypothetical increase of the concentration E_k by a factor λ and a simultaneous decrease of all the rate constants pertaining to the unidirectional rates utilizing the intermediate E_k by the same factor, λ (Kholodenko *et al.*, 1994). Then all the rates of the elementary steps remain unchanged, and so does the overall enzyme rate. Thus, we have

$$\frac{\lambda}{v}\frac{\partial v}{\partial \lambda} = -\sum_i R^v_{k\pm i} + \frac{E_k}{E_T} = 0, \tag{5.461}$$

from which Eq. (5.460) follows immediately.

5.15. CONTROL ANALYSIS OF METABOLIC CHANNELING

Besides homologous enzyme–enzyme interactions (monomer–oligomer associations), heterologous enzyme complexes (i.e., associations of different enzymes) have frequently been detected, in particular in tissues with very high enzyme concentrations (Srivastava and Bernhard, 1986; Srere, 1987). Various experimental data make it very likely that metabolic intermediates are directly transferred between the enzymes in these heterologous complexes. This phenomenon is called metabolic channeling (for a review, see Ovádi, 1991). However, whether this mechanism actually occurs and how important it is still remains in dispute (Gutfreund and Chock, 1991; Wu *et al.*, 1991; Giersch, 1991).

The assembly of enzymes can be transient (reversible, dynamic) or permanent (irreversible, static). In static complexes, the catalytic units may be linked noncovalently (multienzyme aggregates) or covalently (multifunctional proteins). Examples are fatty acid synthase (see Wakil *et al.*, 1983) and the aspartokinase I/homoserine dehydrogenase complex in *E. coli* (see Gottschalk, 1986), respectively. The latter protein is interesting in that it does not catalyze sequential, neighboring reactions as is normally the case in static enzyme complexes, but the first and third reactions of the threonine pathway. There is no sharp conceptual distinction between the phenomena of static channeling and microcompartmentation. The latter term is often used when multienzyme complexes or enzyme arrays attached to membranes or to the cytoskeleton constitute a microenvironment reducing diffusion lengths (see Friedrich, 1984; Welch *et al.,* 1988; Gellerich *et al.,* 1994).

In dynamic channels, the enzymes consecutively associate and dissociate in a way that the metabolic intermediates are "handed over" without the necessity of being released into the aqueous medium. An example of a two-enzyme system with dynamic channeling is shown in Figure 2.1. It is generally accepted that most metabolic channels are not perfect, that is, the individual, nonassociated enzymes are also catalytically active, so that unbound intermediates occur (S_1 in Figure 2.1). It has also been suggested that channels might be leaky (i.e., intermediates may escape into the aqueous medium). In the example shown, leakiness would imply an additional dissociation step from the complex $E_1S_1E_2$ to E_1E_2 and S_1.

From the viewpoint of metabolic control analysis, a static enzyme complex catalyzing several sequential reactions can be treated by considering it as one enzyme catalyzing one overall reaction. Elasticities, control coefficients, and response coefficients for the complex as a whole can be determined. However, things become difficult if the channel is not perfect. As soon as (catalytically active) free enzymes are present in the bulk phase, the pathway flux is a complicated superposition of direct transfer and reactions in the bulk phase. Elasticities of the direct-transfer reaction are then difficult to measure because both the enzyme complex and the free enzymes are present, whereas elasticities are defined for the situation in which the considered reaction proceeds in isolation. The same problem arises for dynamic channels. Therefore, the conceptual distinction between an isolated enzyme and the same enzyme embedded in a pathway takes on a new aspect when channeling is operative. The fact that a system is more than the sum of its constituents is even more relevant here than for unchanneled pathways.

Clearly, in channeled metabolic systems, there is no longer a one-to-one correspondence between enzymes and reactions. Therefore, a clear distinction should be made between control coefficients of reactions and response coefficients with respect to enzyme concentrations. As was shown in Section 5.6.4, the control coefficients of reactions are not unique in the case of dynamic channeling, because they depend on the choice of the perturbation parameter. In contrast, the response coefficients with respect to enzyme concentrations are uniquely defined. Therefore, the attempt made by Sauro and Kacser (1990) to apply the general response equation (5.28) to heterologous enzyme–enzyme interactions is contestable.

It has been shown that the sum of response coefficients for pathway flux with respect to enzyme concentrations exceeds unity when dynamic channeling occurs (Kholodenko and Westerhoff, 1993; Sauro, 1994). It may therefore be sensible to define the response coefficient pertaining to the channel as unity minus this sum.

A possible way of analyzing the control of channeled systems is by considering the elementary steps of enzyme catalysis, extending the analysis of Section 5.14 (Kholodenko and Westerhoff, 1993). Total enzyme concentrations, E_T, cannot be used as perturbation parameters for defining control coefficients at this detailed level of description. This follows from the fact that they are conservation quantities influencing several elementary steps. Furthermore, the rate of an isolated elementary step, w, cannot be expressed as a unique function of total enzyme concentration, because w depends on the distribution of E_T among the particular enzyme intermediates.

As was explained in Sections 5.2 and 5.14, the general definitions (5.3) and (5.5) of control coefficients can also be used for elementary steps. These coefficients *per se* are not of much practical use though, because particular elementary steps are hardly accessible experimentally. To express the overall control exerted by an enzyme, E_i, Kholodenko and Westerhoff (1993) introduced the *impact-control coefficient* as the sum of the control coefficients of all elementary steps that are directly affected by enzyme E_i,

$$ {}^{\mathrm{imp}}C^Y_{E_i} = \sum_{k \in \mathcal{R}i} C^Y_k , \tag{5.462} $$

where $\mathcal{R}_i$ is the set of all E_i-dependent processes. This coefficient, in a sense, evaluates the total impact enzyme E_i has on the steady-state variable Y. A process is called E_i-dependent if its rate depends on the concentrations of the free form of the enzyme E_i or of a complex that involves E_i. In mathematically rigorous terms, a process k is E_i dependent if there is an enzyme subform E^{sub} such that

$$ \frac{\partial w_k}{\partial E^{\mathrm{sub}}} \neq 0, \quad \frac{\partial E_{T,i}}{\partial E^{\mathrm{sub}}} \neq 0, \tag{5.463} $$

where $E_{T,i}$ is the total concentration of enzyme E_i.

Another important quantity expressing the effect of an enzyme E_i is clearly the response coefficient $R^Y_{ET,i}$, which refers to changes in the total concentration $E_{T,i}$. Kholodenko *et al.* (1993b) showed that the impact-control coefficient ${}^{\mathrm{imp}}C^Y_{E_i}$ can be expressed as the sum of the response coefficient $R^Y_{ET,i}$ and terms referring to channeling and conservation relations involving both enzymic species and free intermediates. For proving this relation, they considered a hypothetical perturbation of a given steady state so that

(i) Every concentration involved in the catalytic cycle of a given enzyme E_i is *increased* by a factor λ, which implies that the concentrations of all enzymes forming complexes with E_i are also increased

(ii) The rate constants of all E_i-dependent processes are *decreased* by the same factor λ. By these changes, the total concentration of E_i attains a new value,

$$E_{T,i}(\lambda) = \lambda E_{T,i}. \tag{5.464}$$

The new total concentrations of all enzymes that form complexes with E_i amounts to

$$E_{T,j}(\lambda) = E_{T,j} + (\lambda - 1)E_{ji}^{\text{comp}}, \quad j \neq i, \tag{5.465}$$

where E_{ji}^{comp} is the total concentration of all complexes involving both E_i and E_j before the perturbation. Equation (5.465) is only valid under the assumption that every enzyme may occur in any complex no more than once (i.e., homologous complexes are excluded).

The considered perturbation also changes conservation sums, T_l, that include not only enzymic species E^{sub} involving E_i but also free metabolites, S_i, if such conservation cycles exist. These conservation sums can be decomposed into a part, T_l^{free}, containing free metabolites and a part, T_l^{sub}, involving enzyme subforms. The perturbed conservation sums can then be written as

$$T_l(\lambda) = T_l^{\text{free}} + \lambda T_l^{\text{sub}}. \tag{5.466}$$

To find, in an algebraic way, those conservation quantities T_k affected by changes in the subforms of E_i, stoichiometric analysis can be helpful; for example, by block-diagonalizing the link matrix **L**.

All of the rates w_k of E_i-dependent processes are homogeneous functions of first order of the concentrations of subforms of enzyme E_i, because we exclude dimerization and oligomerization of E_i. They are also homogeneous functions of the rate constants at fixed equilibrium constants of the elementary steps. Therefore, all the rates remain unchanged after the above-mentioned increase in the concentrations of the subforms of enzyme E_i and decrease in the corresponding rate constants (i.e., $\partial J/\partial\lambda = 0$).

Because the steady-state fluxes are functions of the kinetic parameters and the conservation quantities, one can write, by using the chain rule of differentiation,

$$\sum_{\pm k} C_{\pm k}^{J} \frac{\partial \ln(k_{\pm k}/\lambda)}{\partial \ln \lambda} + \sum_{j} R_{E_j}^{J} \frac{\partial \ln E_{T,j}}{\partial \ln \lambda} + \sum_{l} R_{T_l}^{J} \frac{\partial \ln T_l}{\partial \ln \lambda} = 0, \tag{5.467}$$

where the index k refers to all the elementary steps belonging to enzyme E_i ($k \in \mathcal{R}_i$). Taking the derivatives at $\lambda = 1$, we have, for the rate constants,

$$\frac{\partial \ln (k_{\pm k}/\lambda)}{\partial \ln \lambda} = \begin{cases} -1 & \text{if } k \in \mathcal{R}_i \\ 0 & \text{otherwise,} \end{cases} \tag{5.468}$$

and for the conservation relations, due to Eqs. (5.465) and (5.466),

$$\frac{\partial \ln E_{T,j}}{\partial \ln \lambda} = \frac{E_{ji}^{\text{comp}}}{E_{T,j}}, \tag{5.469a}$$

$$\frac{\partial \ln T_l}{\partial \ln \lambda} = \frac{T_l^{\text{sub}}}{T_l}. \tag{5.469b}$$

Because of Eq. (5.468), the first sum on the left-hand side of Eq. (5.467) equals, apart from its sign, the sum of the C_k^J of all E_i-dependent processes. By definition (5.462), this sum equals the impact-control coefficient of enzyme E_i. Therefore, Eqs. (5.467)–(5.469) give

$$^{\text{imp}}C_{E_i}^J = R_{E_i}^J + \sum_{j \neq i} \frac{E_{ji}^{\text{comp}}}{E_{T,j}} + \sum_{l=1} \frac{T_l^{\text{sub}} R_{T_l}^J}{T_l}. \tag{5.470}$$

Kholodenko *et al.* (1993a, 1993b) drew the conclusion that channeled pathways can be more sensitive to regulatory signals than "ideal" ones, because the impact-control coefficient is increased by the two sums on the right-hand side of Eq. (5.470). Indeed, the sum over j refers to enzyme-enzyme interactions as is typical for channeling, but the $R_{E_i}^J$ and the sum over l in Eq. (5.470) for the channeled pathway generally are not the same as those for a comparable non-channeled pathway. Importantly, the impact-control coefficient coincides with the response coefficient of enzyme E_i if this enzyme is not involved in channeling, nor in moiety-conserved cycles of metabolites.

It remains questionable, though, whether the concept of impact-control coefficient is appropriate to describe the control exerted by an enzyme in the situation of metabolic channeling. Changing all E_i-dependent processes by the same fractional amount seems to be impossible in experiment. The evaluation of the sums on the right-hand side of Eq. (5.470) is also problematic. Going down to the level of elementary steps bears the difficulty that the exact number and interconnections of these steps are often unknown.

What is desirable is to define and calculate control coefficients of biochemically meaningful and accessible processes, such as a channeled route as a whole. For example, in Figure 2.1, it would be interesting to have separate control coefficients pertaining to the reactions catalyzed by E_1, E_2, and the E_1E_2 complex. This is, however, difficult because the denominator in the definitions of control coefficients (3.3) and (3.5) refers to isolated reactions, whereas the reaction catalyzed by the E_1E_2 complex cannot be studied in isolation in the case of imperfect channels. It would be a challenge for the modular approach to metabolic control analysis (see Section 5.12) to cope with metabolic channeling.

5.16. COMPARISON OF METABOLIC CONTROL ANALYSIS AND POWER-LAW FORMALISM

Metabolic control analysis is a kind of sensitivity analysis dealing with the effect of perturbations of reaction rates on steady-state variables. Another type of sensitivity analysis based on the power-law approach (cf. Section 2.2.4) was presented by Savageau (1976; see also Savageau *et al.,* 1987b) and applied recently to a model of the tricarboxylic acid cycle (Shiraishi and Savageau, 1993). It makes use of the favorable feature that the power-law rate laws may be transformed into linear equations in the logarithmic concentration space [cf. Eqs. (2.75) and (2.76)]. On the systemic level, however, one arrives at linear equation systems only if a method called aggregation of flux is employed (Savageau *et al.,* 1987a). In that method, rate laws for those processes tending to produce a given substance S_i are first combined to give an aggregate rate, $v^+_{\mathrm{aggr},i}$. Similarly, the kinetic functions of those processes that consume a given substance are summed to give a separate aggregate rate, $v^-_{\mathrm{aggr},i}$ (see Figure 5.16). Instead of the balance equations (2.7), the equations

$$\frac{\mathrm{d}S_i}{\mathrm{d}t} = v^+_{\mathrm{aggr},i} - v^-_{\mathrm{aggr},i} = \alpha_i \prod_{j=1}^{n} S_j^{g_{ij}} - \beta_i \prod_{j=1}^{n} S_j^{h_{ij}}, \quad i = 1, \ldots, n, \tag{5.471}$$

is then used as basis for the system description. The power-law expressions for the aggregate rates are obtained in a similar way as was explained in Section 2.2.4 for isolated reactions. Accordingly, each of the kinetic constants α_i and β_i as well as each of the kinetic orders g_{ij} and h_{ij} now represents the properties of several, aggregated enzymic reactions rather than of only one reaction. At steady state, all time derivatives of concentrations are zero, so that Eq. (5.471) can be transformed to

$$\ln \alpha_i + \sum_j g_{ij} \ln S_j = \ln \beta_i + \sum_j h_{ij} \ln S_j, \quad i = 1, \ldots, n. \tag{5.472}$$

This can be written in matrix notation as

$$\boldsymbol{A} \ln \boldsymbol{S} = \boldsymbol{b}, \tag{5.473}$$

where the matrix **A** and the vector ***b*** contain the following elements:

$$a_{ij} = g_{ij} - h_{ij}, \tag{5.474a}$$

$$b_i = \ln \beta_i - \ln \alpha_i. \tag{5.474b}$$

As the equation system (5.473) is linear in the logarithmic concentrations, it can be treated analytically. However, a number of drawbacks of applying the power-law approach combined with the flux aggregation method should be mentioned.

The method of flux aggregation generates a reaction scheme to which no meaningful stoichiometry matrix can be attached (cf. Figure 5.16).

If the rank of matrix **A** equals *n*, the solution to the equation system (5.473)

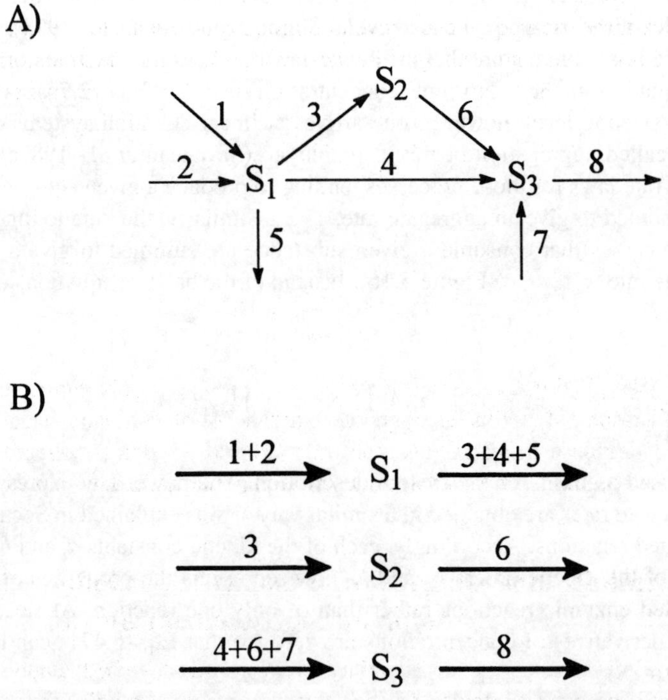

Fig 5.16 Scheme of a reaction system illustrating the method of flux aggregation. (A) Original scheme; (B) "aggregate" scheme. The sums indicate which reactions of the original scheme have been lumped to give the aggregate reactions (thick arrows in B). Note that the aggregate scheme cannot be interpreted as a coherent pathway because the rate of degradation of any metabolite does not occur as the rate of production of another compound and vice versa.

is unique. Otherwise, this equation system is solvable if the rank of the matrix **A** equals the rank of the augmented matrix (a_{ij}, b_i) (cf. Morris, 1992). One then obtains a continuous manifold of solutions. Therefore, the occurrence of isolated multiple steady states with nonzero concentrations cannot be described by the power-law approximation and aggregation of flux, whereas other nonlinear approaches can cope with multistationarity (cf. Section 2.3.3). As this phenomenon plays an important role in biology, the advantage of analytical solvability of the steady-state equations appears not to be very valuable.

Another problem arises when conservation relations or other side constraints are present. As was shown in Section 3.3, the steady state(s) then cannot be calculated from the rate laws alone, but the conservation quantities must be used in the calculation also. To make the power-law approach uniform, Savageau *et al.* (1987a) proposed writing constraints among concentrations in the form of power functions as well.

For example, the conservation relation $ADP + ATP = A =$ const. would be written in the power-law approximation as

$$\left(\frac{ADP}{ADP^0}\right)^{f_1}\left(\frac{ATP}{ATP^0}\right)^{f_2} = 1, \tag{5.475}$$

where

$$f_1 = \frac{ADP^0}{A}, \quad f_2 = \frac{ATP^0}{A}. \tag{5.476}$$

The superscript 0 refers to the reference state of the approximation. Equation (5.475) can be derived by expanding the equation $\ln ATP = \ln(A - ADP) = \ln(A - \exp(\ln ADP))$ into a Taylor series and only considering the terms linear in $(\ln ADP - \ln ADP^0)$.

In the case that constraints are linear conservation relations (as for the above example concerning the conservation of adenine nucleotides), this approach makes things unnecessarily complicated. More importantly, mass conservation is only fulfilled in the reference state, whereas for sensitivity analysis, deviations from the reference state must be studied. Furthermore, whereas conservation relations are a direct consequence of the linear dependencies among balance equations, the side constraints approximated by power laws are not, in general, consistent with the approximate system equations (5.471).

Another drawback is that the method of aggregation of flux entails a questionable reduction in the number of degrees of freedom of the system at steady state, because the stoichiometric relationships are no longer reflected in the system equations (5.471). This concerns, for example, the different possibilities of distribution of flux over the branches in the system (cf. Chapter 3). It is worth mentioning that the aggregation method used in the modular approach of metabolic control analysis (Section 5.12) differs from that underlying Eq. (5.471) because in the former, the number of degrees of freedom in flux and the stoichiometric mass balances are maintained.

It has often been claimed that the kinetic orders g_{ij} and h_{ij} [cf. Eqs. (2.74a,b)] be equivalent to the elasticities defined in Eq. (5.36a) (Savageau *et al.*, 1987b; Cascante *et al.*, 1989a; Sorribas and Savageau, 1989b). Indeed, we have

$$g_{ij} = \frac{\partial \ln v^{+}_{\mathrm{aggr},i}}{\partial \ln S_j}, \tag{5.477a}$$

$$h_{ij} = \frac{\partial \ln v^{-}_{\mathrm{aggr},i}}{\partial \ln S_j}, \tag{5.477b}$$

but $v^{+}_{\mathrm{aggr},i}$ and $v^{-}_{\mathrm{aggr},i}$ have different meanings than v_i in Eq. (5.36a). Because of the method of aggregation of fluxes, they represent aggregate rates of formation and degradation of a substance S_i, whereas v_i in metabolic control analysis denotes the *net* velocity of some reaction, which combines forward and reverse rates of one reaction. Each elasticity ε corresponds to one enzymic reaction, whereas each of the coefficients g_{ij} and h_{ij} generally corresponds to several reactions, which have been aggregated.

Furthermore, taking logarithmic derivatives in Eq. (5.477a) and (5.477b) is a necessary consequence of the power-law formalism, in which rate laws are approximated *ad hoc* by power functions. In contrast, metabolic control analysis is not necessarily based on logarithmic derivatives. One can also use direct derivatives, as in Eq. (5.19). Whether or not normalized quantities are employed is only a question of interpretation (cf. Section 5.7).

Under the condition that matrix $\mathbf{A}$ is nonsingular, Eq. (5.473) can be solved for $\ln \mathbf{S}$ to give

$$\ln \boldsymbol{S} = \mathbf{A}^{-1}\boldsymbol{b}. \tag{5.478}$$

This equation allows one to calculate sensitivities of concentrations with respect to rate constants,

$$\sigma(S_i, a_j) = \frac{\partial \ln S_i}{\partial \ln a_j} = -\frac{\partial \ln S_i}{\partial b_j}, \tag{5.479}$$

$$\sigma(S_i,\beta_j) = \frac{\partial \ln S_i}{\partial \ln \beta_j} = \frac{\partial \ln S_i}{\partial b_j}. \tag{5.480}$$

Here we have replaced the symbol $S(\cdot)$ used for sensitivities in the original work (Savageau, 1976; Savageau *et al.*, 1987b) by $\sigma(\cdot)$, in order to avoid confusion with the symbol for concentrations. By Eq. (5.478), one obtains

$$\sigma(S_i,\beta_j) = -\sigma(S_i,a_j) = (\mathbf{A}^{-1})_{ij}. \tag{5.481}$$

These sensitivities bear a certain analogy to the normalized concentration control coefficients defined in Eq. (5.5). However, the two quantities are not identical, because the sensitivities σ refer to perturbations of the rate constants of aggregate fluxes, whereas control coefficients in metabolic control analysis usually characterize the effect of perturbations of individual enzymes.

The differences between the sensitivities σ and control coefficients become even more obvious by considering the summation theorem derived by Savageau *et al.* (1987b):

$$\sum_{j=1}^{n} [\sigma(S_i,\beta_j) + \sigma(S_i,a_j)] = 0. \tag{5.482}$$

This theorem is not equivalent to the summation theorem (5.42), because each term $\sigma(S_i,\beta_j) + \sigma(S_i,a_j)$ is zero on its own, so that this equation does not properly reflect the contribution of all reactions in the control of the concentration S_i. This is because the particular equations constituting Eq. (5.471) are not coupled with each other *via* the rate constants a_i and β_i. Therefore, an equal fractional increase of only one pair of rate constants, a_i and β_i, leaves the steady state of all S_j unchanged. In contrast, the summation theorem (5.42) is related to the situation that all rate constants of the system are changed by the same fractional amount. This discrepancy is due to the flux aggregation method, which decouples the metabolites. It is no longer considered that consumption of some substance coincides with formation of another (see Figure 5.16). It is an oversimplification to treat the rate constants a_i and β_i (and likewise the kinetic orders) to be independent of the other rate constants a_j and β_j. This was acknowledged by Sorribas and Savageau (1989a) but was not taken into account in their general formalism.

Sensitivities of *rates* with respect to rate constants have also been defined. Because at steady state the total flux feeding into a substance equals the total flux consuming this substance, such sensitivities are, in the flux aggregation approach, only meaningful when defined for unidirectional rates. From Eq. (5.471), one obtains

$$\sigma(v_{\text{aggr},i}^+, \alpha_j) = \frac{\partial \ln v_{\text{aggr},i}^+}{\partial \ln \alpha_j} = \delta_{ij} + \sum_{k=1}^{n} g_{ik}\sigma(S_k, \alpha_j) , \tag{5.483}$$

$$\sigma(v_{\text{aggr},i}^+, \beta_j) = \frac{\partial \ln v_{\text{aggr},i}^+}{\partial \ln \beta_j} = \sum_{k=1}^{n} g_{ik}\sigma(S_k, \beta_j) , \tag{5.484}$$

and similar equations for $S(v_{\text{aggr},i}^-, \beta_j)$. Summation of Eqs. (5.483) and (5.485) over j yields

$$\sum_{j=1}^{n} [\sigma(v_{\text{aggr},i}^+, \alpha_j) + \sigma(v_{\text{aggr},i}^+, \beta_j)] = 1, \tag{5.485}$$

because of Eq. (5.481). At variance with the summation theorem (5.43) of metabolic control analysis, exactly one term of the sum (5.485) equals unity, and all others are zero.

Starting from the equation

$$\mathbf{A}^{-1}\mathbf{A} = \mathbf{I} \tag{5.486}$$

and using Eqs. (5.474), (5.479), and (5.480), the relation

$$\sum_j \left[\sigma(S_i, \alpha_j) g_{jk} + \sigma(S_i, \beta_j) h_{jk} \right] = -\delta_{jk} \tag{5.487}$$

can be derived. For the sensitivities of rates, one can deduce

$$\sum_j \left[\sigma(v_i^+, \alpha_j) g_{jk} + \sigma(v_i^+, \beta_j) h_{jk} \right] = 0. \tag{5.488}$$

Although Eqs. (5.487) and (5.488) exhibit a certain formal analogy to the connectivity theorems (5.53a) and (5.53b), respectively, they are not identical to the latter, nor generalized versions, as the coefficients have a meaning different from the coefficients in metabolic control analysis. For a further discussion on the role of the theorems in both approaches, see the work of Cornish-Bowden (1989).

Power-law approaches have been used not only for sensitivity analysis but also for simulation of biochemical systems far from a chosen reference state (Shiraishi and Savageau, 1993; Torres, 1994). The criticism put forward above concerning small deviations from the reference state is all the more valid for such simulations.

5.17. COMPUTATIONAL ASPECTS

It is an important achievement of metabolic control analysis to have provided a means to quantify the control properties of enzymic reactions embedded in arbitrarily complex metabolic networks. In the previous sections on applications, we selected simple examples and concentrated on analytical solutions for didactic reasons. For complex metabolic networks, the treatment by "pencil and paper" soon becomes impossible, although the basic equations of metabolic control analysis are linear.

A number of powerful computer programs performing calculations in metabolic control analysis on IBM-PC compatibles and to some extent also on UNIX computers have been developed (Hofmeyr and Van der Merwe, 1986; Cornish-Bowden and Hofmeyr, 1991; Letellier *et al.*, 1991; Sauro and Fell, 1991; Sauro, 1993; Mendes, 1993; Thomas and Fell, 1993; Ehlde and Zacchi, 1993).

The program CONTROL developed by Letellier *et al.* (1991) uses the matrix formalism of metabolic control analysis as introduced by Reder (1988). The program is written in Turbo-Pascal and offers two submenus. The first serves to calculate the values of all (normalized or non-normalized) flux control coefficients and concentration control coefficients of a metabolic network from the elasticity coefficients, the values of which must be put in together with the stoichiometry matrix of the network. Information about the rate laws thus enters the computation only via the elasticities. In the second submenu, the link matrix and null-space matrix are calculated in the form given in eqs. (3.7) and (3.47), respectively. The generalized summation theorems (5.44) and connectivity theorems (5.51) are displayed in a form with these matrices specified but the control and elasticity coefficients unspecified (given as symbols).

The program package SCAMP (Sauro and Fell, 1991; Sauro, 1993) running under MS-DOS and on the Atari is a control analysis program and, moreover, a general metabolic simulator. It can be used to make time-dependent simulations by numerically integrating systems of ordinary differential equations. SCAMP also has options to detect and analyze steady states. It makes the conservation relations and all the coefficients defined in metabolic control analysis available. Rate laws can be defined by the user or chosen from a database. The program works by reading an ASCII file of instructions (a command file) detailing the model in a specific command language. The structure of the metabolic network must be given in the form of reaction equations (such as $glucose—S1 for the transformation of glucose treated as an external metabolite into an internal metabolite S_1). SCAMP then translates the command file into an intermediate code that is executed by a run-time interpreter. It is able to generate the stoichiometry matrix and the governing differential equations from the reaction equations and rate laws. The user can select to have some or all control coefficients calculated by numerical modulation or by the matrix method outlined in Section 5.2, and to have elasticities calculated by modulation or by symbolic differentiation. For both simulations and steady-state analysis, additional quantities can be monitored, for example the sum of some control coefficients, or other user-defined quantities or functions. Predefined changes to parameters, for example, after a certain time of simulation, can be made by if/then functions. A routine for graphical output is included.

The program MetaModel (Cornish-Bowden and Hofmeyr, 1991) running on IBM-PC compatibles serves similar purposes as the package SCAMP, but it is menu-operated and therefore more user-friendly. The user is not obliged to learn a specific command language. However, it does not include as many facilities to calculate arbitrary quantities as SCAMP. All rate laws except for a predefined, "minimal" Michaelis–Menten kinetics must be defined by the user. For steady-state calculations, conservation equations have also to be indicated by the user so that the input is somewhat redundant.

The program GEPASI (Mendes, 1993) is also menu-operated, taking advantage of the front-end facilities of MS-Windows, such as menus, dialogue boxes, push buttons, and the help engine. The reactions can be endowed with rate laws chosen from a menu or defined by the user. The input of the values of kinetic parameters is done in a window separate from that for the input of structural data because one is likely to input many different sets of parameters for the same reaction scheme. As in SCAMP, the algorithm used for the integration of the ordinary differential equations is the LSODA (Petzold, 1983), which automatically detects whether or not the system is stiff and uses an appropriate method accordingly. Concentrations, fluxes, elasticities, control coefficients, and response coefficients can be calculated. They can conveniently be plotted versus time or in a two- or three-dimensional phase space, whereby the program GNUPLOT is used. Results can also be written in an output file.

Thomas and Fell (1993) presented the C program MetaCon (under MS-DOS), which is, in essence, an automation of a matrix method developed by Fell and coworkers (Fell and Sauro, 1985; Sauro *et al.,* 1987; Small and Fell, 1989). In the present book, we review that matrix method only in part (on discussing the branch-point relationships in Section 5.4.3.2), because it is equivalent to the method developed by Reder (1988) (cf. Section 5.2) This equivalence was demonstrated by Thomas and Fell (1993) themselves.

The input of the reaction scheme in MetaCon proceeds in a similar way as in SCAMP, by parsing (reading) an input file and creating the corresponding stoichiometry matrix. The elasticities are written in the input file as values or symbolic expressions. MetaCon allows a combination of symbolic (algebraic) and numeric information in a much more extended way than other programs. Depending on the amount of data that can be provided as input, the (normalized) control coefficients can evaluate to a number or can be expressed as algebraic expressions containing enzyme-kinetic constants, equilibrium constants, fluxes, and so forth. A unique feature of MetaCon is a routine to calculate the sensitivities of control coefficients with respect to all elasticities, fluxes, and metabolite concentrations when they appear in the expressions for the control coefficients. In addition, if the elasticities are, in turn, defined in terms of kinetic constants and metabolite concentrations, the sensitivities of the control coefficients to these can also be determined. Knowledge about the effect of deviations in the values of these quantities on control coefficients is particularly valuable when they are inaccurately known. The program uses formulas derived by Small and Fell (1990) and Thomas and Fell (1994) to calculate these sensitivities. MetaCon includes some tests for model validity and integrity. For example, it allows one to check whether all columns of the stoichiometry matrix contain at least one nonzero entry, and to check whether the rank of **N** is smaller than the number of reactions, which is a necessary condition for a nontrivial steady state to exist. The program also produces a message if the null-space matrix contains a row of zeros, which would then correspond to a strictly detailed balanced reaction (see Section 3.3.2).

All programs mentioned in this section are in the public domain. The packages SCAMP, MetaModel, GEPASI, and MetaCon are continuously updated and are available at an ftp server on the Internet (address 161.73.104.10, directory pub/software).

It is worth noting that the approaches presented in Sections 5.8–5.15 are also amenable to automation on computer, which opens interesting programming tasks in the future.

6

Application of Optimization Methods and the Interrelation with Evolution

In the preceding sections, our interest was focused on the mathematical description of the behavior of *variables* of metabolic systems, that is, concentrations of pathway intermediates and fluxes, either in stationary states or in time-dependent states. Other quantities such as the kinetic constants of enzymes or the stoichiometric coefficients which define the topology of enzymic networks are considered as given parameters (i.e., they are inputs of the models). Any explanation for the observed values was not attempted. For traditional simulation models as well as in the context of metabolic control analysis, this distinction between variables and parameters is reasonable. Variations in the concentrations or fluxes may be experimentally observed in short time intervals, whereas the topology of the networks and the kinetic properties may change only very slowly or are even fixed during the life span of an organism.

In the present chapter we draw attention to the fact that, in contrast to chemical systems of an inanimate nature, biochemical systems of living cells are the outcome of evolution. In the light of the Darwinian theory one may state for biochemical systems that during evolution (i) new types of reactions were recruited by the cells leading to an increase in the complexity of biological organization and (ii) existing enzymic systems have adapted to environmental conditions. Both processes have been driven by mutation and natural selection. It seems, therefore, plausible to assume that contemporary metabolic systems have developed by stepwise improvement of their functioning.

Obviously, it would be a formidable task to follow in detail the origination and further development of metabolism during billions of years where living conditions have permanently changed and from which only few traces exist. On the other hand, it may be worth trying to explain the structural features of contemporary enzymic reaction systems on the basis of optimization principles. Certainly, evolution did not lead to a "global optimal state," but it is an experimental fact that mutations or other changes in the structure of present-day metabolism lead in most cases to a worse functioning of the cells (cf. Belfiore, 1980). The concept of optimization is also relevant in the design and improvement of bioreactors.

A crucial point is to formulate appropriate performance functions whose maximum (or minimum) values might correspond to the outcome of the evolution of cellular metabolism. In the literature, the following optimization principles are considered: (a) maximization of reaction rates and steady-state fluxes, (b) minimization of the concentrations of metabolic intermediates, (c) minimization of transient times (for a review, cf. Heinrich *et al.,* 1991). Investigations concerning optimal stoichiometries (Meléndez-Hevia and Isidoro, 1985; Meléndez-Hevia and Torres, 1988; Meléndez-Hevia *et al.,* 1994) and maximization of thermodynamic efficiencies (Stucki, 1980) have also been implemented. Because many properties of cellular reaction systems may influence the fitness of the whole organism, the optimization problem may be considered as a multiobjective one.

In the following quantitative treatments we assume that during evolution of cellular metabolism, some state function Φ was maximized by variations of the system parameters,

$$\Phi(\boldsymbol{p}) = \max. \tag{6.1a}$$

Minimization problems may be transformed into such a maximization principle by considering $-\Phi = \max$. The parameters may enter the performance function Φ directly or via parameter-dependent concentrations or fluxes so that Eq. (6.1a) may be written in more detail as

$$\Phi(\boldsymbol{S}(\boldsymbol{p}), \boldsymbol{J}(\boldsymbol{p}), \boldsymbol{p}) = \max. \tag{6.1b}$$

In optimization studies concerning metabolic systems, one has to take into account certain constraints which may be of different type. First, there are a number of physical constraints limiting the range of variations of kinetic parameters, for example, for the following reasons: (a) any parameter configuration has to meet the thermodynamic equilibrium condition which is independent of the properties of the catalyst, (b) there are upper limits for the elementary rate constants due to physico-chemical constraints, for example, diffusional limitations, and (c) the stoichiometry of metabolic systems has to fulfill certain physical requirements such as mass conservation. Second, there are biological constraints which are often called *cost functions* (Reich, 1983; Rosen, 1986) and which are more difficult to express in clear-cut mathematical terms. Various cost functions possibly relevant for the evolutionary optimization of metabolism have been proposed: (a) the total enzyme content of a cell or a given pathway and (b) the total energy utilization (Reich, 1983; Stucki, 1980; Heinrich *et al.,* 1987). Mathematically, these types of constraints may often be taken into account by the use of the method of Lagrange multipliers, that is, by considering the extremum properties of the function

$$\Phi^* = \Phi - \sum_k \lambda_k(\chi_k - \chi_k^0), \tag{6.2}$$

where λ_k are the Lagrange multipliers and $\chi_k = \chi_k(\boldsymbol{p})$ denote the parameter-dependent cost functions, the values of which are prescribed to be χ_k^0.

Studies have been made on the optimum properties of single enzymes as well as on the mutual interdependence of the enzymes within metabolic pathways. The inclusion of systemic properties into optimization analysis may lead to considerable mathematical difficulties arising from the nonlinearities in the system equations for metabolic networks.

Taking into account not only one but several optimization principles leads to a *multicriteria optimization problem*

$$\Phi_j = \max, \quad j = 1, \ldots, m, \tag{6.3}$$

where the various performance functions may be gathered in a vector $\boldsymbol{\Phi} = (\Phi_1, \ldots, \Phi_m)$. Obviously, the situation may occur that the principles contradict each other, which means that the optimal state is characterized by a trade-off between different performance functions. The role of trade-offs in the evolutionary adaptation of biochemical networks has been stressed also by several other authors (Majewski and Domach, 1985, 1990a; Liljenström and Blomberg, 1987).

Multicriteria optimization is related to the concept of semiordered sets. Traditional optimization approaches in biology start from the assumption that biological systems could be compared according to a total ordering, that is, for any two systems X and Y, exactly one of the relations "greater than," "less than," and "equal to" holds true. Here, "greater than" means that X is better fit than Y so that X will survive when competing with Y. However, there are many instances where biological systems cannot be compared in this way, in particular, if two systems under study do not interfere at all with each other. Moreover, two systems cannot be compared when the ranking in fitness varies with circumstances.

It appears that four different relations should be distinguished,

$$\mathrm{X} > \mathrm{Y}, \qquad \mathrm{Y} > \mathrm{X}, \qquad \mathrm{X} = \mathrm{Y}, \qquad \mathrm{X} \mathbin{?} \mathrm{Y}, \tag{6.4}$$

where the latter relation signifies that X cannot be compared with Y. Upon inclusion of the plausible axiom of transitivity,

$$(X > Y, Y > Z) \Rightarrow X > Z, \tag{6.5}$$

relations (6.4) give rise to a semiordering structure (cf. Rédei, 1967).

The possible relevance of multicriteria optimization principles in biology also follows from the fact that they often give rise to connected or disconnected *manifolds of solutions.* These could account for the rather large variation of biochemical data found in organisms of different species and even of one and the same species. Disconnected solution sets show a conspicuous correspondence with the fact that two given biological species are not generally connected by a continuous line of intermediary forms.

The adequacy of optimization approaches depends essentially on the formulation of appropriate objective functions used to evaluate the fitness of a biological system. Generally, it seems to be difficult to derive the objective functions from more fundamental principles, such as the laws of physics. Accordingly, it is appropriate to derive them from heuristic arguments, and their validity should be judged by comparing theoretically predicted optimum properties with those of real systems. For the optimization of metabolic conversions in bioreactors, the proper choice of the objective functions is less problematic, because they are related to the specific goal of the biotechnological process.

6.1. OPTIMIZATION OF THE CATALYTIC PROPERTIES OF SINGLE ENZYMES

6.1.1. Basic Assumptions

It has often been stressed that evolutionary pressure on the enzyme function was mainly directed toward maximization of catalytic activity,

$$v = \max \tag{6.6}$$

(Fersht, 1974; Crowley, 1975; Albery and Knowles, 1976a, 1976b; Cornish-Bowden, 1976a; Brocklehurst, 1977; Heinrich and Hoffmann, 1991; Pettersson, 1992). This hypothesis is strongly supported by the fact that the rates of enzymatically catalyzed reactions are typically 10^6–10^{12}-fold higher than those of the corresponding uncatalyzed reactions (cf. Voet and Voet, 1990). Obviously, such high reaction rates may only be achieved if the kinetic properties of the enzymes fulfill certain requirements. It has been stated, for example, that enzymes with optimal catalytic activity are characterized by Michaelis constants close to the concentrations of their substrates *in vivo* (Hochachka and Somero, 1973; Cornish-Bowden, 1976a). Other authors came to the conclusion that the K_m values tend to be large relative to the respective substrate concentrations (e.g., Crowley, 1975).

These early studies were mainly based on the most simple enzyme mechanism depicted in Scheme 1 (Section 2.2.2), with the special assumption that the release of product from the enzyme-intermediate complex (step 2) is irreversible.

In the following, we consider an enzymatic reaction which involves two reversible binding processes of the substrate S and product P to the enzyme E and a reversible transformation of two enzyme-intermediate complexes (Scheme 2 in Section 2.2.2; in the present chapter, the substrate and product are denoted as S and P, respectively). The kinetic properties of the enzyme may be described on the basis of Eq. (2.20) which involves the phenomenological parameters K_{m1} and K_{m2} (here denoted by K_{mS} and K_{mP}), V_m^+ and V_m^-. Evolutionary variations of these parameters are interrelated due to their dependence on the rate constants $k_{\pm i}$ of the elementary steps [Eqs. (2.27a)–(2.27d)] and, in particular, due to the Haldane relation $V_m^+ K_{mP}/V_m^- K_{mS} = q =$ const. [Eq. (2.26)]. Therefore, the analysis of evolutionary optimization of the catalytic properties of enzymes on the basis of the principle (6.6) should focus first on variations of the $k_{\pm i}$ values. Thereafter, conclusions concerning optimal values of K_m and V_m may be derived.

The steady-state reaction rate for the enzymatic process depicted in Scheme 2 may be expressed as

$$v = \frac{E_T}{D}(S \cdot q - P)\,, \tag{6.7a}$$

with the thermodynamic equilibrium constant

$$q = \frac{k_1 k_2 k_3}{k_{-1} k_{-2} k_{-3}} \tag{6.7b}$$

and the denominator

$$D = \frac{1}{k_{-3}} + \frac{k_3}{k_{-2}k_{-3}} + \frac{k_2 k_3}{k_{-1}k_{-2}k_{-3}} + \left(\frac{k_1}{k_{-1}k_{-3}} + \frac{k_1 k_2}{k_{-1}k_{-2}k_{-3}} + \frac{k_1 k_3}{k_{-1}k_{-2}k_{-3}}\right)S + \left(\frac{k_2}{k_{-1}k_{-2}} + \frac{1}{k_{-1}} + \frac{1}{k_{-2}}\right)P. \tag{6.7c}$$

We are interested in those values of the elementary rate constants maximizing the absolute value $|v|$ of the reaction rate under the constraints of fixed values of the concentrations of the reactants and of the equilibrium constant q. Without loss of generality, it is assumed that $q \geq 1$. For $q < 1$, the optimal rate constants may be obtained from the solution derived for $q > 1$ by the transformations $v \to -v$ and $q \to 1/q$, and by interchanging the meaning of the symbols k_1 and k_{-3}, k_{-1}

and k_3, k_2 and k_{-2}, as well as of S and P. According to Eq. (6.7) the reaction rate v is a homogeneous function of first degree of the elementary rate constants $k_{\pm i}$, that is

$$v(\alpha k_i, \alpha k_{-i}) = \alpha v(k_i, k_{-i}) \tag{6.8}$$

with an arbitrary value of $a > 0$. For that reason, the rate v could be increased in an unlimited way when no constraints for the rate constants of the elementary reactions are imposed. According to quantum-mechanical and diffusional constraints, it is reasonable to take into account upper bounds on the individual rate constants upon optimizing the reaction rate, that is,

$$k_{\pm i} \leq k_{\pm i,\max}\,. \tag{6.9}$$

Due to Eq. (6.8) and condition (6.9), states of maximal activity have the property that one or more kinetic constants assume their maximal values. Because for q = const., the numerator in Eq. (6.7a) is independent of the rate constants, optimal states are characterized by those values of the rate constants minimizing the denominator D.

For the mechanism depicted in Scheme 2, three groups of kinetic constants may be distinguished: (a) the second-order rate constants k_1 and k_{-3} characterizing the binding of the substrate and product, respectively, to the enzyme, (b) the first-order rate constants k_2 and k_{-2} characterizing the isomerization step, and (c) the first-order rate constants k_{-1} and k_3 characterizing the dissociation of reactants from enzyme-intermediate complexes. Accordingly, we consider three different upper limits for the rate constants

$$k_1, k_{-3} \leq k_{\mathrm{b}}, \tag{6.10a}$$

$$k_2, k_{-2} \leq k_{\mathrm{m}}, \tag{6.10b}$$

$$k_{-1}, k_3 \leq k_{\mathrm{r}}. \tag{6.10c}$$

These conditions indicate that no distinction is made for the allowed ranges of the rate constants belonging to the same group.

In all what follows, dimensionless values for the rate constants $k_{\pm i}$, the concentrations S and P as well as for the enzymic activity v will be used. In order to avoid new symbols we simply redefine the previously used quantities,

$$\frac{k_a}{k_{\mathrm{b}}} \to k_a, \qquad \frac{k_\beta}{k_{\mathrm{m}}} \to k_\beta, \qquad \frac{k_\gamma}{k_{\mathrm{r}}} \to k_\gamma \tag{6.11a}$$

($a = 1, -3$, $\beta = 2, -2$, and $\gamma = -1, 3$) and

$$\frac{k_b S}{k_r} \rightarrow S, \qquad \frac{k_b P}{k_r} \rightarrow P, \qquad \frac{v}{k_r E_T} \rightarrow v. \tag{6.11b}$$

Using normalized rate constants, condition (6.9) assumes the form

$$k_i, k_{-i} \leq 1. \tag{6.12}$$

To simplify the mathematical treatment, we confine ourselves to the special case $k_r/k_m = 1$; that is, no distinction is made between the upper bounds for the two different types of first-order rate constants. Then, by using the normalized quantities, the rate equation keeps the form (6.7) except that the factor E_T is omitted (for a more general treatment, cf. Heinrich and Hoffmann, 1991).

6.1.2. Optimal Values of Elementary Rate Constants

Examination of Eq. (2.28), of which Eq. (6.7) is a special case for $r = 3$, shows that the forward rate constants k_i enter the denominator only together with k_{-i} and $k_{-(i-1)}$ in the form $k_i/(k_{-i}k_{-(i-1)})$, whereas all k_{-i} enter the denominator also in the form $1/k_{-i}$ (note the cyclic notation $k_{-3} = k_{-0}$). This gives rise to the following.

Theorem 6A. *For ordered enzyme reaction mechanisms, a state with nonmaximal values of k_i and k_{-i} or k_i and $k_{-(i-1)}$ cannot be optimal.*

This theorem follows from the fact that such a state can be improved by increasing k_i and k_{-i} or k_i and $k_{-(i-1)}$ by the same factor. This change affects neither the equilibrium constant nor the terms $k_i/(k_{-i}k_{-(i-1)})$ but decreases the terms $1/k_{-i}$ in the denominator of Eq. (2.28). For three-step mechanisms, 10 different optimal solutions L_j are therefore possible for a given value of $q \geq 1$:

(a) Three solutions with a submaximal value of one backward rate constant,

$$L_1: k_{-1} < 1; \qquad L_2: k_{-2} < 1; \qquad L_3: k_{-3} < 1 \tag{6.13a}$$

(b) Three solutions with submaximal values of two backward rate constants,

$$L_4: k_{-1}, k_{-3} < 1; \qquad L_5: k_{-1}, k_{-2} < 1; \qquad L_6: k_{-2}, k_{-3} < 1 \tag{6.13b}$$

(c) Three solutions with submaximal values of one backward rate constant and one forward rate constant,

$$L_7: k_1, k_{-2} < 1; \qquad L_8: k_2, k_{-3} < 1; \qquad L_9: k_3, k_{-1} < 1 \tag{6.13c}$$

(d) One solution with all backward rate constants being submaximal,

$$L_{10}\colon k_{-1}, k_{-2}, k_{-3} < 1 \tag{6.13d}$$

The denominators D_j in Eq. (6.7) for the various solutions L_j may be expressed in terms of S, P, q, and the respective submaximal rate constants given in Eqs. (6.13a)–(6.13d). For example,

$$D_1 = 2 + q + 3qS + (1 + 2q)P, \tag{6.14a}$$

$$k_{-1} = q^{-1}, \tag{6.14b}$$

$$D_9 = 1 + q + qS + P + k_3 + \frac{2(S + P)}{k_{-1}}, \tag{6.15a}$$

$$\frac{k_3}{k_{-1}} = q, \tag{6.15b}$$

and

$$D_{10} = \left(\frac{1}{k_{-3}} + \frac{1}{k_{-2}k_{-3}} + \frac{1}{k_{-1}k_{-2}k_{-3}}\right) + \left(\frac{1}{k_{-1}k_{-3}} + \frac{2}{k_{-1}k_{-2}k_{-3}}\right)S + \left(\frac{1}{k_{-1}} + \frac{1}{k_{-2}} + \frac{1}{k_{-1}k_{-2}}\right)P, \tag{6.16a}$$

$$k_{-1}k_{-2}k_{-3} = q^{-1}. \tag{6.16b}$$

All kinetic constants which, for given D_j, do not enter relations (6.13)–(6.16) assume their maximal values for the indicated solutions; that is, their normalized values are equal to unity.

Under consideration of the interrelations among kinetic parameters due to the fixed equilibrium constant [e.g., Eqs. (6.15b) and (6.16b)], it is seen that the denominators D_4 to D_{10} may attain local minima with respect to variations of the kinetic parameters involved. For example, after elimination of k_3 in Eq. (6.15a) by Eq. (6.15b), D_9 becomes minimum for values of k_{-1} that fulfill the condition

$$\frac{\partial D_9}{\partial k_{-1}} = q - \frac{2(S + P)}{k_{-1}^2} = 0. \tag{6.17}$$

From this equation, one derives

$$k_{-1} = \sqrt{\frac{2(S + P)}{q}}, \quad k_3 = \sqrt{2q(S + P)}. \tag{6.18}$$

In a similar way, one may determine the kinetic constants which minimize the denominators $D_4, \ldots D_8$. The results are listed in Table 6.1 where the rows L_j contain the parameters which minimize D_i. The local minimum of D_{10} is determined by the conditions

$$\frac{\partial D_{10}}{\partial k_{-1}} = (1 + k_{-2})q - \frac{P}{k_{-1}^2}\left(1 + \frac{1}{k_{-2}}\right) = 0, \tag{6.19a}$$

$$\frac{\partial D_{10}}{\partial k_{-2}} = qk_{-1} + qS - \frac{P}{k_{-2}^2}\left(1 + \frac{1}{k_{-1}}\right) = 0. \tag{6.19b}$$

These equations may be transformed into the equation system

$$k_{-1}^4 + k_{-1}^3 - \frac{Pk_{-1}}{q} - \frac{S \cdot P}{q} = 0, \tag{6.20a}$$

$$k_{-2} = \frac{P}{qk_{-1}^2}. \tag{6.20b}$$

After solving the fourth-order equation (6.20a), the rate constants k_{-2} and k_{-3} can be obtained from Eq. (6.20b) and the equation $k_{-3} = (qk_{-1}k_{-2})^{-1}$.

As is seen from Table 6.1, the optimal solutions L_j ($j \geq 4$) depend on the concentrations S and P. Therefore, conditions (6.12) impose various constraints on the allowed (S,P) values, depending on the type of the solution. For example, solution L_9 given in Eq. (6.18) only exists if

$$S + P \leq \frac{1}{2q}, \tag{6.21a}$$

$$S \geq 0, \quad P \geq 0, \tag{6.21b}$$

and solution L_{10} determined by Eq. (6.20) only exists if

$$S \leq \frac{2q}{P} - 1, \tag{6.22a}$$

$$S \geq \frac{P}{q}, \tag{6.22b}$$

$$S \leq P(qP^2 + qP - 1). \tag{6.22c}$$

Inequalities (6.21) and (6.22) and analogous relations for the solutions $L_4, \ldots,$ L_8 define, within the space of reactant concentrations, different subregions where the solutions L_j ($4 \leq j \leq 10$) lead to rate constants fulfilling condition (6.12). The solutions L_1, L_2, and L_3 are independent of S and P and are, therefore, possible for all reactant concentrations. Some of these regions will overlap. There-

Table 6.1 Optimal Solutions for the Rate Constants for the Enzymic Reaction Depicted in Scheme 2 as Functions of the Concentration of the Product for $q \geq 1$

Solution	k_1	k_{-1}	k_2	k_{-2}	k_3	k_{-3}
L_1	1	$\frac{1}{q}$	1	1	1	1
L_2	1	1	1	$\frac{1}{q}$	1	1
L_3	1	1	1	1	1	$\frac{1}{q}$
L_4	1	$\sqrt{\frac{P}{q}}$	1	1	1	$\sqrt{\frac{1}{Pq}}$
L_5	1	$\sqrt{\frac{S + P}{q(1 + P)}}$	1	$\sqrt{\frac{1 + P}{q(S + P)}}$	1	1
L_6	1	1	1	$\sqrt{\frac{2P}{q(1 + S)}}$	1	$\sqrt{\frac{1 + S}{2Pq}}$
L_7	$\sqrt{\frac{2q(1 + P)}{S}}$	1	1	$\sqrt{\frac{2(1 + P)}{Sq}}$	1	1
L_8	1	1	$\sqrt{\frac{2q(1 + S)}{P}}$	1	1	$\sqrt{\frac{2(1 + S)}{Pq}}$
L_9	1	$\sqrt{\frac{2(S + P)}{q}}$	1	1	$\sqrt{2q(S + P)}$	1
L_{10}	$k_1 = k_2 = k_3 = 1,\ k_{-1}^4 + k_{-1}^3 - \frac{Pk_{-1}}{q} - \frac{SP}{q} = 0,\ k_{-2} = \frac{P}{qk_{-1}^2},\ k_{-3} = \frac{1}{qk_{-1}k_{-2}}$					

fore, to make the solutions L_j unique functions of the reactant concentrations, one has to determine for given (S, P) values that solution which gives the highest enzymic activity. This may be achieved by introducing the optimal rate constants for L_j (Table 6.1) into the corresponding expressions D_j and by comparing the resulting minimal denominators $\hat{D}_j$. For $j \leq 9$ the minimal denominators $\hat{D}_j$ are listed in Table 6.2. The determination of $\hat{D}_{10}$ requires numerical solution of Eq. (6.20).

In this way one arrives at a unique subdivision of the (S, P)-plane into subregions R_j such that within region R_j solution L_j applies. In Figure 6.1 these subregions are depicted for $q = 2$. From the results listed in Table 6.1 one may derive that the optimal values of the rate constants change continuously as the values of S and P vary even if a boundary between neighboring regions R_j is

Table 6.2 Denominators $\hat{D}$ of the Kinetic Equation (6.7) Corresponding to the Optimal Solutions L_j for $j = 1, \ldots, 9$

Solution	$\hat{D}(S,P)$
L_1	$\hat{D}_1 = 2 + q + 3qS + (1 + 2q)P$
L_2	$\hat{D}_2 = (1 + 2q)(1 + S + P)$
L_3	$\hat{D}_3 = 3(q + qS + P)$
L_4	$\hat{D}_4 = q + 3qS + P = 4\sqrt{qP}$
L_5	$\hat{D}_5 = 1 + q + 2qS + qP + 2\sqrt{q(S + P)(1 + P)}$
L_6	$\hat{D}_6 = 2q(1 + S) + P + 2\sqrt{2qP(1 + S)}$
L_7	$\hat{D}_7 = 1 + 2qS + P + 2\sqrt{2qS(1 + P)}$
L_8	$\hat{D}_8 = q(1 + S) + 2P + 2\sqrt{2qP(1 + S)}$
L_9	$\hat{D}_9 = 1 + q + qS + P + 2\sqrt{2q(S + P)}$

crossed. From Table 6.1 and Figure 6.1, the following properties of the optimal solutions may be derived:

(a) At very low substrate and product concentrations, optimal enzymic activity is achieved by improving the binding of S and P to the enzyme (solution L_9: *high (S, P)-affinity solution*).

(b) When the substrate is present at a high concentration, it is weakly bound to the enzyme in the optimal state (solution L_7: *low S-affinity solution*). An analogous statement applies to the product (solution L_8: *low P-affinity solution*).

(c) k_2 is always maximum, except for region R_8 where the reaction proceeds backward.

(d) At variance with previous assumptions (e.g., Albery and Knowles, 1976a, 1976b) optimal enzymic activity is not compulsorily achieved by maximal values of the second-order rate constants. As for k_1 this is the case for L_7 and as for k_{-3} for L_3, L_4, L_6, L_8, and L_{10}.

(e) Independent of the equilibrium constant q of the overall reaction, the *internal equilibrium constant* $q_{\text{int}} = k_2/k_{-2}$ equals unity for solutions L_1, L_3, L_4, and L_9. $q_{\text{int}} \cong 1$ is valid for all near-equilibrium enzymes.

The optimal values for the elementary rate constants are not only functions of S and P but they also depend on the equilibrium constant q. For $q \rightarrow 1$, solutions L_1, L_2, and L_3 become identical, whereas the regions R_4, R_5, R_6, and R_{10} disappear.

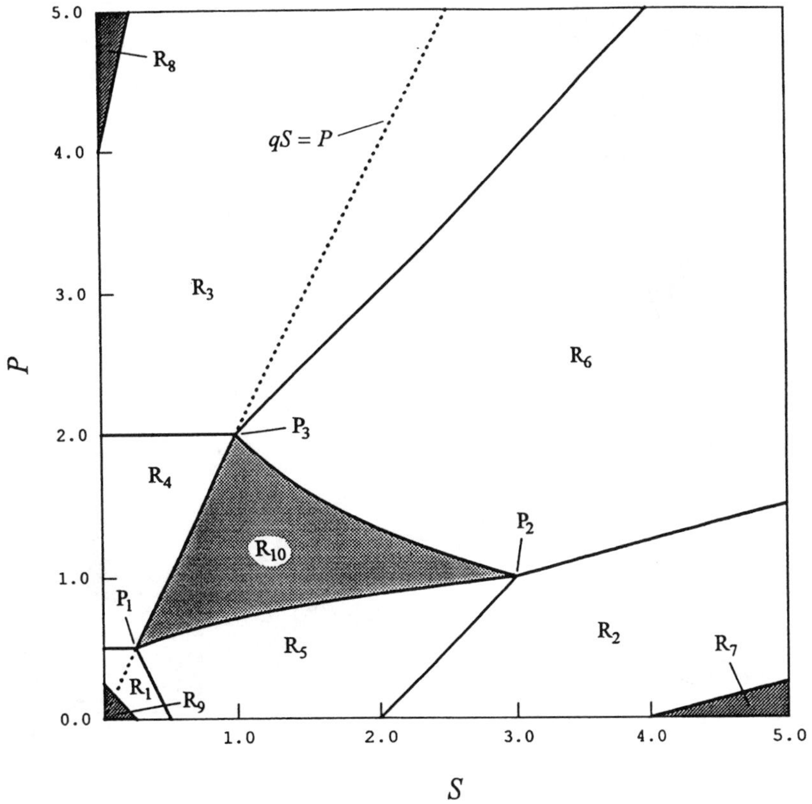

Figure 6.1 Subdivision of the (S, P)-plane into subregions R_i corresponding to the 10 solutions for optimal rate constants of the reversible three-step kinetic mechanism depicted in Scheme 2 for $q = 2$. The vertices P_1, P_2, and P_3 of the central region R_{10} have the coordinates $(1/q^2, 1/q)$, $(2q - 1, 1)$, and $(1, q)$, respectively. Along the dotted line, $qS = P$ holds.

Therefore, the case $q = 1$ is fully characterized by solution L_1 ($k_{\pm i} = 1$) and the solutions L_7, L_8, and L_9 (cf. Figure 6.2). Region R_{10}, which is determined by conditions (6.22a)–(6.22c), increases strongly in size with increasing values of the equilibrium constant q. One may conclude that for irreversible reactions ($q \to \infty$) solution L_{10} becomes valid for all positive values of S and P. According to the central location of region R_{10} within the space of reactant concentrations, the corresponding solution L_{10} has been called the *central solution* (Wilhelm *et al.*, 1994).

For the reactant concentrations $S = P = 1$ which always belong to region

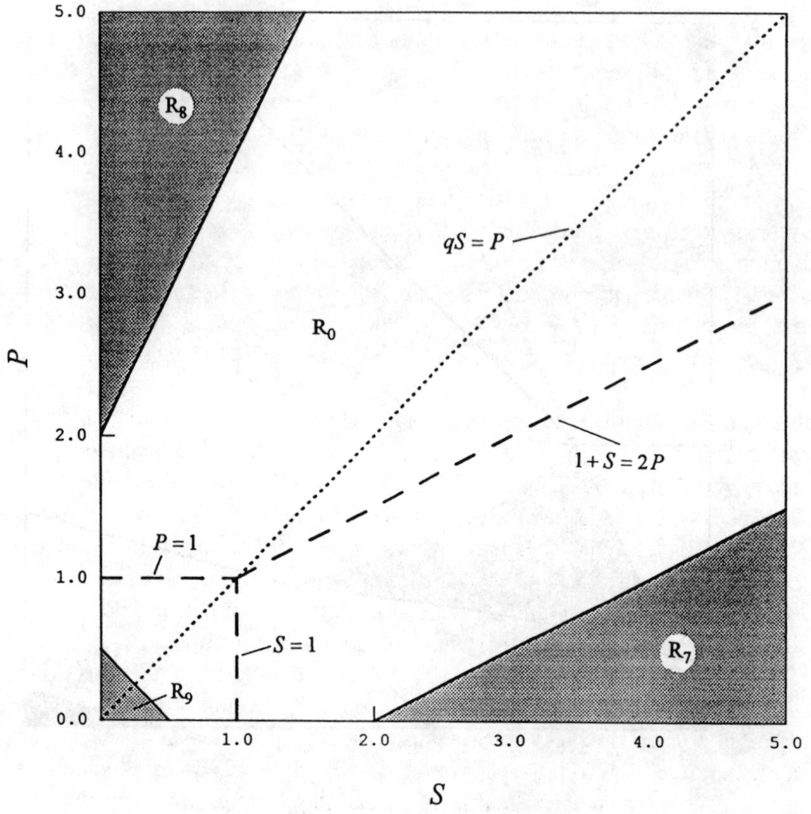

Figure 6.2 Subdivision of the (*S*, *P*)-plane into subregions R_i corresponding to the optimal solutions for the rate constants of the reversible three-step kinetic mechanism depicted in Scheme 2 for $q = 1$. Subregions R_4, R_5, and R_6 degenerate to the broken lines, where $P = 1$, $S = 1$, and $1 + S = 2P$, respectively. R_{10} degenerates to the point $S = P = 1$. The optimal solutions for regions R_1, R_2, and R_3 (combined to region R_0) are identical.

R_{10}, the fourth-order equation (6.20a) can be solved analytically. One obtains the two real solutions $k_{-1} = \sqrt[3]{1/q}$ and $k_{-1} = -1$, and the two complex solutions $k_{-1} = (-1/2 \pm i\sqrt{3/2})\sqrt[3]{1/q}$. As only the positive real solution is relevant, we conclude, with the help of Eqs. (6.16b) and (6.20b),

$$k_{-1} = k_{-2} = k_{-3} = \sqrt[3]{\frac{1}{q}}, \qquad k_1 = k_2 = k_3 = 1. \tag{6.23}$$

This solution shows some correspondence to that proposed by Stackhouse *et al.* (1985). In their *descending staircase* model, it was suggested that in the optimal state each of the three catalytic steps contributes equally to the equilibrium constant q.

It is important to note that the existence of 10 different solutions L_j is a direct consequence of the fact that upper limits are not only introduced for the second-order rate constants but also for the first-order rate constants [cf. Eq. (6.10)]. This may be seen as follows. In the limit $k_r = k_m \rightarrow \infty$, for finite values of the second-order rate constant k_b one gets infinitesimally small values of the normalized reactant concentrations (cf. Eq. (6.11b)). Accordingly, only one solution, namely L_9, applies as $S + P < 1/2q$ [cf. Eq. (6.21a)]. In terms of non-normalized quantities, solution L_9 reads

$$k_1 = k_{-3} = k_b, \tag{6.24a}$$

$$k_2 = k_{-2} = k_m, \tag{6.24b}$$

$$k_{-1} = \sqrt{\frac{2k_b k_m}{q}(S + P)}, \tag{6.24c}$$

$$k_3 = \sqrt{2qk_b k_m(S + P)}. \tag{6.24d}$$

From these equations, it follows that for fixed non-normalized reactant concentrations the limit $k_r = k_m \rightarrow \infty$ will lead to infinite values of the first-order rate constants k_{-1}, k_3 and k_2, k_{-2}, whereas the second-order rate constants k_1 and k_{-3} remain finite. Taking into account the rate equation (6.7) for solution L_9 [Eq. (6.18)], one obtains, with D_9 from Table 6.2, in the limit $k_r = k_m \rightarrow \infty$,

$$v = \frac{k_b E_T(qS - P)}{1 + q}. \tag{6.25}$$

This expression is identical with the formula proposed by Albery and Knowles (1976a, 1976b) and Pettersson (1989) for the rate equation of a *perfect catalyst.* However, in addition to solution L_9, we have derived another nine solutions which are to be considered as not less "perfect" if the normalized reactant concentrations S and P are not small compared to $1/2q$.

The procedure for calculating kinetic parameters in states of maximal activity outlined above may be generalized to enzymic reactions involving arbitrary numbers of elementary steps (Wilhelm *et al.,* 1994). For example, using the rate equation (2.28) one obtains for an ordered uni-uni reaction with n elementary steps the following high-affinity solution

$$k_{-1} = \sqrt{\frac{(n-1)(S+P)}{q(n-2)}}, \qquad k_n = \sqrt{\frac{q(n-1)(S+P)}{n-2}} \tag{6.26}$$

($k_1 = k_{-n} = k_{\pm i} = 1, i = 2, \ldots, n-1$) which is a generalization of solution L_9 obtained for $n = 3$. The general solution for $S = P = 1$ reads

$$k_{-i} = \sqrt[n]{\frac{1}{q}}, \quad k_i = 1, \quad i = 1, \ldots, n. \tag{6.27}$$

Other generalizations concern enzymic reactions with more than one substrate or product (Wilhelm *et al.*, 1994). For example, for a bi-uni reaction where the enzyme catalyzes the interconversion of two substrates S_1 and S_2 into one product P by an ordered four-step mechanism, optimization of the elementary rate constants results, according to Theorem 6A, in 31 different solutions. Most remarkably, one obtains also in this case a *central solution* characterized by maximal values for all forward rate constants and submaximal values for all backward rate constants. In the space of the reactant concentrations S_1, S_2, and P, this central solution for bi-uni reactions applies in a three-dimensional central region which has the topology of a tetrahedron and which increases in size with increasing values of the thermodynamic equilibrium constant q (cf. the property of the central region R_{10} for the three-step mechanism of uni-uni reactions).

Example. Let us consider the hydrolysis of pyrophosphate to inorganic phosphate, that is, the reaction $PP_i \rightarrow P_i + P_i$ which is catalyzed by the enzyme inorganic pyrophosphatase (EC 3.6.1.1). The detailed catalytic mechanism with four reaction steps and the participation of magnesium is depicted in Figure 6.3.

The following first-order and second-order rate constants of the elementary steps have been reported (Baykov *et al.*, 1990, 1993):

$$k_1 = 14 \times 10^7/\text{M s}, \quad k_2 = 400/\text{s}, \quad k_3 = 390/\text{s}, \quad k_4 = 600/\text{s}, \tag{6.28a}$$

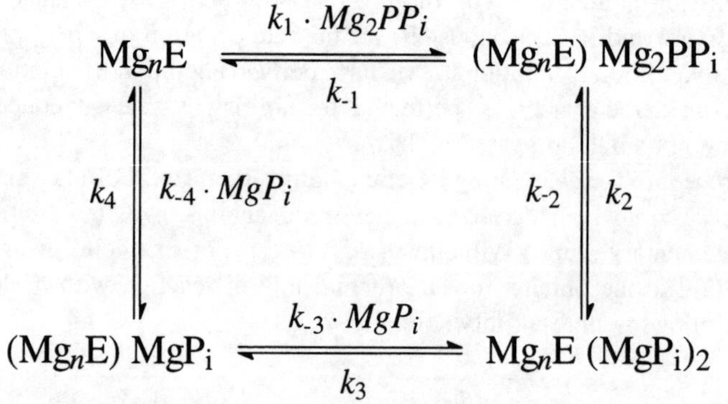

Figure 6.3 Reaction scheme of inorganic pyrophosphatase.

$$k_{-1} = 12/\text{s}, \qquad k_{-2} = 81/\text{s}, \tag{6.28b}$$
$$k_{-3} = 1.3 \times 10^5/\text{M s}, \qquad k_{-4} = 20 \times 10^5/\text{M s}.$$

The equilibrium constant corresponding to these data reads

$$q = \frac{k_1 k_2 k_3 k_4}{k_{-1} k_{-2} k_{-3} k_{-4}} \cong 51.85 \text{ M}. \tag{6.29}$$

Let us assume, for simplicity's sake, that the highest value in the group of the first-order rate constants ($k_2,k_3,k_4,k_{-1},k_{-2}$) and the highest value in the group of the second-order rate constants (k_1,k_{-3},k_{-4}) approximately represent in each case the upper bound for the rate constants (i.e., $k_{\text{monomol}}^{\max} = k_4$ and $k_{\text{bimol}}^{\max} = k_1$). Normalization of the data given in Eqs. (6.28a) and (6.28b) leads to

$$k_1 = 1, \qquad k_2 = 0.67, \qquad k_3 = 0.65, \qquad k_4 = 1, \tag{6.30a}$$

$$k_{-1} = 0.02, \qquad k_{-2} = 0.135, \qquad k_{-3} = 0.0009, \qquad k_{-4} = 0.014, \tag{6.30b}$$

with a normalized equilibrium constant $q = 1.28 \times 10^7$. For such a high equilibrium constant, the model predicts optimal elementary rate constants which belong to the central solution for the four-step mechanism where all forward rate constants are maximal, and all backward rate constants assume submaximal values. Because in Eqs. (6.30a) and (6.30b) all normalized forward rate constants are close to unity and all normalized backward rate constants are much smaller, the data given in Eqs. (6.30a) and (6.30b) correspond rather well to the theoretical expectations. Furthermore, the internal rate constants of the catalytic step amounts to $q_{\text{int}} = k_2/k_{-2} \cong 4.94$ which is much smaller than the normalized thermodynamic equilibrium constant q ($q_{\text{int}}/q \cong 3.86 \times 10^{-7}$). This supports the hypothesis that the internal equilibrium constants of enzyme catalyzed reactions are close to unity [see above and the works of Burbaum *et al.* (1989), Pettersson (1991), and Wilhelm *et al.* (1994)]. One may conclude, therefore, that the kinetic design of pyrophosphatase has been selected with respect to flux maximization.

6.1.3. Optimal Michaelis Constants

The kinetic equation (6.7) may be rewritten in the form of the reversible Michaelis–Menten equation given in Eq. (2.20). The relations (2.27c) and (2.27d) for the Michaelis constants remain valid using normalized quantities if these constants are scaled in the same way as the reactant concentrations [cf. Eq. (6.11b)]. Optimal values for these phenomenological parameters are obtained by introducing $k_{\pm i}$ from Table 6.1 into expressions (2.27c) and (2.27d). For simplicity's sake we consider only the Michaelis constants for two special cases:

(a) Case $S = P = 1$: For these values of the normalized reactant concentrations, solution L_{10} applies and from expressions (2.27c) and (2.27d) one obtains with Eq. (6.23)

$$K_{mS} = \frac{1 + q^{1/3} + q^{2/3}}{q^{1/3}(1 + 2q^{1/3})}, \tag{6.31a}$$

$$K_{mP} = \frac{1 + q^{1/3} + q^{2/3}}{2 + q^{1/3}} \tag{6.31b}$$

and from that

$$\frac{\partial K_{mS}}{\partial q} < 0, \tag{6.32a}$$

$$\frac{\partial K_{mP}}{\partial q} > 0. \tag{6.32b}$$

From these relations it follows that for solution L_{10} higher equilibrium constants imply lower K_{mS} and higher K_{mP} values. Using $S = P = 1$, one derives with Eqs. (6.31a) and (6.31b)

$q = 1$:

$$\frac{S}{K_{mS}} = 1, \tag{6.33a}$$

$$\frac{P}{K_{mP}} = 1, \tag{6.33b}$$

$q \gg 1$:

$$\frac{S}{K_{mS}} \cong 2, \tag{6.33c}$$

$$\frac{P}{K_{mP}} \cong q^{-1/3}. \tag{6.33d}$$

Relations (6.33a) and (6.33c) bear the interesting fact that the optimal Michaelis constant K_{mS} of the substrate is of the same order of magnitude as the substrate concentration S, irrespective of the equilibrium constant q. This is not the case for the relation between K_{mP} and P, at least for the considered solution (L_{10}).

(b) Case $q = 1$: As outlined above, this case is fully characterized by solutions L_7, L_8 and L_9 as well as $L_1 = L_2 = L_3$. Using the expressions listed in Table 6.1 one gets with Eqs. (2.27c) and (2.27d)

L_j ($j \leq 6, j = 10$):

$$K_{mS} = 1, \qquad K_{mP} = 1, \tag{6.34}$$

L_7:

$$K_{mS} = \sqrt{\frac{S}{2(1 + P)}}, \qquad K_{mP} = 1 \tag{6.35}$$

L_8:

$$K_{mS} = 1, \qquad K_{mP} = \sqrt{\frac{P}{2(1 + S)}} \tag{6.36}$$

L_9:

$$K_{mS} = K_{mP} = \sqrt{2(S + P)} \tag{6.37}$$

Equations (6.34)–(6.37) give a strong support of the hypothesis that higher reactant concentrations imply higher Michaelis constants. Let us first consider solution L_9 which for $q = 1$ applies for $2(S + P) < 1$. In this concentration range, K_{mS} and K_{mP} are monotonic increasing functions of the reactant concentrations. In particular, in the limiting case $S,P \to 0$ one obtains $K_{mS},K_{mP} \to 0$. On the other hand, solutions L_7 and L_8 which are applicable for high substrate concentrations and high product concentrations, respectively, have high K_{mS} and K_{mP} values [cf. Eqs. (6.35) and (6.36)]. For all solutions the following relations hold:

$$\frac{\partial K_{mS}}{\partial S} \geq 0, \qquad \frac{\partial K_{mP}}{\partial P} \geq 0. \tag{6.38}$$

The results given in Eqs. (6.34)–(6.37) may be visualized in a space with the Michaelis constants as coordinates (Figure 6.4).

The solutions L_j ($j \leq 6, j = 10$) are represented by the point ($K_{mS} = 1, K_{mP} = 1$), whereas solutions L_7, L_8, and L_9 are represented by lines. It is seen that solution L_9 which is valid for low concentrations of S and P is characterized by low values of both K_{mS} and K_{mP}. Solutions L_7 and L_8 applicable for high concentrations of S and P are characterized by high values of K_{mS} and K_{mP}, respectively.

6.2. OPTIMIZATION OF MULTIENZYME SYSTEMS

6.2.1. Maximization of Steady-State Flux

The maximization of catalytic efficiencies as studied for single enzymes remains relevant also in the context of enzymic networks. Here, the difficulty arises that the concentrations of the intermediates (i.e., the substrates and products of the participating enzymes) are not fixed but depend on the kinetic parameters,

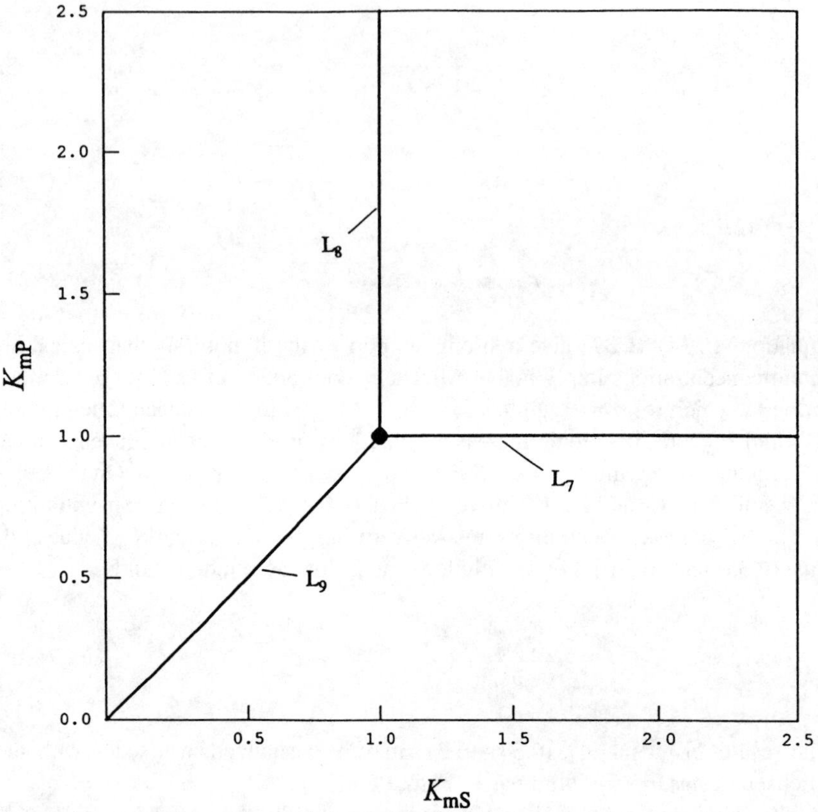

Figure 6.4 Michaelis constants K_{mS} and K_{mP} in optimal states of the three-step mechanism depicted in Scheme 2 for $q = 1$ according to Eqs. (6.34)–(6.37). The point $K_{mS} = K_{mP} = 1$ characterizes the solutions L_1 to L_6 and L_{10}.

which have changed during biological evolution. Moreover, due to the nonlinearity of most rate equations, the mathematical treatment is hampered by the fact that there are generally no explicit expressions for the parameter dependence of the performance function Φ. This holds true even for unbranched chains (Scheme 11, Section 5.4.3.1), if they include saturable enzymes. The calculation of kinetic parameters in states of maximal steady-state activity (J = max) could be based on Eq. (5.82). With $S_{n+1} = P_2 =$ const. one arrives at an implicit nonlinear equation, which cannot, in general, be solved for J [see comments on Eqs. (5.82) and (5.83)]. For simplicity's sake we here consider only the case of nonsaturated enzymes where J may be expressed analytically as a function of the first-order rate constants k_{+j} and k_{-j} of the participating reactions [cf. Eqs. (5.85) and

(5.88)]. If each reaction is described by a three-step mechanism as depicted in Scheme 2 (Section 2.2.2), one obtains with Eqs. (2.27a)–(2.27d)

$$k_j = \frac{V_j^+}{K_j^+} = \frac{k_{1,j}k_{2,j}k_{3,j}E_j}{k_{2,j}k_{3,j} + k_{-1,j}k_{3,j} + k_{-1,j}k_{-2,j}}, \tag{6.39a}$$

$$k_{-j} = \frac{V_j^-}{K_j^-} = \frac{k_{-1,j}k_{-2,j}k_{-3,j}E_j}{k_{2,j}k_{3,j} + k_{-1,j}k_{3,j} + k_{-1,j}k_{-2,j}}, \tag{6.39b}$$

where E_j denotes the concentration of the enzyme catalyzing reaction j. $k_{\pm i,j}$ are the rate constants of the elementary reactions of enzyme E_j. As is often done (Albery and Knowles, 1976a, 1976b), we introduce apparent second-order rate constants κ_j and κ_{-j}, so that Eqs. (6.39a) and (6.39b) can be written as

$$k_j = \kappa_j E_j, \qquad k_{-j} = \kappa_{-j}E_j. \tag{6.40}$$

To distinguish between the contributions of the enzyme concentration and the (intrinsic) second-order rate constants to the catalytic efficiency of the particular enzymes, we first consider a reference state where all enzyme concentrations are equal, $E_j = \tilde{E}$. For this state, the characteristic times read [cf. Eq. (4.1)]

$$\tilde{\tau}_j = \frac{1}{\tilde{E}(\kappa_j + \kappa_{-j})}. \tag{6.41}$$

In other states, with the enzyme concentrations E_j, the characteristic times are

$$\tau_j = \tilde{\tau}_j \frac{\tilde{E}}{E_j}. \tag{6.42}$$

With the equilibrium constants

$$q_j = \frac{k_j}{k_{-j}} = \frac{\kappa_j}{\kappa_{-j}}, \tag{6.43}$$

it follows from Eqs. (6.40)–(6.43) that

$$k_j = \frac{q_j E_j}{\tilde{\tau}_j(1 + q_j)\tilde{E}}, \quad k_{-j} = \frac{E_j}{\tilde{\tau}_j(1 + q_j)\tilde{E}}. \tag{6.44}$$

Equation (5.88) for the steady-state flux may be rewritten in terms of the relaxation times, with the help of Eq. (6.44),

$$J = \left(P_1 \prod_{j=1}^{r} q_j - P_2\right)\left(\tilde{E} \sum_{m=1}^{r} \frac{\tilde{\tau}_m}{E_m}(1 + q_m) \prod_{j=m+1}^{r} q_j\right)^{-1} \tag{6.45}$$

In the following, we are interested in those enzyme concentrations maximizing the steady-state flux J under the constraint

$$\sum_{j=1}^{r} E_j \leq E_{\text{tot}}, \tag{6.46}$$

which expresses the fact that the total enzyme concentration for a metabolic pathway is limited by the capacity of the living cell to synthesize proteins (Waley, 1964). Because expression (6.45) is a homogeneous function of first degree of the enzyme concentrations, their total must equal E_{tot} in optimal states. Using the method of Lagrange multipliers, the spectrum of optimal enzyme concentrations is determined by the condition

$$\frac{\partial}{\partial E_j}\left[J - \lambda\left(\sum_{m=1}^{r} E_m - E_{\text{tot}}\right)\right] = \frac{\partial J}{\partial E_j} - \lambda = 0. \tag{6.47}$$

Introducing expression (6.45) into Eq. (6.47) yields

$$\frac{E_j}{E_k} = \sqrt{\frac{\tilde{\tau}_j(1 + q_j)}{\tilde{\tau}_k(1 + q_k)} \prod_{l=j+1}^{k} q_l} \quad \text{with } k > j. \tag{6.48}$$

For the special case of three enzymes, a similar equation had been derived by Waley (1964). Equation (6.48) leads to the conclusion that the concentrations of slow enzymes (i.e., of enzymes with long characteristic times $\tilde{\tau}$ in the reference state) are in states of maximal steady-state activity generally higher than those of fast enzymes. In other words, poor catalysts should be present in high concentrations. However, the optimal distribution of enzyme concentrations also depends on the equilibrium constants. In the special case that all the enzymes have the same catalytic efficiency (i.e., $\tilde{\tau}_j = \tilde{\tau}$), Eq. (6.48) predicts for all $q_j > 1$ a monotonic decrease of the enzyme concentrations from the beginning toward the end of the chain. If all the enzymes have the same intrinsic properties ($\tilde{\tau}_j = \tilde{\tau}$, $q_j = q$), one derives from Eq. (6.48) the relation

$$E_j = E_k\sqrt{q^{k-j}}. \tag{6.49}$$

Introducing this relation into the condition of fixed total enzyme concentration, with the formula of geometric progressions one obtains

$$E_j = \frac{E_{\text{tot}}(q^{1/2} - 1)q^{(r-j)/2}}{q^{r/2} - 1}. \tag{6.50}$$

In Figure 6.5 the optimal enzyme concentrations are depicted as functions of the positions j and the equilibrium constant q according to formula (6.50). It is clearly seen that for $q > 1$ the enzyme concentrations decrease monotonically toward the end of the chain. This decrease is the stronger the higher the equilibrium constant is. In the limiting case $q = 1$, Eq. (6.50) describes a uniform distribution $E_j = E_{\text{tot}}/r$.

Inserting the distribution (6.48) into Eq. (6.45), one arrives at an expression for the optimal flux which reads (also in the general case that the equilibrium constants are not equal to each other)

$$J = E_{\text{tot}}\left(P_1 \prod_{j=1}^{r} q_j - P_2\right)\left[\tilde{E}\left(\sum_{j=1}^{r} \sqrt{\tilde{\tau}_j(1 + q_j) \prod_{m=j+1}^{r} q_m}\right)^2\right]^{-1}. \tag{6.51}$$

From this equation, in the limits $q_j \to \infty$ and $q_j \to 1$ one derives

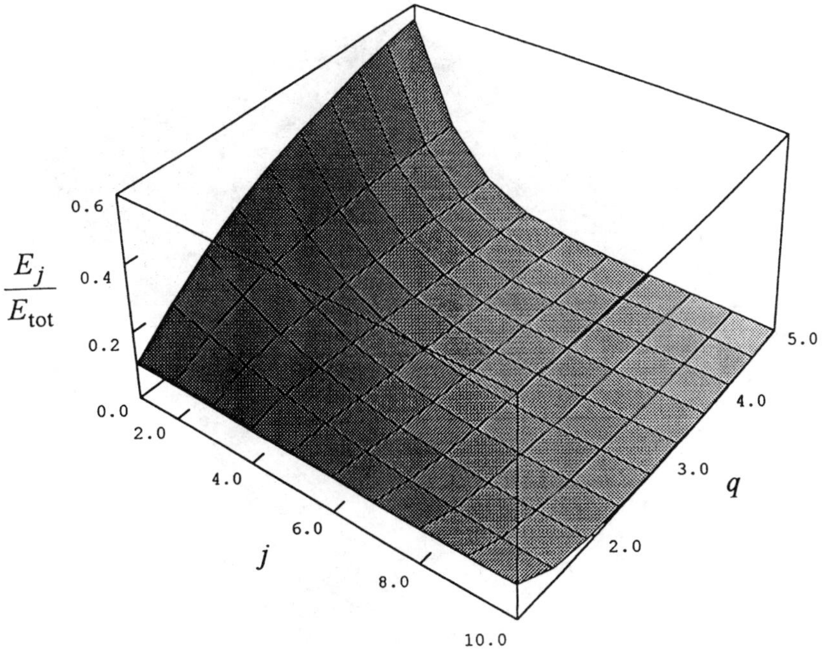

Figure 6.5 Optimal enzyme concentrations E_j/E_{tot} for an unbranched reaction chain with $r = n + 1 = 10$ in states of maximal steady-state flux as functions of the equilibrium constants for the case $q_j = q$, $\tilde{\tau}_j = \tilde{\tau}$), according to Eq. (6.50)

$$J = \begin{cases} \dfrac{E_{tot}}{\tilde{E}} \dfrac{P_1}{\tilde{\tau}} & \text{for } q_i \to \infty \\ \\ \dfrac{E_{tot}}{\tilde{E}} \left(\dfrac{P_1 - P_2}{\tilde{\tau} r^2} \right) & \text{for } q_i \to 1. \end{cases} \tag{6.52}$$

In Figure 6.6, the optimal flux is represented as a function of the equilibrium constant and the chain length. From Eq. (6.51) for the special case $q_j = q$ and Eq. (6.52), one may conclude that for $1 \leq q < \infty$ the optimal flux decreases with increasing chain length (cf. Figure 6.6). This decrease is the stronger the lower the equilibrium constant is, as long as $q > 1$.

In the present case, where the reactions are described by linear rate equations, the optimal distribution of enzyme concentrations is independent of concentrations P_1 and P_2 of the initial substrate and the end product, respectively, of the pathway. This is no longer the case if saturation kinetics of the individual enzymes is taken into account (cf. Heinrich *et al.,* 1987; Heinrich and Hoffmann, 1991).

It is worth mentioning that optimization of the steady-state flux (J = max)

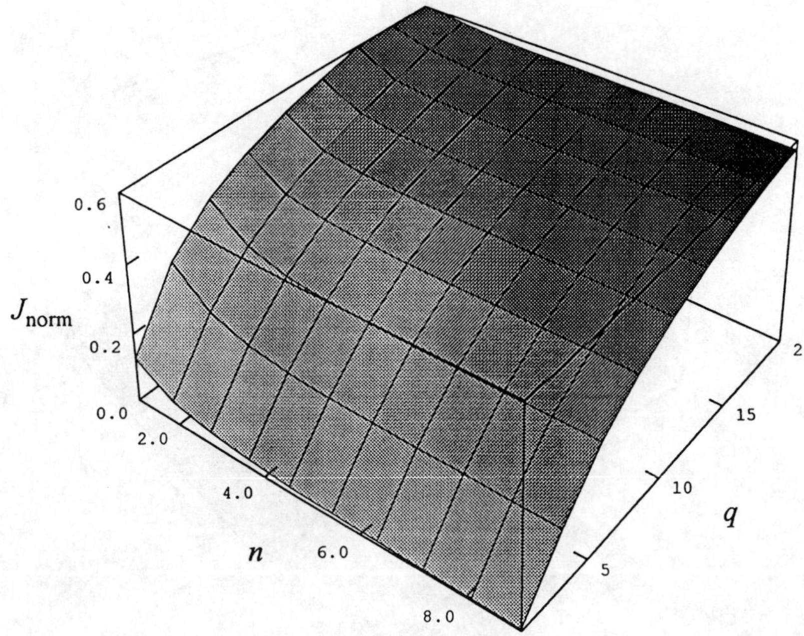

Figure 6.6 Optimal steady-state flux of an unbranched reaction chain as a function of the chain length and the equilibrium constant $q_j = q$ of the participating reactions for the case $\tilde{\tau}_j = \tilde{\tau}$, $P_2 = 0$ according to Eq. (6.51). The figure shows the normalized optimal flux $J_{norm} = J \cdot \tilde{E}\tilde{\tau}/E_{tot}P_1$.

under the constraint of fixed total enzyme concentration (E_{tot} = const.) is mathematically equivalent to the problem of minimizing the total enzyme concentration at fixed steady-state flux (Heinrich *et al.*, 1987; Brown, 1991). In the latter case, the method of Lagrange multipliers leads to the variational equation

$$\frac{\partial}{\partial E_j}\left(\sum_{m=1}^{r} E_m - \lambda'(J - J^0)\right) = 1 - \lambda' \frac{\partial J}{\partial E_j} = 0, \tag{6.53}$$

which may be transformed into Eq. (6.47) by choosing $\lambda' = 1/\lambda$. Therefore, the solution of Eq. (6.53) leads to the enzyme distribution of optimal enzyme concentrations as given in Eq. (6.48).

Now we show that there is a close relation between flux maximization and the extremal properties of other quantities.

Maximization of total entropy production: For the unbranched chain, the entropy production reads

$$\sigma = \sum_{j=1}^{r} \frac{v_j A_j}{T} = J \cdot R \ln\left(\prod_{j=1}^{r} q_j \frac{P_1}{P_2}\right), \tag{6.54}$$

where A_j denotes the affinity of reaction j [cf. Eq. (2.16)]. As P_1 and P_2 as well as all q_j are considered to be constant, the principles J = max and σ = max are, for unbranched chains, equivalent to each other. The principle of maximal entropy production (i.e., the establishment of system states far from thermodynamic equilibrium) was suggested to play an important role in the evolution of biochemical systems (Nicolis and Prigogine, 1977, pp. 442–445).

Maximization of growth rate: The cellular growth rate can be expressed as

$$G = \frac{1}{V}\frac{dV}{dt}, \tag{6.55}$$

where V denotes the cellular volume (Kacser and Beeby, 1984). To apply the principle G = max, one may again consider the unbranched chain depicted in Scheme 11 (Section 5.4.3.1) by taking into account that enzymes are not only the catalysts of metabolic systems but also some of their net products. In a simple model, one may assume a proportionality between the rate of enzyme production and the steady-state flux J, that is, $dE_{tot}/dt \propto J$. Neglecting the synthesis of structural proteins and using the assumption that the cell volume is proportional to the protein content of the cell, the growth rate may be expressed as

$$G(E_1, \ldots, E_r) = \frac{J}{\sum_{j=1}^{r} E_j}. \tag{6.56}$$

The effect of changes in the profile of enzyme concentrations along the pathway on the growth rate is described by the variational equation

$$\frac{\partial G}{\partial E_j} = \frac{1}{E_{\text{tot}}} \frac{\partial J}{\partial E_j} - \frac{J}{E_{\text{tot}}^2} . \tag{6.57}$$

The optimal distribution of the enzyme concentrations is determined by $\partial G/\partial E_j = 0$, that is, by

$$\frac{\partial J}{\partial E_j} = \frac{J}{E_{\text{tot}}} . \tag{6.58}$$

As this condition is equivalent to Eq. (6.47), the enzyme distribution maximizing the exponential growth rate and that maximizing the steady-state flux under the constraint of fixed total enzyme concentration are the same (see also Reich, 1983, 1985).

The problem of identifying optimal enzyme concentrations bears some relationship to *metabolic control analysis* which may be seen as follows. From Eq. (6.47) one derives

$$C_j^J = \frac{E_j}{J} \frac{\partial J}{\partial E_j} = \lambda \frac{E_j}{J} , \tag{6.59}$$

which implies

$$C_j^J = \frac{E_j}{E_{\text{tot}}} , \tag{6.60}$$

due to the summation theorem of flux control coefficients [cf. Eq. (5.43)]. This relation means that in states of maximal steady-state activity the normalized control coefficients C_j^J and the optimal enzyme concentrations in unbranched pathways show the same distribution (see also Heinrich and Holzhütter, 1985; Brown, 1991).

6.2.2. Influence of Osmotic Constraints and Minimization of Intermediate Concentrations

In the context of evolutionary optimization of metabolic pathways considerations about the limited solvent capacity and the osmotic balance of living cells may play an important role (Atkinson, 1969; Savageau, 1976; Heinrich *et al.*, 1987; S. Schuster and Heinrich, 1991). Because most molecules in the living cell contain polar groups or are electrically charged, they fix cell water by hydration. In view of the huge number of different substances in the cell, it was argued that the total solute concentration should be low enough in order to allow sufficiently fast diffusion. For all cells having no cell wall, an additional constraint for concentrations results from the fact that these cells must be in osmotic equilibrium with the extracellular medium. This constraint could not be fulfilled if the numerous intracellular substances had too large concentrations. The fact that enzyme concentrations are usually very low in comparison with their substrates might also be rationalized by solvation and osmotic pressure arguments, although the metabolic effort needed for enzyme synthesis is probably the limiting factor (cf. Brown, 1991).

Let us again consider the unbranched pathway depicted in Scheme 11 (Section 5.4.3.1). It is easy to see that these osmotic conditions are not always fulfilled for a distribution of enzyme concentrations as given in Eq. (6.48). If all equilibrium constants are greater than unity and all $\tilde{\tau}_j$ are of the same order of magnitude, the decrease of the optimal enzyme concentrations toward the end of the chain will result in a strong accumulation of intermediate concentrations. In particular, one derives from Eq. (6.48) that in the limit $q_j \to \infty$ one obtains $E_1 = E_{\text{tot}}$ and $E_j \to 0$ for $j \geq 2$, which implies infinite steady-state concentrations S_i. Besides an upper limit for the total enzyme concentration, one may, therefore, take into account an upper limit for the total concentration of intermediates, that is,

$$\Omega = \sum_{i=1}^{n} S_i \leq \Omega^0. \tag{6.61}$$

Ω represents the total osmolarity of the pathway under the assumption that all osmotic coefficients are equal to unity. Using this condition in the form of an equality constraint the variational equation for the determination of optimal enzyme concentrations reads

$$\frac{\partial}{\partial E_j}\left[J - \lambda_1\left(\sum_{m=1}^{r} E_m - E_{\text{tot}}\right) - \lambda_2\left(\sum_{i=1}^{n} S_i - \Omega^0\right)\right] = 0 \tag{6.62}$$

with the Lagrange multipliers λ_1 and λ_2. Although solutions of this equation can generally be found only numerically (Heinrich *et al.*, 1987), treatment of the limit $q_j \to \infty$ is rather easy. In this case, Eq. (6.44) implies

$$k_j = \frac{E_j}{\tilde{\tau}\tilde{E}}, \qquad k_{-j} = 0. \tag{6.63}$$

Therefore, the steady-state flux and the sum of the steady-state concentrations may be expressed as

$$J = \frac{E_1 P_1}{\tilde{\tau}_1 \tilde{E}}, \tag{6.64a}$$

$$\Omega = P_1 \sum_{i=1}^{n} \frac{\tilde{\tau}_{i+1} E_1}{\tilde{\tau}_1 E_{i+1}}, \tag{6.64b}$$

respectively. The optimal enzyme concentrations are determined by the following variational equation:

$$\frac{\partial}{\partial E_j}\left(\frac{E_1 P_1}{\tilde{\tau}_1 \tilde{E}}\right)\left(1 - \lambda_2 \tilde{E} \sum_{i=1}^{n} \frac{\tilde{\tau}_{i+1}}{E_{i+1}}\right) = 0, \tag{6.65}$$

which contains only the Lagrange multiplier for the osmotic constraint because the condition of fixed total enzyme concentration may be taken into account by the relation

$$E_1 = E_{\text{tot}} - \sum_{m=2}^{r} E_m. \tag{6.66}$$

From these equations, one derives for $j \geq 2$

$$\left(-\frac{P_1}{\tilde{\tau}_1 \tilde{E}}\right)\left(1 - \lambda_2 \tilde{E} \sum_{i=1}^{n} \frac{\tilde{\tau}_{i+1}}{E_{i+1}}\right) + \lambda_2 \frac{P_1 E_1}{\tilde{\tau}_1} \frac{\tilde{\tau}_j}{E_j^2} = 0. \tag{6.67}$$

This equation implies that

$$E_j = a\sqrt{\tilde{\tau}_j} \quad \text{for } j \geq 2. \tag{6.68}$$

Once again one may conclude that poor catalysts with long characteristic times $\tilde{\tau}$ should be present in high concentrations [cf. Eq. (6.48)]. After determining the common factor a as well as E_1 by the two constraints, one arrives eventually at the following optimal distribution for the enzyme concentrations:

$$E_1 = \frac{E_{\text{tot}} \Omega^0 \tilde{\tau}_1}{\Omega^0 \tilde{\tau}_1 + P_1 n^2 \langle\sqrt{\tilde{\tau}}\rangle^2}, \tag{6.69a}$$

$$E_j = \frac{E_{\text{tot}} P_1 n \langle\sqrt{\tilde{\tau}}\rangle \sqrt{\tilde{\tau}_j}}{\Omega^0 \tilde{\tau}_1 + P_1 n^2 \langle\sqrt{\tilde{\tau}}\rangle^2} \quad \text{for } j \geq 2, \tag{6.69b}$$

with

$$\langle\sqrt{\tilde{\tau}}\rangle = \frac{1}{n} \sum_{j=1}^{n} \sqrt{\tilde{\tau}_{j+1}}. \tag{6.69c}$$

Equations (6.69a) and (6.69b) may be considerably simplified when all the enzymes have the same intrinsic properties ($\tilde{\tau}_j = \tau$). Then, one obtains

$$E_1 = \frac{\Omega^0 E_{\text{tot}}}{\Omega^0 + n^2 P_1}, \tag{6.70a}$$

$$E_j = \frac{n P_1 E_{\text{tot}}}{\Omega^0 + n^2 P_1}, \quad j \geq 2. \tag{6.70b}$$

Equations (6.69) and (6.70) indicate that osmotic constraints may have a strong influence on the optimal distribution of enzyme concentrations. In the case $q_j \to \infty$, for example, vanishing enzyme concentrations are excluded for finite values of Ω^0. Nevertheless, the former result for the case without osmotic constraints ($E_1 \to E_{\text{tot}}$, $E_j \to 0$, $j \geq 2$) may be derived from Eq. (6.69) in the limit $\Omega^0 \to \infty$. Furthermore, in the present case the optimal enzyme concentrations depend on the concentration P_1 of the pathway substrate. From Eqs. (6.69a) and (6.69b), it is easy to see that for

$$\frac{\Omega^0}{P_1} < \frac{n}{\tilde{\tau}_1} \langle \sqrt{\tilde{\tau}} \rangle \sqrt{\tilde{\tau}_j}\,, \tag{6.71}$$

the concentrations E_j with $j \geq 2$ become even higher than the concentration of the first enzyme.

The solution of the variational equation (6.62) for the optimal distribution of enzyme concentration is identical to that obtained from

$$\frac{\partial}{\partial E_j}\left[\sum_{i=1}^{n} S_i - \lambda_1\left(\sum_{m=1}^{r} E_m - E_{\text{tot}}\right) - \lambda_2'(J - J^0)\right] = 0, \tag{6.72}$$

which results from the extremum principle $\Omega = \min$ under the constraints $E_{\text{tot}} = \text{const.}$ and $J = J^0 = \textit{const.}$

Let us now consider the principle $\Omega = \min$ in a more qualitative way and without the constraint $E_{\text{tot}} = \text{const.}$ For a given metabolic pathway it is relatively easy to distinguish between *biologically important* substances which must be present in certain amounts (e.g., storage metabolites and structural components) and metabolites which serve only as reaction intermediates. Evolutionary pressure is likely to diminish only the levels of these intermediates (see Srivastava and Bernhard, 1986; Ovádi, 1991; Mendes *et al.,* 1992). This leads to the extremum principle

$$\Omega = \sum_{i=1}^{n} g_i S_i = \min, \tag{6.73}$$

where, in an extension of Eq. (6.61), arbitrary osmotic coefficients g_i are considered, which are positive functions of the metabolite concentrations. The following assumptions are made:

$$\frac{\partial(g_i S_i)}{\partial S_i} \geq 0, \tag{6.74a}$$

$$\frac{\partial g_i}{\partial S_j} = 0 \quad \text{for } j \neq i. \tag{6.74b}$$

Condition (6.74a) is obviously satisfied for ideal solutions, where $g_i = 1$. It is generally also fulfilled for dilute, nonideal solutions (cf. Moelwyn-Hughes, 1964). The reaction system should be delimited in such a way that all "biologically important" substances are external metabolites, so that they do not enter the sum in relation (6.73).

We again restrict the analysis to steady states, so that $J = J^0 = \text{const.}$ is included as a side condition to the minimization problem (6.73). Further plausible side conditions are to fix the concentrations of external metabolites and the equilibrium constants of reactions. The rate laws are supposed to be comprised in the generalized mass-action kinetics (2.15).

Let us again consider *unbranched reaction chains* as represented in Scheme 11 (Section 5.4.3.1). Without loss of generality, we can assume that the external pools and equilibrium constants have such values that the steady-state flux is positive. With the help of the generalized rate equation (2.15), the steady-state flux can be written as

$$J = v_j = F_j(k_{+j}S_{j-1} - k_{-j}S_j), \quad j = 1, \ldots, r = n + 1 \tag{6.75}$$

($S_0 = P_1$ and $S_{n+1} = P_2$). In order that this flux is positive, the following inequalities have to be fulfilled:

$$\frac{J}{F_j k_{+j}} = \left(S_{j-1} - \frac{S_j}{q_j}\right) > 0. \tag{6.76a}$$

This implies the condition

$$S_{j-1} \geq \frac{S_j}{q_j}. \tag{6.76b}$$

The case that the concentration values lie on the boundary of the admissible region [i.e., that equality in one of the relations (6.76b) applies], occurs if, and only if, a reaction j is in quasi-equilibrium, that is, if $|J/F_j k_{+j}| \ll 1$. Considering inequalities (6.76a) consecutively for $j = r$ backward up to $j = 1$, one derives that all S_i are simultaneously minimized if all reactions but the first are very fast. In this state, the sum given in relation (6.73) is minimized also because this quantity is a monotonic increasing function of all concentrations S_i, due to relation (6.74a). Therefore, the solution of the optimization problem reads

$$F_1k_1 = J^0\left[P_1 - P_2\left(\prod_{j=1}^{r} q_j\right)^{-1}\right]^{-1}, \quad F_jk_j \to \infty, j \geq 2, \tag{6.77}$$

where only the kinetic parameters of the first enzyme depend on the concentrations of the external parameters. This result shows some correspondence to the result expressed in relation (6.69) which states that for very low Ω^0, the concentration E_1 of the first enzyme becomes lower than the concentrations of all other enzymes E_j ($j \geq 2$). Furthermore, it is in agreement with the frequently observed feature that the first step of a pathway is a nonequilibrium reaction (Savageau, 1976; Easterby, 1981; Dibrov *et al.*, 1982).

Also for *branched pathways* of monomolecular reactions, the solutions to the minimization problem under study have the property generally not to depend on the details of the functions $g_i(S_i)$, provided that condition (6.74) is satisfied. They depend, however, on the concentrations of external metabolites. The solutions are characterized by the fact that all reactions attain quasi-equilibrium except for reactions behind initial substrates of the system and one reaction behind each ramification point (see Heinrich *et al.*, 1991).

The extremum principle (6.73) can be rephrased as a multicriteria minimization problem,

$$S_i = \min, \quad i = 1, \ldots, n, \tag{6.78}$$

which is a *vector-optimization problem* because the S_i can be gathered in a vector, $\mathbf{S}$. For the concepts and methods of multicriteria optimization, the reader is referred to the works of Zeleny (1974) and Sawaragi *et al.* (1985). For the present case a *nondominated solution* (also called a compromise solution), $\mathbf{S}^*$, to the problem (6.78) has the property that there is no other concentration vector for which no concentration is higher and at least one concentration is smaller then in $\mathbf{S}^*$. It can be shown that the set of all nondominated solutions to the multicriteria minimization problem (6.78) coincides with the set of solutions to the minimization problem (6.73) for all positive functions $g_i(S_i)$ fulfilling condition (6.74a) (S. Schuster and Heinrich, 1991). The most important conclusion one can draw from this optimization study is which reactions are at quasi-equilibrium. Because for reasons of monotonicity the solution is always situated on the boundary of the admissible region for concentrations, where some reactions are infinitely fast, the optimal state is always characterized by a distinct decomposition of the network into near-equilibrium and nonequilibrium reactions.

The outlined treatment was also applied to systems of more complex stoichiometry, including a model of glycolysis, the pentose monophosphate shunt, and the glutathione system in human erythrocytes (S. Schuster *et al.*, 1991). The solution set is then composed of four faces of the concentration polyhedron. All of these faces have in common that the enzymes hexokinase and 2,3-bisphosphoglycerate phosphatase are slow, which is in accordance with reality.

6.2.3. Minimization of Transient Times

A necessary condition for the occurrence of steady states is their stability. However, this property may not suffice for the maintenance of such states under the influence of permanent larger fluctuations. In addition to stability, rapid relaxation toward the original steady state after fluctuations or to new steady states after changes in the environmental conditions is thus of importance for the biological function of a metabolic pathway (Rosen, 1967; Majewski and Domach, 1985). Accordingly, the minimization of transient times can be postulated as an optimality criterion relevant in biological evolution.

Using definition (5.278) for transient times, Cleland (1979) studied the minimization principle

$$\tau = \min \tag{6.79}$$

subject to the condition that the total mass concentration, M, of the pathway enzymes is bounded above,

$$M = \sum_j \mu_j E_j \leq M^0 = \text{const.} \tag{6.80}$$

with μ_j denoting the molar mass of the jth enzyme. The side condition (6.80) proposed also by Kuchel (1985) may be more realistic than relation (6.46) because it takes into account that the metabolic effort necessary for the synthesis of a protein is more closely related to its mass concentration than to its molar concentration. It would be straightforward to replace, in the maximization of flux (Section 6.2.1), side condition (6.46) by relation (6.80).

We now consider the optimization principle (6.79) for an unbranched reaction chain and assume that the reactions are irreversible and that the enzymes operate in the linear region. Furthermore, the turnover numbers k_{cat} are considered to be all equal, so that $V_i^+/V_j^+ = E_i/E_j$. Due to $J = S_{j-1}V_j^+/K_j^+$, Eq. (5.278) then simplifies to

$$\tau = \frac{1}{k_{cat}} \sum_{j=2}^{r} \frac{K_j^+}{E_j}. \tag{6.81}$$

Note that the sum in Eq. (6.81) runs from 2 to $r = n + 1$, because the first enzyme does not affect the total transient time when the concentration of the pathway substrate is constant. Accordingly, the first enzyme may be omitted in the cost function (6.80). Equation (6.81) shows that the transient time is a monotonic decreasing function of all enzyme concentrations. Therefore, in the side condition (6.80), the equality sign applies for optimal states. The optimal parameter distribution can be found by the equation

$$\frac{\partial}{\partial E_i}\left[\frac{1}{k_{cat}} \sum_{j=2}^{r} \frac{K_j^+}{E_j} - \lambda\left(\sum_{j=2}^{r} (\mu_j E_j - M^0)\right)\right] = 0, \tag{6.82}$$

where λ denotes a Lagrangian multiplier. This gives

$$\frac{E_i}{E_j} = \sqrt{\frac{K_i^+ \mu_j}{K_j^+ \mu_i}}. \tag{6.83}$$

Taking into account the side condition (6.80), one obtains

$$E_i = \frac{M^0 \sqrt{K_i^+/\mu_i}}{\sum_{j=2}^{r} \sqrt{K_j^+ \mu_j}}. \tag{6.84}$$

This result shows that enzymes with high molar masses should have small concentrations, with the relationship $E_j \propto \sqrt{1/\mu_i}$. Moreover, poorly binding enzymes must be present in high amounts to achieve a short transient time of the pathway. Equation (6.84) was first derived by Cleland (1979), who dealt with the question of under what conditions coupled enzyme assays attain the steady state very rapidly and only require small amounts of enzymes.

The optimization principle (6.79) subject to the constraint (6.80) is equivalent to the principle of minimizing the total mass concentration of the enzymes, M, with the side condition $\tau = \tau^0 = const.$ (Kuchel, 1985). The solution of that problem again leads to Eq. (6.83), while instead of Eq. (6.84), the following formula is obtained:

$$E_i = \frac{1}{\tau k_{cat}} \sqrt{\frac{K_i^+}{\mu_i}} \sum_{j=2}^{r} \sqrt{K_j^+ \mu_j}. \tag{6.85}$$

In the minimization of transient times, it is also sensible to compare systems with the same steady-state flux. This leads to the side condition J = const. It pertains to situations where the flux is determined by the biological function the reaction chain fulfills. The minimization problem (6.79) can then even be solved in the case that all reactions but the first are reversible. Owing to the equation $\tau = \Omega/J$ [Eq. (5.278)], the extremum principle (6.79) is then equivalent to the minimization of intermediate concentrations. Conversely, with the side condition $\Omega =$ const., it gives the same solutions as the maximization of flux under that side condition (cf. Section 6.2.2).

States of minimal transient times for unbranched reaction chains with reversible reactions have also been calculated on the basis of definition (5.279) instead of definition (5.278) (S. Schuster and Heinrich, 1987). Applying Eqs. (5.283a) and (5.283b), one derives, after some algebra, for the transient time of the last intermediate, S_n, in unbranched chains after perturbation of the concentration S_a,

$$\tau_n^{(a)} = \sum_{j=a+1}^{n} \frac{1}{k_j}\left[\left(1 + \sum_{i=2}^{j-1} \prod_{l=2}^{i} q_l\right)\left(\prod_{l=2}^{j-1} q_l\right)^{-1}\right], \tag{6.86}$$

where it is assumed that the first reaction is irreversible. The quantity $\tau_n^{(a)}$ may be regarded as the propagation time of a perturbation of the ath intermediate to the end of the chain. The minimization principle

$$\tau_n^{(a)} = \min \tag{6.87}$$

was investigated under the constraint of constant flux. As it was assumed that the first reaction were irreversible, the steady-state flux reads $J = k_1P_1$. One obtains the solution $k_1 = P_1/J$, $k_i \to \infty$ for all $i > a$, and arbitrary rate constants for the reactions with $1 < j \leq a$. If, in addition, the side condition $\Omega =$ const. is imposed, a similar solution obtains, where the rate constants for the reactions from 1 to a have to be chosen so as to satisfy the condition of fixed total osmolarity (see S. Schuster and Heinrich, 1987).

As outlined in Section 4.1, an alternative approach to comprehend relaxation processes is on the basis of the eigenvalues of the Jacobian. When the system is stable, all eigenvalues have negative real parts, and the long-term behavior is determined by that eigenvalue the real part of which has the smallest absolute value. Denoting this eigenvalue by λ^*, we can consider $[-\mathrm{Re}(\lambda^*)]^{-1}$ as a characteristic time of the pathway. It is therefore sensible to study the extremum principle

$$-\mathrm{Re}(\lambda^*) = \max. \tag{6.88}$$

For nonlinear systems, λ^* is a complicated function of the kinetic parameters, which cannot normally be given in closed form. Qualitative assertions about the solutions can nevertheless be made if only the side condition that all fluxes in the system are fixed is included, so that the steady-state condition is satisfied. Re(λ^*) can then tend to minus infinity, namely in all situations where rank(**N**) reactions reach quasi-equilibrium, whereas the remaining r − rank(**N**) reactions have such parameters that the independent fluxes attain the prescribed values. In such situations, any intermediate is connected by a chain of fast reactions with an external metabolite, so that fluctuations can be propagated very fast to the outside of the system (S. Schuster, 1989). For example, in unbranched reactions chains, all reactions but one have to be at quasi-equilibrium in order to maximize −Re(λ^*), so that r different solution arise. If, in addition, the side condition of fixed total osmolarity is included, optimal solutions can only be calculated numerically. One obtains optimal states where two adjacent reactions are slow and the others are very fast (S. Schuster and Heinrich, 1987).

Summarizing the above results, we can state that minimization of transient times without side conditions limiting the enzyme concentrations generally gives rise to pronounced time hierarchy (i.e., to a distinct separation of slow and fast reactions). Some reactions remain slow to meet the constraint of fixed fluxes. Time hierarchy is actually a ubiquitous phenomenon in living cells (cf. Chapter 4). Accordingly, the extremum principles studied above may be wellsuited to account for this phenomenon.

6.3. OPTIMAL STOICHIOMETRIES

In the previous sections of this chapter the *optimization of kinetic parameters* has been considered. It leads to maximal reaction rates or to minimal values of transition times and of total osmolarity of metabolic systems. It may be argued that this kind of evolutionary optimization was nothing else than a fine-tuning which guaranteed the efficient interplay of enzymes within the pathways whose basic structure had evolved in a much earlier stage of evolution. The question arises of whether the special *topology of enzymatic systems* expressed by the molecular interactions may also be described as a result of an evolutionary optimization process. We are far from understanding in detail the origination of the different metabolic pathways observed in contemporary living cells. However, biochemists have rather clear ideas concerning the *temporal order* of the emergence of the main biochemical pathways (cf. Wald, 1964; Hochachka and Somero, 1973; Holms, 1986). The main assumption is that there was a close mutual interaction between the evolution of the metabolic machinery and the composition of the earth's atmosphere. It is generally believed that life started when molecular oxygen was still absent in the atmosphere. Accordingly, *anaerobic fermentation* (i.e., *glycolysis*) was the first source of metabolic energy and, in a sense, fermentation has played, until the present stage of evolution, the central role in metabolism. Besides the production of two ATP molecules per one molecule of glucose degraded, alcoholic fermentation yields two molecules of carbon dioxide whose concentration probably had also been very low in the early atmosphere. The second big achievement in the evolution of metabolism may have been the establishment of the *hexosemonophosphate pathway* which is also able to take place under anaerobic conditions. It produces NADPH which may be used for the reductive synthesis of organic compounds from glucose, under participation of ATP derived from glycolysis, and it is also accompanied by the release of carbon dioxide. After this stage, *photosynthesis* became possible, where the energy of sunlight is used to produce glucose from water and carbon dioxide. The basic steps of this process resemble those of the hexosemonophosphate pathway running in reverse order. Most importantly, photosynthesis involves the release of molecular oxygen into the atmosphere. This paved the way for the development of *cellular respiration,* that is, the complete oxidation of glucose by molecular oxygen to carbon dioxide and water *via* the *citric acid cycle.* In combination with *oxidative phosphorylation,* 38 molecules of ATP may be synthesized from ADP and inorganic phosphate by the degradation of one molecule of glucose. Respiration is thus much more efficient than anaerobic fermentation.

On a more detailed level, the problem of the development of specific molecular interactions, as expressed by the stoichiometry of present-day metabolism, was probably closely related to that of the optimization of kinetic properties of enzymes. It was proposed that the evolution of metabolic pathways had involved the specialization of a smaller set of enzymes with less developed regulatory mechanisms and a much broader substrate specificity than the enzymes of present-day metabolism (Ycas, 1974; Jensen, 1976; Kacser and Beeby, 1984). Such diversity may be regarded as necessary to make a metabolic system possible despite the limited gene content of primitive cells. Probably, the translation process itself evolved from a less accurate mechanism.

The investigation of optimal stoichiometries is of importance not only for the understanding of biological evolution but also for optimization studies in biotechnology. For example, the computer-aided detection of elementary modes and the generation of alternative biosynthetic routes can be of significant value in the improvement of biotechnological procedures (cf. Mavrovouniotis *et al.,* 1990). For theoretical concepts in evolutionary biotechnology, see Eigen and Gardiner (1984) and P. Schuster (1995).

6.3.1. Optimal Properties of the Pentose Phosphate Pathway

The relation between optimal kinetic and stoichiometric properties has been stressed in the pioneering work of Meléndez-Hevia and Isidoro (1985) (see also Meléndez-Hevia and Torres, 1988; Meléndez-Hevia *et al.,* 1994). Analyzing the stoichiometric structure of the nonoxidative phase of the pentose phosphate pathway, they came to the conclusion that the reduction of the number of reaction steps in the transformation of an initial substrate S into an end product P may be considered as a general principle of evolutionary optimization of metabolic pathways. In fact, as has been shown in Section 6.2.1, the optimal flux through an unbranched chain of reactions will decrease with the increasing number of intermediate products if the total amount of available enzyme is limited [cf. Eq. (6.52)]. Concerning the pentose phosphate pathway, the question of whether Nature managed the conversion of six pentoses into five hexoses in a minimum number of reaction steps was raised.

For the solution of this problem, a *game of combinatorial optimization* obeying the following rules was proposed (Meléndez-Hevia and Isidoro, 1985): (a) The various sugars are only characterized by the numbers of their carbon atoms; (b) at the beginning there are six sugars containing five carbons each; (c) each reaction step involves the transfer of two carbons [transketolase reaction (EC 2.2.1.1)] or three carbons [transaldolase (EC 2.2.1.2) or aldolase (EC 4.1.2.13) reactions] from one sugar to another; (d) any compound cannot contain less than three carbons; (e) the goal is to produce five sugars with six carbons each by a minimum number of steps.

The optimal strategy for this game is shown in Table 6.3. Identifying steps 1A, 1B, 3A, and 3B with the reactions of transketolase, steps 2A and 2B with the reactions of transaldolase, and step 4 with fructose-1,6-bisphosphate aldolase, it is seen that the solution given in Table 6.3 is exactly the same as the sequence of reactions taking place in the nonoxidative phase of this pathway (Figure 6.7) (Horecker *et al.,* 1954; Wood and Katz, 1958).

A similar optimization procedure has been applied to the nonreductive phase of the Calvin cycle (Meléndez-Hevia, 1990). Here the goal is the to convert 12 sugars with 3 carbons into 6 of 5 carbons and 1 of 6 carbons. It has been shown that the *simplest combinatorial solution* of this problem is identical to the actual reaction sequence in the Calvin cycle.

6.3.2. Optimal Location of ATP-Consuming and ATP-Producing Reactions in Glycolysis

In Section 5.4.4, we have dealt with glycolysis by confining ourselves to its control properties. Now we try to gain some further insight into this pathway by consideration of evolutionary optimization principles. A remarkable feature of the stoichiometry of glycolysis is that it involves *ATP-consuming reactions,* despite the fact that its main biological function consists in the *production of ATP.* It is, furthermore, striking that the two ATP-consuming reactions, hexokinase (HK) and phosphofructokinase (PFK), are located in the upper part, whereas the two ATP-producing reactions, phosphoglycerate kinase (PGK) and pyruvate kinase (PK), belong to the lower part of this pathway (cf. Figure 3.1). Certainly, there

Table 6.3 Optimal Strategy of the "Combinatorial Game" Proposed by Meléndez-Hevia and Isidoro (1985) for the Explanation of the Stoichiometry of the Nonoxidative Phase of the Pentose Phosphate Pathway

	Sugar Molecules						
Step	1	2	3	4	5	6	Step
	5	5	5	5	5	5	
1A	TK_1						
	3	7	5	5	5	5	
2A	TA						
	6	4	5	5	5	5	
3A		TK_2					
	6	6	3	5	5	5	
					TK_1		1B
	6	6	3	5	7	3	
					TA		2B
	6	6	3	5	4	6	
				TK_2			3B
	6	6	3	3	6	6	
4			Ald				4
	6	6	0	6	6	6	

are various constraints concerning the *chemical possibilities* of converting glucose into lactate, which are in favor of this special stoichiometric design. Beyond, it seems worthwhile to consider also the possible *kinetic advantages* of such a distribution of ATP-consuming and ATP-producing steps. Accordingly, we deal with the kinetic effect of changes in the number and location of ATP-consuming and ATP-producing reactions on the energy yield of glycolysis.

To allow general conclusions we do not incorporate too many details of pres-

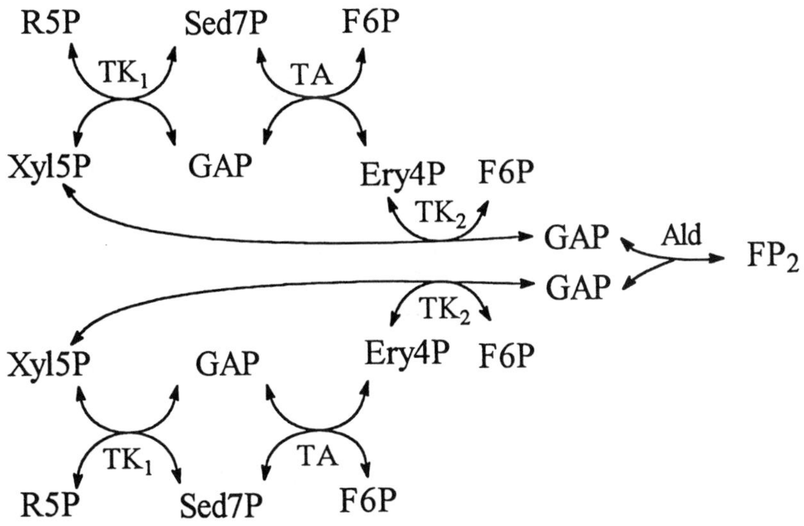

Figure 6.7 Reaction scheme of the nonoxidative part of the pentose phosphate pathway. Abbreviations: TA, transaldolase; TK_1, transketolase 1; TK_2, transketolase 2; Ald, aldolase; Ery4P, erythrose-4-phosphate; F6P, fructose-6-phosphate; FP_2, fructose-1,6-bisphosphate; GAP, glyceraldehyde-3-phosphate; R5P, ribose-5-phosphate; Sed7P, seduheptulose-7-phosphate; Xyl5P, xylulose-5-phosphate.

ent-day glycolysis. We start with the analysis of an unbranched pathway and consider, thereafter, the effect of branching as observed in glycolysis at the aldolase reaction. Further chemical constraints which may have been important during development of the structural design of glycolysis are neglected.

Our analysis is based on Eq. (5.88) which determines the steady-state flux J through an unbranched chain of r reactions as depicted in Scheme 11 (Section 5.4.3.1). This formula may be applied also for chains with bimolecular reactions involving cofactors, if they are considered as external reactants. In this case, one has to replace the kinetic constants k_i and k_{-i} by apparent first-order rate constants $\tilde{k}_i$ and $\tilde{k}_{-i}$ which are obtained as products of the corresponding second-order rate constants κ_i and κ_{-i} and the concentrations of those external reactants participating in the corresponding reaction steps. Throughout this section, we will consider the concentration of free inorganic phosphate to be constant and incorporate it into the rate constants.

We denote the ATP-producing sites as *P-sites* and the ATP-consuming sites as *C-sites,* cf. Figure 6.8. Both types of reactions are called *coupling sites.* Reactions which are involved neither in ATP production nor in ATP consumption are denoted as *O-sites.* The coupling sites are described by the following rate equations

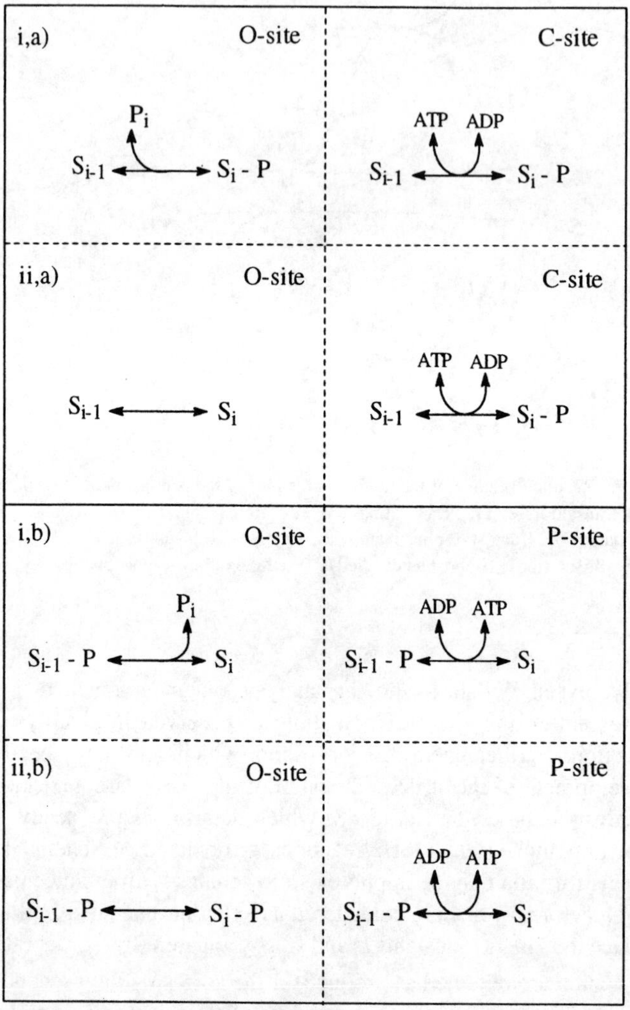

Figure 6.8 Scheme representing the possible replacement of O-sites (left) by C-sites or P-sites (right) according to the possibilities that (i) the composition of substrates and products remains unchanged, whereby a phosphate takes part as a reactant (case i,a) or product (case i,b) in the O-site; (ii,a) introduction of a C-site implies addition of a phosphate group to the reaction product; and (ii,b) introduction of a P-site implies removal of a phosphate group from the product.

C-sites: $$v_i = \kappa_i ATP \cdot S_{i-1} - \kappa_{-i} ADP \cdot S_i \tag{6.89a}$$

P-sites: $$v_i = \kappa_i ADP \cdot S_{i-1} - \kappa_{-i} ATP \cdot S_i \tag{6.89b}$$

Denoting by a and b the number of C-sites and P-sites, respectively, the ATP-production rate is related to the glycolytic flux in the following way:

$$J_{ATP} = (b - a)J. \tag{6.90}$$

To identify the optimal structural design according to the principle $J_{ATP} = \max$, the kinetic properties of chains with different numbers and different locations of coupling sites are compared. Clearly, the first step cannot produce ATP and the final step cannot be a C-site, due to the composition of glucose and lactate. For simplicity's sake, no further restrictions concerning the allowed number and distribution of coupling sites are made, except for $a + b \leq r$.

Concerning the replacement of one type of reaction site by another type, two situations can be distinguished. (i) When the composition of all intermediates is considered to be fixed, an O-site can only be replaced by a C-site (case i,a) if it is linked with phosphorylation anyway, and by a P-site (case i,b) if it is a dephosphorylation step. (ii) A second possibility is to allow changes in the composition of the intermediates, which makes possible to compare alternative paths of a given overall transformation. This means that introduction of a C-site (case ii,a) or a P-site (case ii,b) brings about that the substrate or the product of the reaction differs from that at the corresponding O-site by one phosphate group (see Figure 6.8). For example, the glyceraldehyde-phosphate dehydrogenase reaction, in which a phosphate takes part as a reactant, is an O-site according to case (i,a). The 2,3-bisphosphoglycerate phosphatase is an O-site according to case (i,b). The phosphoglucoisomerase and triose-phosphate isomerase reactions are examples of O-sites according to case (ii,b).

Coupling of the ith reaction to ATP consumption or ATP production will change the thermodynamic properties. For the equilibrium constant of the uncoupled reaction, q_i, we have $q_i = k_i/k_{-i}$. The equilibrium constant of the coupled reaction reads $q_i' = \kappa_i/\kappa_{-i}$. When possibility (i) mentioned above is considered for the replacement of O-sites, q_i' and q_i are related to each other by

C-sites: $$q_i' = q_i\hat{K}^{-1}, \tag{6.91a}$$

P-sites: $$q_i' = q_i\hat{K}, \tag{6.91b}$$

where $\hat{K} \cong 4.4 \cdot 10^{-6}$ denotes the equilibrium constants for the interconversion of ADP into ATP, which is actually an apparent equilibrium constant because the concentration of inorganic phosphate is considered constant. A relation similar to (6.91) also applies to possibility (ii), with the transformation factor, $\hat{K}$, now depending not only on the free-energy differences of ADP and ATP but also on the free-energy differences between the phosphorylated and unphosphorylated substrate or product. Because part of the free-energy of ATP goes into the phosphorylated sugar, $\hat{K}$ will be closer to unity than in case (i). For example, the standard free-energy difference, ΔG^0, between glucose-6-phosphate and glucose is 14 kJ/mole, and ΔG^0 between ADP and ATP is about -30 kJ/mole [depending on various factors such as pH and the magnesium concentration, cf. Gnaiger and Wyss (1994)].

By necessity, changes in the equilibrium constants as given in Eq. (6.91) are brought about by changes in the forward and backward rate constants. We use the following relations between the first-order rate constants of the uncoupled reactions and the apparent first-order rate constants of the coupled reactions,

$$\text{C-sites:} \qquad \tilde{k}_i = \frac{k_i}{\alpha_1}, \qquad \tilde{k}_{-i} = k_{-i}\beta_1, \tag{6.92a}$$

$$\text{P-sites:} \qquad \tilde{k}_i = k_i\alpha_2, \qquad \tilde{k}_{-i} = \frac{k_{-1}}{\beta_2}, \tag{6.92b}$$

where for C-sites, $\tilde{k}_i = \kappa_i ATP$, $\tilde{k}_{-i}ADP$, and for P-sites, $\tilde{k}_i = \kappa_i ADP$, $\tilde{k}_{-i} = \kappa_{-i}ATP$. In these equations we neglect the possible dependencies of the factors $\alpha_{1/2}$ and $\beta_{1/2}$ on the special properties of reaction i.

The combination of Eqs. (6.91) and (6.92) leads to

$$\alpha_1\beta_1 = \alpha_2\beta_2 = K = \hat{K}\frac{ADP}{ATP}. \tag{6.93}$$

Physiological values of the concentration ratio ADP/ATP are in the range 0.1–0.3 in various cells and organelles. Therefore, it follows from Eq. (6.93) that not only $\hat{K}$ but also K is much smaller than unity.

We use the plausible assumption that coupling of a reaction to a highly exergonic reaction increases reaction rate, whereas coupling to an endergonic reaction slows down the reaction. Accordingly,

$$K < \alpha_1, \alpha_2, \beta_1, \beta_2 < 1. \tag{6.94}$$

Thermodynamically, a chain with a C-sites and b P-sites may be characterized by the overall affinity

$$A = RT\ln\left(\frac{P_1}{P_2}K^{(b-a)}Q\right), \tag{6.95a}$$

$$Q = \prod_{i=1}^{r} q_i, \tag{6.95b}$$

which depends on the number but not on the location of C-, O-, and P-sites in the chain. Because $K < 1$, the overall affinity decreases as the excess number, $d = b - a$, of ATP-producing sites increases. The glycolytic flux is positive as long as the overall affinity is positive, which is fulfilled for

$$d < -\frac{\ln(P_1 Q/P_2)}{\ln K} = d_{\max}. \tag{6.96}$$

The maximal excess number $d_{\max}$ of ATP-producing sites may be expressed by standard free-energy changes. If $P_1 \cong P_2$, one obtains the maximal excess number as the ratio of the standard free-energy change of the uncoupled interconversion of glucose into two molecules of lactate and the standard free-energy change of ATP hydrolysis,

$$d_{\max} \cong \frac{\Delta G^{\circ}_{\text{glyc}}}{\Delta G^{\circ}_{\text{ATP}}}. \tag{6.97}$$

With $\Delta G^{\circ}_{\text{glyc}} = -197$ kJ/mole and $\Delta G^{\circ}_{\text{ATP}} = -30.5$ kJ/mole (Lehninger, 1982), one derives $d_{\max} \cong 6.5$.

The main conclusions concerning the optimal kinetic properties of ATP-producing reaction chains may be derived from the following two theorems.

Theorem 6B. (1) *The replacement of an O-site by a C-site (i.e., $a \to a + 1$) at any reaction increases the glycolytic rate J.* (2) *The replacement of an O-site by a P-site (i.e., $b \to b + 1$) decreases J.*

This theorem points to the kinetic effects of a *change of the number* of coupling sites. The kinetic effects of a *variation of the location* of coupling sites at fixed numbers a and b are described by

Theorem 6C. *J as well as J_{ATP} are increased first by an exchange of a P-site at reaction i for an O-site at reaction m with $i < m$ and second by an exchange of a C-site at reaction j for an O-site at reaction m with $m < j$, provided that the affinity A and the excess number d of ATP-producing sites are positive.*

We now prove these theorems under the simplification that $\alpha_1 = \alpha_2 = \alpha$ and $\beta_1 = \beta_2 = \beta$. The extension to the general case is straightforward.

Proof of Theorem 6B. Let us consider the replacement of an O-site by a C-site at reaction i (Part 1 of Theorem 6B). According to Eqs. (5.88) and (6.91)–(6.93), the flux $J(\mathrm{O}_i)$ with an O-site at reaction i and the flux $J(\mathrm{C}_i)$ with a C-site at reaction i read, respectively,

$$J(\mathrm{O}_i) = (P_1 Q - P_2)\left(\sum_{j=1}^{i-1} \frac{Q_{j,r}}{k_j} + \frac{Q_{i,r}}{k_i} + \sum_{j=i+1}^{r} \frac{Q_{j,r}}{k_j}\right)^{-1}, \tag{6.98}$$

$$J(\mathrm{C}_i) = (P_1 K^{-1} Q - P_2)\left(K^{-1} \sum_{j=1}^{i-1} \frac{Q_{j,r}}{k_j} + K^{-1} a \frac{Q_{i,r}}{k_i} + \sum_{j=i+1}^{r} \frac{Q_{j,r}}{k_j}\right)^{-1}, \tag{6.99}$$

where $Q_{j,r} = \prod_{m=j}^{r} q_m$. Strictly speaking, Eq. (6.98) applies to the situation that all reactions are O-sites. They can, however, also be applied to the case $a,b > 0$ when all earlier replacements have already been taken into account by including the values K, a, and β in the values of (apparent) kinetic and equilibrium constants. From Eqs. (6.98) and (6.99) it follows directly that $J(\mathrm{C}_i) > J(\mathrm{O}_i)$ if and only if

$$P_1 Q_{1,r} \frac{Q_{i,r}}{k_i} K^{-1}(1 - a) + P_1 Q_{1,r} \sum_{j=i+1}^{r} \frac{Q_{j,r}}{k_j} (K^{-1} - 1) + P_2 \sum_{j=1}^{i-1} \frac{Q_{j,r}}{k_j} (K^{-1} - 1) + P_2 \frac{Q_{i,r}}{k_i} (K^{-1} a - 1) > 0. \tag{6.100}$$

Condition (6.100) holds true under consideration of relation (6.94) which completes the proof. Part 2 of Theorem 6B can be proved in an analogous way.

Proof of Theorem 6C. For fixed numbers of P- and C-sites, the numerator of Eq. (5.88) is independent of the distribution of these sites along the chain. To investigate the influence of the location of P-sites on J and J_{ATP} (first statement of Theorem 6C), we compare, therefore, the denominators D of Eq. (5.88) for the following two situations: (a) P-site at reaction i and O-site at reaction m [denominator $D(\mathrm{P}_i,\mathrm{O}_m)$] and (b) O-site at reaction i and P-site at reaction m [denominator $D(\mathrm{O}_i,\mathrm{P}_m)$] where in both cases $i < m$. One obtains

$$D(\mathrm{P}_i, \mathrm{O}_m) = K \sum_{j=1}^{i-1} \frac{Q_{j,r}}{k_j} + K a^{-1} \frac{Q_{i,r}}{k_i} + \sum_{j=i+1}^{r} \frac{Q_{j,r}}{k_j}, \tag{6.101a}$$

$$D(\mathrm{O}_i, \mathrm{P}_m) = K \sum_{j=1}^{m-1} \frac{Q_{j,r}}{k_j} + K a^{-1} \frac{Q_{m,r}}{k_m} + \sum_{j=m+1}^{r} \frac{Q_{j,r}}{k_j}. \tag{6.101b}$$

From Eqs. (6.101a) and (6.101b) it follows that

$$D(P_i,O_m) - D(O_i,P_m) = \frac{Q_{i,r}}{k_i}K(a^{-1} - 1) + \sum_{j=i+1}^{m-1} \frac{Q_{j,r}}{k_j}(1 - K) + \frac{Q_{m,r}}{k_m}(1 - Ka^{-1}) > 0, \quad (6.102)$$

where condition (6.94) has been taken into account. For a positive overall affinity A, Eq. (6.102) implies $J(O_i,P_m) > J(P_i,O_m)$ and for $b - a > 0$, $J_{ATP}(O_i,P_m) > J_{ATP}(P_i,O_m)$ also, which completes the proof. The second statement of Theorem 6C can be proved in an analogous way.

It follows from Theorem 6C that J_{ATP} becomes maximum when all P-sites are located at the lower end of the chain and all C-sites are located at the upper end of the chain. According to Eq. (5.88), the optimal ATP-production rate reads, therefore,

$$J_{ATP}(a, b) = \frac{b - a}{D_a + D_o + D_b}\left(P_1K^{b-a}\prod_{j=1}^{r} q_j - P_2\right) \quad (6.103a)$$

with

$$D_a = \sum_{j=1}^{a} \frac{a}{k_j} K^{b-a+j-1}Q_{j,r}, \quad (6.103b)$$

$$D_o = K^b \sum_{j=a+1}^{r-b} \frac{Q_{j,r}}{k_j}, \quad (6.103c)$$

$$D_b = \sum_{j=r-b+1}^{r} \frac{1}{k_j a} K^{r-j}Q_{j,r}. \quad (6.103d)$$

Using Eqs. (6.103a)–(6.103d) it is now shown that an optimum for the ATP-production rate J_{ATP} is not only obtained by proper localization of C- and P-sites at the two ends of the chain but also by variation of their numbers a and b. We consider the special case of equal values for all thermodynamic equilibrium constants and all forward and backward rate constants of the uncoupled reactions (i.e., $q_i = q$, $k_i = k$, $k_{-i} = k/q$). With these conditions, expressions (6.103b)–(6.103d) permit explicit evaluation by means of the formula for geometric progressions. One obtains

$$kD_a = aK^{b-1}q^{r-a+1}\left(\frac{(q/K)^a - 1}{q/K - 1}\right), \quad (6.104a)$$

$$kD_o = K^b q^{b+1}\left(\frac{q^{r-b-a} - 1}{q - 1}\right), \quad (6.104b)$$

$$kD_b = q\frac{K}{a}\left(\frac{(qK)^b - 1}{qK - 1}\right). \quad (6.104c)$$

Figures 6.9A and 6.9B show the glycolytic rate J and the ATP-production rate J_{ATP}, respectively, as functions of the number of coupling sites for a chain with 10 reactions for special values of the thermodynamic parameters Q and K as well as of the coupling parameters α and β. The curves are calculated on the basis of Eqs. (6.90), (6.103a) and (6.104a)–(6.104c). The starting points of the curves at low b values are given by the condition that we only consider chains where the number of C-sites does not exceed the number of P-sites (i.e., $b \geq a$). The end points at high b values are determined by the limited total number of sites (i.e., $a + b \leq r$). More rigorously, one should take into account that the total numbers of steps where substrates are phosphorylated and dephosphorylated must be equal. This leads, with $b \geq a$, to the condition $a + b + (b - a) = 2b \leq r$, because $b - a$ is the number of O-sites (i.e., sites where inorganic phosphate can be incorporated).

As is seen in Figure 6.9A, the glycolytic rate J decreases for all possible values of a monotonically with the number b of P-sites. This property follows directly from Theorem 6B, Part 2. For low values of a ($a < 2$) the flux J may become negative at very high numbers of P-sites ($b - a > d_{max}$) [cf. Eq. (6.96)]. For small values of b with $b > a$, the flux J is rather insensitive to variations of b. This is in accordance with the result that for $q_i > 1$, flux control in unbranched chains is mainly exerted by the first enzymic steps, that is, a change of the kinetic properties of reactions at the end of the chain (resulting from the incorporation of P-sites) has little effect on the steady-state flux (cf. Section 5.4.3.1). The ATP-production rate J_{ATP} shown in Figure 6.9B displays a maximum at variations of the number b of P-sites as long as the number a of C-sites is not too high. This is explained by the fact that the two factors in Eq. (6.90), $b - a$ and J, change in opposite directions upon variation of b. In particular, the increase of J_{ATP} results from the insensitivity of J to variations of b for low b values. At higher values of b, the decrease of J overcompensates the increase of b. The flux J shown in Figure 6.9A increases with the number a of ATP-consuming sites at the upper end of the chain. This results from Theorem 6B, Part 1. This effect is most pronounced at the transition from $a = 0$ to $a = 1$ which makes the first reaction quasi-irreversible due to $K \ll 1$. Because steps behind quasi-irreversible reactions in unbranched chains exert minor flux control, further replacement of O-sites by C-sites at subsequent reactions yields less effect. Upon transition from $a = 0$ to $a = 1$, the ATP-production rate as a function of b retains the property of exhibiting a maximum. This maximum is higher for $a = 1$ (located at $b = 5$ at the chosen parameter values) than for $a = 0$ ($b = 4$).

The curves shown in Figures 6.9A and 6.9B are calculated for $Q = 10^3$ and $K = 0.3455$. According to Eq. (6.96), these parameter values correspond to a realistic value of $d_{max} = 6.5$. However, the standard free-energy changes listed below Eq. (6.97) result in a much higher overall equilibrium constant Q and in a much lower equilibrium constant K for the synthesis of ATP from ADP and

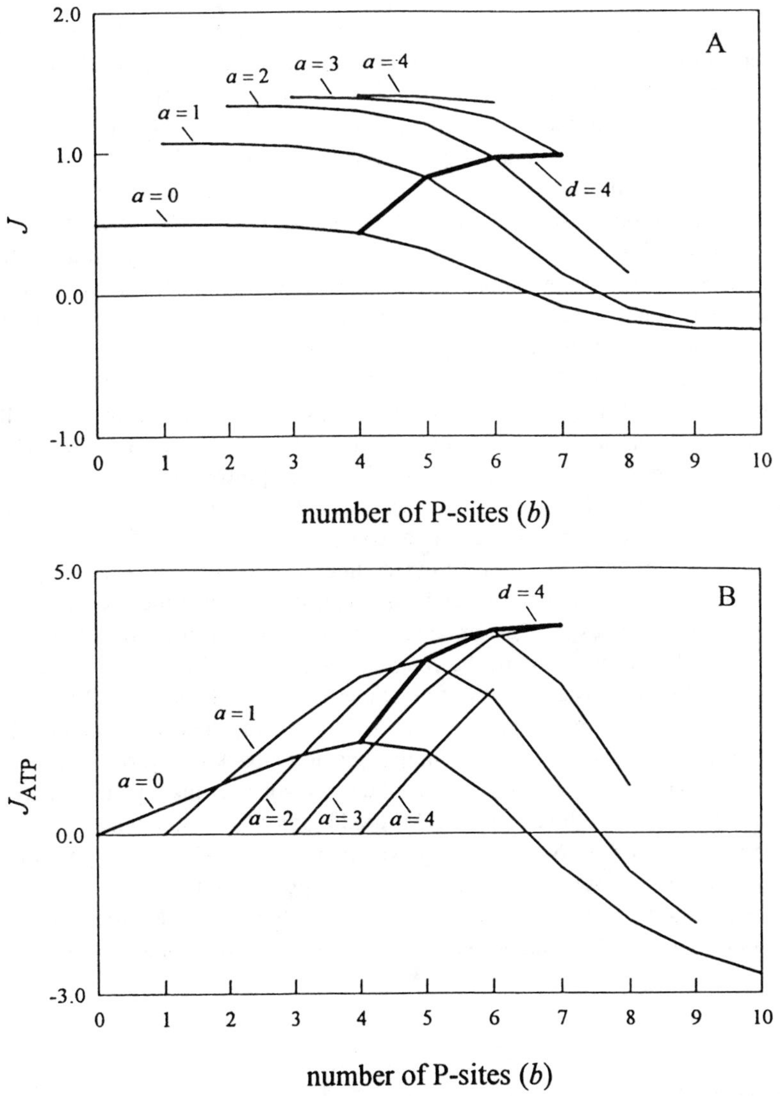

Figure 6.9 Glycolytic rate J (A) and ATP-production rate J_{ATP} (B) as functions of the number b of P-sites located at the end of the chain for various values of the number a of C-sites at the upper end of the chain for $r = n + 1 = 10$. Parameter values: $Q = 10^3$; $K = 0.3455$; $\alpha = \beta = K^{0.5}$; $P_1 = P_2 = 1$. The thermodynamic limit according to Eq. (6.96) is $d_{max} = 6.5$. The thick lines connect points for the same excess number of P-sites ($d = 4$).

inorganic phosphate. In a plot of the ATP-production rate for very high values of Q and such values of K that again $d_{max} = 6.5$, it can be seen that J_{ATP} retains the property of exhibiting a maximum with the variation of the number b of P-sites (Heinrich *et al.*, 1996). For different values $a \neq 0$, the maxima with respect to variation of b are virtually the same and are much higher than for $a = 0$. One may conclude, therefore, that for realistic values of thermodynamic parameters, one ATP-consuming site at the first step of the chain would be sufficient to guarantee high ATP production. However, the existence of two ATP-consuming sites at the upper end of glycolysis (which is not less optimal thermodynamically than the case $a = 1$) may be explained on the basis of the chemical fact that two phosphate groups are necessary for a "symmetric pathway" in the degradation of the triose phosphates in the lower part of glycolysis (see below).

Effect of branching: In the above analysis, the fact that real glycolysis is characterized by a splitting of C_6 compounds into two C_3 compounds at the aldolase reaction has been neglected. To introduce this feature the branching model depicted in Figure 6.10 may be considered. There, a splitting of the compound S_m into the compounds S_{m+1} and S^*_{m+1} occurs at the step $m + 1$. The latter two compounds can be interconverted in an isomerization reaction.

According to the results derived for the unbranched chain, it is meaningful to assume that all C-sites may be located only in the upper part of the chain (reactions 1 to m) and all P-sites only in the lower part of the chain (reactions $m + 2$ to $r = n + 1$). The steady state of the chain is characterized by $J_2 = 2J_1$, where J_1 and J_2 are the steady-state fluxes of reactions 1 to $m + 1$ and $m + 2$ to $n + 1$, respectively. Therefore, we now have $d = 2b - a$ for the excess number. Due to the fact that reaction $m + 1$ is bimolecular in the backward direction, a quadratic equation results for the glycolytic flux. By solving this equation, one can express the glycolytic flux and the ATP-production flux as functions of a and b (Heinrich *et al.*, 1996). Note that only even numbers a of C-sites may be considered because otherwise the degradation pathways of S_{m+1} and S^*_{m+1} could not be the same. It turns out that the conclusions concerning the optimal number of ATP-consuming and ATP-producing steps derived for the linear model remain valid

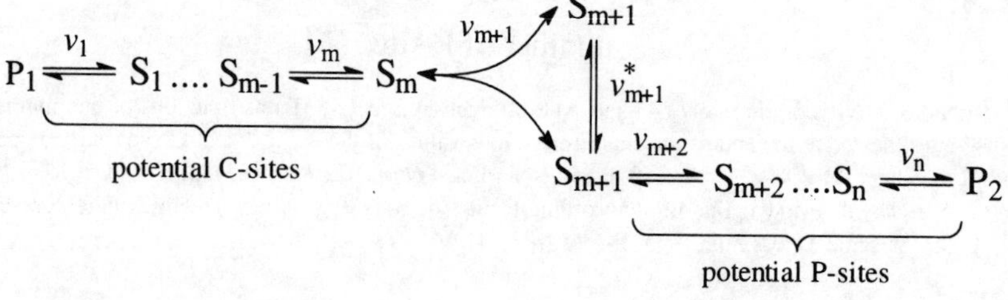

Figure 6.10 Branched reaction scheme representing potential ATP-producing pathways.

upon consideration of branching. In particular, replacement of an O-site by a C-site in the first part of the chain results in an increase of the ATP-production rate. One may conclude that there is no essential kinetic difference in the linear and branching models with respect to the efficiency of ATP production. However, the branched system seems to be more effective from a chemical point of view. For example, an excess number $d = 2$ as observed in glycolysis can be realized in the branched model with $a = 2$ and $b = 2$, whereas in the linear model, there are the possibilities $a = 1, b = 3$ and $a = 2, b = 4$, that is, in the linear model in each case more ATP-producing reactions are necessary then in the branching model.

The main result of the present investigation is that the optimization of kinetic properties favors pathways where the first steps are exergonic or coupled to exergonic processes (as ATP hydrolysis) and the subsequent steps are endergonic or coupled to endergonic processes (as ATP production). This result is in accordance not only with glycolysis but also with other metabolic systems. For example, the citric acid cycle starts with two exergonic reactions: (a) the citrate synthase reaction (EC 4.1.3.7) which involves hydrolysis of the energy-rich thioester bond of Acetyl-CoA and (b) the isocitrate dehydrogenase (EC 1.1.1.42). The subsequent reactions yield the energy-rich compound GTP and the redox equivalents NADH and $FADH_2$. The last reaction of the cycle, the malate dehydrogenase reaction (EC 1.1.1.37), is very endergonic. Another example is gluconeogenesis, which starts by circumventing the pyruvate kinase step by two steps: the pyruvate carboxylase (EC 6.4.1.1) and the phosphoenolpyruvate carboxykinase (EC 4.1.1.32) which both involve hydrolysis of either ATP or GTP. Similarly, the fatty acid oxidation is initiated by the fatty acid activation in an ATP-dependent acylation reaction to form fatty acyl-CoA. Further fatty acid oxidation yields NADH and $FADH_2$ which are reoxidized through oxidative phosphorylation to form ATP.

Coupling of the first step in glycolysis to ATP consumption, which is favorable with respect to enhancement of ATP-production rate, makes this step quasi-irreversible. In this situation, the flux control coefficients of all subsequent reactions are virtually zero, as has been shown in Section 5.4.3.1. This would imply the problem that regulation by the demand for the end product would not be effective. This may be the reason why contemporary glycolysis and other pathways are characterized by a large number of internal regulators (enzyme activation or inhibition by substances other than substrates or products) which allow that also other reactions exert flux control. For example, in glycolysis inhibition of the hexokinase by glucose 6-phosphate and (in erythrocytes) by 2,3-bisphosphoglycerate leads to nonvanishing control coefficients of the enzymes located downstream of hexokinase despite the fact that the hexokinase-reaction is quasi-irreversible (cf. Section 5.4.4).

In our analysis, we have neglected feedback loops. One may expect, however, that the incorporation of such mechanisms would have a similar effect as the product inhibition resulting from reversible reactions. Because we have used, in Figure 6.9, a value for the overall equilibrium constants, Q, that is lower than the real value, we have taken into account a certain degree of reversibility. It is worth mentioning that incorporation of variable concentrations of ATP and ADP and of nonglycolytic ATP-consuming processes would give rise to further feedback effects, which result from stoichiometric coupling (cf. Section 5.4.4.3).

In future studies, it would also be worth combining the analysis of the stoichiometric design with the optimization of kinetic parameters (e.g., enzyme concentrations) as presented in Section 6.2.

6.3.3. Concluding Remarks

The theoretical investigation of optimal stoichiometries of metabolic pathways is still at the very beginning. Probably, the problem may be tackled in the future by application of mathematical methods developed in theoretical chemistry for predicting the conceivable existence of chemical objects for a given collection of atoms as well as for generating reaction pathways by computers (Bauer *et al.*, 1988).

Furthermore, the methods outlined in Chapter 3 may be useful for solving problems in the evolutionary optimization of the stoichiometry of metabolic pathways. A given distribution of steady-state fluxes may be considered as a linear superposition of fundamental flux modes which are independent of the kinetic details of the participating reactions; that is, they exclusively reflect the stoichiometric properties of the pathway. These are the basis vectors $\boldsymbol{k}$ of the null-space of the stoichiometry matrix **N** or, more specifically, the elementary modes of a pathway (see Section 3.2.4). The dimension of this null-space is closely related to the number of branches of a metabolic network and may be used, therefore, to characterize the number of different metabolic functions of the network. For example, an unbranched pathway, as depicted in Scheme 11 (Section 5.4.3.1), is characterized by only one $\boldsymbol{k}$ vector [$\boldsymbol{k} = (1, \ldots, 1)^{\mathrm{T}}$] which is in line with the fact that there is only one steady-state flux and only one end product, independent of the total number r of participating reactions. In other words, for unbranched chains the number, f, of metabolic functions equals one ($f = 1$). In contrast, a branched pathway as depicted in Scheme 7 (Section 3.2.4) is characterized by two basis vectors $\boldsymbol{k}_1$ and $\boldsymbol{k}_2$ and accordingly by two independent steady-state fluxes which may be independently regulated by variation of the kinetic parameters. This dimension of the null-space is invariant against changes of the number of reactions participating in the three branches of the given reaction scheme. Therefore, one may attribute to the given scheme of reactions two different metabolic functions ($f = 2$). Extending this consideration to more complex networks, one may argue that, due to the limited enzyme content of a cell, evolution was characterized by an increasing number of metabolic functions relative to the total number of reactions. This would lead to the optimization principle

$$\hat{f} = \frac{f}{r} = \max. \tag{6.105}$$

For any metabolic pathway a good estimate for the number of different metabolic functions could be obtained from the rank of the stoichiometric matrix using the formula $f = r - \text{rank}(\mathbf{N})$, as f equals the number of different basis vectors of the null-space. From this it follows that $\hat{f}$, which may characterize the *degree of functionalization* of metabolic networks, is bounded between zero and unity ($0 \leq \hat{f} \leq 1$). Another possibility is to define f as the number of elementary modes, which in a sense may also serve as a measure of the number of different functions. However, in the usual case that the number of elementary modes is greater than the dimension of the null-space, they are linearly dependent and cannot, hence, be regulated independently.

Whereas it is rather easy to calculate f for systems of moderate complexity, the dimension of the null-space for large networks existing in real cells is still unknown. In view of the proposed principle (6.105) it would be an intriguing task to derive, in a first step, an estimate for $\hat{f}$ of living cells by calculating the dimension of the null-space of stoichiometry matrices on a large scale, by taking into account as many reactions as possible from the biochemical transformations documented in the literature.

References

Acerenza, L. and Kacser, H. (1990) Enzyme kinetics and metabolic control. A method to test and quantify the effect of enzymic properties on metabolic variables, *Biochem. J.*, **269**, 697–707.

Acerenza, L., Sauro, H.M. and Kacser, H. (1989) Control analysis of time-dependent metabolic systems, *J. Theor. Biol.*, **137**, 423–444.

Albe, K.R., Butler, M.H. and Wright, B.E. (1990) Cellular concentrations of enzymes and their substrates, *J. Theor. Biol.*, **143**, 163–195.

Alberts, B., Bray, D., Lewis, J., Raff, M., Roberts, K. and Watson, J.D. (1983) *Molecular Biology of the Cell*, Garland Publishing, New York.

Alberty, R.A. (1994) Constraints in biochemical reactions, *Biophys. Chem.*, **49**, 251–261.

Albery, W.J. and Knowles, J.R. (1976a) Free-energy profile for the reaction catalyzed by triosephosphate isomerase, *Biochemistry*, **15**, 5627–5631.

Albery, W.J. and Knowles, J.R. (1976b) Evolution of enzyme function and the development of catalytic efficiency, *Biochemistry*, **15**, 5631–5640.

Anderson, K.S., Miles, E.W. and Johnson, K.A. (1991) Serine modulates substrate channeling in tryptophan synthase. A novel intersubunit triggering mechanism, *J. Biol. Chem.*, **266**, 8020–8033.

Andronov, A.A., Vitt, A.A. and Khaikin, S.E. (1966) *Theory of Oscillators*, Pergamon Press, Oxford.

Aris, R. (1965) Prolegomena to the rational analysis of systems of chemical reactions, *Arch. Rational Mech. Anal.*, **19**, 81–99.

Ataullakhanov, F.I., Vitvitsky, V.M., Zhabotinsky, A.M., Pichugin, A.V., Kholodenko, B.N. and Ehrlich, L.I. (1981) Regulation of glycolysis in human erythrocytes. The mechanism of ATP concentration stabilization, *Acta Biol. Med. Germ.*, **40**, 991–997.

Atkinson, D. E. (1968) The energy charge of the adenylate pool as a regulatory parameter. Interaction with feedback modifiers, *Biochemistry*, **7**, 4030–4034.

Atkinson, D.E. (1969) Limitation of metabolite concentrations and the conservation of solvent capacity in the living cell, *Curr. Top. Cell. Regul.*, **1**, 29–43.

Atkinson, D.E. (1986) Dynamic interactions between metabolic sequences, in: Damjanovich, S., Keleti, T. and Trón, L. (eds.) *Dynamics of Biochemical Systems*, Academiai Kiádo, Budapest, pp. 129–141.

Baconnier, P.F., Pachot, P. and Demongeot, J. (1993) An attempt to generalize the control coefficient concept, *J. Biol. Syst.*, **1**, 335–347.

Bak, T.A. (1959) *Contributions to the Theory of Chemical Kinetics. A Study of the Connection between Thermodynamics and Chemical Rate Processes.* Munksgaard, København.

Battelli, F. and Lazzari, C. (1985) On the pseudo-steady-state approximation and Tikhonov theorem for general enzyme systems, *Math. Biosci.*, **75**, 229–246.

Battelli, F. and Lazzari, C. (1986) Singular perturbation theory for open enzyme reaction networks, *IMA J. Math. Appl. Med. Biol.*, **3**, 41–51.

Bauer, J., Fontain, E. and Ugi, I. (1988) Computer-assisted bilateral solution of chemical problems and generation of reaction networks, *Anal. Chim. Acta*, **210**, 123–134.

Bauer, S.H. (1990) Comments on current aspects of chemical kinetics, *Int. J. Chem. Kinet.*, **22**, 113–133.

Baykov, A.A., Shestakov, A.S., Kasho, V.N., Vener, A.V. and Ivanov, A.H. (1990) Kinetics and thermodynamics of catalysis by the inorganic pyrophosphatase of Escherichia coli in both directions, *Eur. J. Biochem.*, **194**, 879–887.

Baykov, A.A., Shestakov, A.S. and Kasho, V.N. (1993) Kinetics and thermodynamics of catalysis by inorganic pyrophosphatase, *Braz. J. Med. Biol. Res.*, **26**, 347–350.

Belfiore, F. (1980) *Enzyme Regulation and Metabolic Diseases*, S. Karge, Basel.

Berridge, M.J. (1989) Cell signalling through cytoplasmic calcium oscillations, in: Goldbeter, A. (ed.) *Cell to Cell Signalling: From Experiments to Theoretical Models*, Academic Press, London, pp. 449–459.

Betts, G.F. and Srivastava, D.K. (1991) The rationalization of high enzyme concentration in metabolic pathways such as glycolysis, *J. Theor. Biol.*, **151**, 155–167.

Betz, A. and Selkov, E. (1969) Control of phosphofructokinase [PFK] activity in conditions simulating those of glycolysing yeast extract, *FEBS Lett.*, **3**, 5–9.

Bjornbom, P.H. (1977) The relation between the reaction mechanism and the stoichiometric behavior of chemical reactions, *AIChE J.*, **23**, 285–288.

Blangy, D., Buc, H. and Monod, J. (1968) Kinetics of the allosteric interactions of phosphofructokinase from Escherichia coli, *J. Mol. Biol.*, **31**, 13–35.

Bloomfield, V., Peller, L. and Alberty, R.A. (1962) Multiple intermediates in steady-state enzyme kinetics. II. Systems involving two reactants and two products, *J. Am. Chem. Soc.*, **84**, 4367–4374.

Bodenstein, M. (1913) Eine Theorie der photochemischen Reaktionsgeschwindigkeiten, *Z. Phys. Chem.*, **85**, 329–397.

Bohnensack, R. (1981) Control of energy transformation in mitochondria. Analysis by a quantitative model, *Biochim. Biophys. Acta*, **634**, 203–218.

Bohnensack, R. (1985) Mathematical modeling of mitochondrial energy transduction, *Biomed. Biochim. Acta*, **44**, 853–862.

Brand, M.D. (1994) The stoichiometry of proton pumping and ATP synthesis in mitochondria, *The Biochemist*, **16**, 20–24.

Brand, M.D., Hafner, R.P. and Brown, G.C. (1988) Control of respiration in non-phosphorylating mitochondria is shared between the proton leak and the respiratory chain, *Biochem. J.*, **255**, 535–539.

Brocklehurst, K. (1977) Evolution of enzyme catalytic power, *Biochem. J.*, **163**, 111–116.

Brown, G.C. (1991) Total cell protein concentration as an evolutionary constraint on the metabolic control distribution in cells, *J. Theor. Biol.*, **153**, 195–203.

Brown, G.C. (1992) Control of respiration and ATP synthesis in mammalian mitochondria and cells, *Biochem. J.*, **284**, 1–13.

Brown, G.C. and Cooper, C.E. (1993) Control analysis applied to single enzymes: can an isolated enzyme have a unique rate-limiting step?, *Biochem. J.*, **294**, 87–94.

Brown, G.C. and Cooper, C.E. (1994) The analysis of rate limitation within enzymes: relations between flux control coefficients of rate constants and unidirectional rates, rate constants and thermodynamic parameters of single isolated enzymes, *Biochem. J.*, **300**, 159–164.

Brown, G.C., Hafner, R.P. and Brand, M.D. (1990) A 'top-down' approach to the determination of control coefficients in metabolic control theory, *Eur. J. Biochem.*, **188**, 321–325.

Brumen, M. and Heinrich, R. (1984) A metabolic osmotic model of human erythrocytes, *BioSystems*, **17**, 155–169.

Burbaum, J.J., Raines, R.T., Albery, W.J. and Knowles, J.R. (1989) Evolutionary optimization of the catalytic effectiveness of an enzyme, *Biochemistry*, **28**, 9293–9305.

Burns, J.A., Cornish-Bowden, A., Groen, A.K., Heinrich, R., Kacser, H., Porteous, J.W., Rapoport, S.M., Rapoport, T.A., Stucki, J.W., Tager, J.M., Wanders, R.J.A. and Westerhoff, H.V. (1985) Control analysis of metabolic systems, *Trends Biochem. Sci.*, **10**, 16.

Caplan, S.R. (1981) Reciprocity or near-reciprocity of highly coupled enzymatic processes at the multidimensional inflection point, *Proc. Nat. Acad. Sci. USA*, **78**, 4314–4318.

Cárdenas, M.L. and Cornish-Bowden, A. (1989) Characteristics necessary for an interconvertible enzyme cascade to generate a highly sensitive response to an effector, *Biochem. J.*, **257**, 339–345.

Casas, J.L., Garcia-Canovas, F., Tudela, J. and Acosta, M. (1993) A kinetic study of simultaneous suicide inactivation and irreversible inhibition of an enzyme. Application to 1-aminocyclopropane-1-carboxylate (ACC) synthase inactivation by its substrate S-adenosylmethionine, *J. Enzym. Inhib.*, **7**, 1–14.

Cascante, M., Franco, R. and Canela, E.I. (1989a) Use of implicit methods from general sensitivity theory to develop a systematic approach to metabolic control. I. Unbranched pathways, *Math. Biosci.*, **94**, 271–288.

Cascante, M., Franco, R. and Canela, E.I. (1989b) Use of implicit methods from general sensitivity theory to develop a systematic approach to metabolic control. II. Complex systems, *Math. Biosci.*, **94**, 289–309.

Cavallotti, P., Celeri, G. and Leonardis, B. (1980) Calculation of multicomponent multiphase equilibria, *Chem. Eng. Sci.*, **35**, 2297–2304.

Cavieres, J.D. (1977) The sodium pump in human red cells, in: Ellory, J.C. and Lew, V.L. (eds.) *Membrane Transport in Red Cells*, Academic Press, London, pp. 1–37.

Chance, B., Williams, G.R., Holmes, W.F. and Higgins, J. (1955) Respiratory enzymes in oxidative phosphorylation. V. A mechanism for oxidative phosphorylation, *J. Biol. Chem.*, **217**, 439–451.

Chance, B., Holmes, W., Higgins, J. and Connelly, C.M. (1958) Localization of interaction sites in multi-component transfer systems: theorems derived from analogues, *Nature*, **182**, 1190–1193.

Chance, B., Estabrook, R.W. and Ghosh, A. (1964a) Damped sinusoidal oscillations of cytoplasmic reduced pyridine nucleotide in yeast cells, *Proc. Nat. Acad. Sci. USA*, **51**, 1244–1251.

Chance, B., Ghosh, A., Higgins, J.J. and Maitra, P.K. (1964b) Cyclic and oscillatory responses of metabolic pathways involving chemical feedback and their computer representations, *Ann. NY Acad. Sci.*, **115**, 1010–1024.

Cheung, C.-W., Cohen, N.S. and Raijman, L. (1989) Channeling of urea cycle intermediates *in situ* in permeabilized hepatocytes, *J. Biol. Chem.*, **264**, 4038–4044.

Chock, P.B., Rhee, S.G. and Stadtman, E.R. (1980) Interconvertible enzyme cascades in cellular regulation, *Ann. Rev. Biochem.*, **49**, 813–843.

Chou, K.-C. (1993) Graphic rule for non-steady-state enzyme kinetics and protein folding kinetics, *J. Math. Chem.*, **12**, 97–108.

Christensen, H., Martin, M.T. and Waley, S.G. (1990) ß-Lactamases as fully efficient enzymes. Determination of all the rate constants in the acyl-enzyme mechanism, *Biochem. J.*, **266**, 853–861.

Clarke, B.L. (1981) Complete set of steady states for the general stoichiometric dynamical system, *J. Chem. Phys.*, **75**, 4970–4979.

Clarke, B.L. (1988) Stoichiometric network analysis, *Cell Biophys.*, **12**, 237–253.

Cleland, W.W. (1979) Optimizing coupled enzyme assays, *Anal. Biochem.*, **99**, 142–145.

Corio, P.L. and Johnson, B.G. (1991) Conditions for reaction mechanisms, *J. Phys. Chem.*, **95**, 4166–4171.

Cornish-Bowden, A. (1976a) The effect of natural selection on enzymic catalysis, *J. Mol. Biol.*, **101**, 1–9.

Cornish-Bowden, A. (1976b) *Principles of Enzyme Kinetics*, Butterworths, London.

Cornish-Bowden, A. (1989) Metabolic control theory and biochemical systems theory: different objectives, different assumptions, different results, *J. Theor. Biol.*, **136**, 365–377.

Cornish-Bowden, A. (1991) Failure of channelling to maintain low concentrations of metabolic intermediates, *Eur. J. Biochem.*, **195**, 103–108.

Cornish-Bowden, A. (1994) Product inhibition in mechanisms in which the free enzyme isomerizes, *Biochem. J.*, **301**, 621–623.

Cornish-Bowden, A. (1995) Metabolic control analysis in theory and practice, *Adv. Molec. Cell Biol.*, **11**, 21–64.

Cornish-Bowden, A. and Cárdenas, M.L. (eds.) (1990) *Control of Metabolic Processes*, Plenum Press, New York.

Cornish-Bowden, A. and Cárdenas, M.L. (1993) Channelling can affect concentrations of metabolic intermediates at constant net flux: artefact or reality?, *Eur. J. Biochem.*, **213**, 87–92.

Cornish-Bowden, A. and Hofmeyr, J.-H.S. (1991) MetaModel: a program for modelling and control analysis of metabolic pathways on the IBM PC and compatibles, *CABIOS*, **7**, 89–93.

Cornish-Bowden, A. and Hofmeyr, J.-H.S. (1994) Determination of control coefficients in intact metabolic systems, *Biochem. J.*, **298**, 367–375.

Cornish-Bowden, A. and Wharton, C.W. (1988) *Enzyme Kinetics*, IRL Press, Oxford.

Crowley, P.H. (1975) Natural selection and the Michaelis constant, *J. Theor. Biol.*, **50**, 461–475.

Cuthbertson, K.S.R. (1989) Intracellular calcium oscillators, in: Goldbeter, A. (ed.) *Cell to Cell Signalling: From Experiments to Theoretical Models*, Academic Press, London, pp. 435–447.

Dibrov, B.F., Zhabotinsky, A.M. and Kholodenko, B.N. (1982) Dynamic stability of steady states and static stabilization in unbranched metabolic pathways, *J. Math. Biol.*, **15**, 51–63.

Dupont, G. and Goldbeter, A. (1989) Theoretical insights into the origin of signal-induced calcium oscillations, in: Goldbeter, A. (ed.) *Cell to Cell Signalling: From Experiments to Theoretical Models*, Academic Press, London, pp. 461–474.

Dupont, G. and Goldbeter, A. (1992) Oscillations and waves of cytosolic calcium: insights from theoretical models, *BioEssays*, **14**, 485–493.

Dupont, G. and Goldbeter, A. (1993) One-pool model for Ca^{2+} oscillations involving Ca^{2+} and inositol 1,4,5-trisphosphate as co-agonists for Ca^{2+} release, *Cell Calcium*, **14**, 311–322.

Dupont, G., Berridge, M.J. and Goldbeter, A. (1991) Signal-induced Ca^{2+} oscillations: properties of a model based on Ca^{2+}-induced Ca^{2+} release, *Cell Calcium*, **12**, 73–85.

Easterby, J.S. (1973) Coupled enzyme assays: a general expression for the transient, *Biochim. Biophys. Acta*, **293**, 552–558.

Easterby, J.S. (1981) A generalized theory of the transition time for sequential enzyme reactions, *Biochem. J.*, **199**, 155–161.

Edelstein-Keshet, L. (1988) *Mathematical Models in Biology*, Random House, New York.

Ehlde, M. and Zacchi, G. (1993) Derivation of a general matrix method for the calculation of control coefficients, in: Schuster, S., Rigoulet, M., Ouhabi, R. and Mazat, J.-P. (eds.) *Modern Trends in Biothermokinetics*, Plenum Press, New York, pp. 243–252.

Eigen, M. and Gardiner, W. (1984) Evolutionary molecular engineering based on RNA replication. *Pure Appl. Chem.*, **56**, 967–978.

Endo, M., Tanaka, M. and Ogawa, Y. (1970) Calcium induced release of calcium from the sacroplasmatic reticulum of skinned skeletal muscle fibres, *Nature*, **228**, 34–36.

Engel, J., Fechner, M., Sowerby, A.J., Finch, S.A.E. and Stier, A. (1994) Anisotropic propagation of Ca^{2+} waves in isolated cardiomyocytes, *Biophys. J.*, **66**, 1756–1762.

Érdi, P. and Tóth, J. (1989) *Mathematical Models of Chemical Reactions. Theory and Applications of Deterministic and Stochastic Models*, Manchester University Press, Manchester, U.K.

Eschrich, K., Schellenberger, W. and Hofmann, E. (1980) In vitro demonstration of alternate stationary states in an open enzyme system containing phosphofructokinase, *Arch. Biochem. Biophys.*, **205**, 114–121.

Eschrich, K., Schellenberger, W. and Hofmann, E. (1990) A hysteretic cycle in glucose 6-phosphate metabolism observed in a cell-free yeast extract, *Eur. J. Biochem.*, **188**, 697–703.

Fabiato, A. and Fabiato, F. (1975) Concentrations induced by a calcium-triggered release of calcium from the sarcoplasmatic reticulum of single skinned cardiac cell, *J. Physiol. Lond.*, **249**, 469–495.

Farquhar, G.D. (1979) Models describing the kinetics of ribulose biphosphate carboxylase-oxygenase, *Arch. Biochem. Biophys.*, **193**, 456–468.

Feinberg, M. (1989) Necessary and sufficient conditions for detailed balancing in mass action systems of arbitrary complexity, *Chem. Eng. Sci.*, **44**, 1819–1827.

Feistel, R. and Ebeling, W. (1989) *Evolution of Complex Systems*, KluwerAcademic Publishers, Dordrecht.

Fell, D.A. (1990) Substrate cycles: theoretical aspects of their role in metabolism, *Comm. Theor. Biol.*, **6**, 1–14.

Fell, D.A. (1992) Metabolic control analysis: a survey of its theoretical and experimental development, *Biochem. J.*, **286**, 313–330.

Fell, D.A. (1993) The analysis of flux in substrate cycles, in: Schuster, S., Rigoulet, M., Ouhabi, R. and Mazat, J.-P. (eds.) *Modern Trends in Biothermokinetics*, Plenum Press, New York, pp. 97–101.

Fell, D.A. and Sauro, H.M. (1985) Metabolic control and its analysis. Additional relationships between elasticities and control coefficients, *Eur. J. Biochem.*, **148**, 555–561.

Fell, D.A. and Sauro, H.M. (1990) Metabolic control analysis. The effects of high enzyme concentrations, *Eur. J. Biochem.*, **192**, 183–187.

Fell, D.A. and Small, J.R. (1986) Fat synthesis in adipose tissue. An examination of stoichiometric constraints, *Biochem. J.*, **238**, 781–786.

Fell, D.A. and Snell, K. (1988) Control analysis of mammalian serine biosynthesis. Feedback inhibition on the final step, *Biochem. J.*, **256**, 97–101.

Fersht, A.R. (1974) Catalysis, binding and enzyme-substrate complementarity, *Proc. R. Soc. Lond. B*, **187**, 379–407.

Fitton, V., Rigoulet, M., Ouhabi, R. and Guérin, B. (1994) Mechanistic stoichiometry of yeast mitochondrial oxidative phosphorylation, *Biochemistry*, **33**, 9692–9698.

Freedman, J.C. and Hoffman, J.F. (1979) Ionic and osmotic equilibria of human red blood cells treated with nystatin, *J. Gen. Physiol.*, **74**, 157–185.

Frenzen, C.L. and Maini, P.K. (1988) Enzyme kinetics for a two-step enzymic reaction with comparable initial enzyme-substrate ratios, *J. Math. Biol.*, **26**, 689–703.

Frieden, C. (1964) Treatment of enzyme kinetic data. 1. The effect of modifiers on the kinetic parameters of single substrate enzymes, *J. Biol. Chem.*, **239**, 3522–3531.

Friedrich, P. (1984) *Supramolecular Enzyme Organization*, Pergamon Press, Oxford.

Galazzo, J. and Bailey, J. (1990) Fermentation pathway kinetics and metabolic flux control in suspended and immobilized Saccharomyces cerevisiae, *Enzyme Microbiol. Technol.*, **12**, 162–172.

Gantmacher, F.R. (1959) *The Theory of Matrices*, Chelsea Publishing Co., New York.

Gary-Bobo, C.M. and Solomon, A.K. (1968) Properties of hemoglobin solutions in red cells, *J. Gen. Physiol.*, **52**, 825–853.

Gavalas, G.R. (1968) *Nonlinear Differential Equations of Chemically Reacting Systems*, Springer-Verlag, Berlin.

Gellerich, F.N., Kunz, W.S. and Bohnensack, R. (1990) Estimation of flux control coefficients from inhibitor titrations by non-linear regression, *FEBS Lett.*, **274**, 167–170.

Gellerich, F.N., Laterveer, F.D., Gnaiger, E. and Nikolay, K. (1994) Effect of macromolecules on ADP transport into mitochondria, in: Gnaiger, E., Gellerich, F.N. and Wyss, M. (eds.) *What is Controlling Life?*, Innsbruck University Press, Innsbruck, pp. 181–185.

Gerisch, G. and Hess, B. (1974) Cyclic-AMP-controlled oscillations in suspended Dictyostelium cells: their relation to morphogenetic cell interactions, *Proc. Nat. Acad. Sci. USA*, **71**, 2118–2122.

Ghosh, A. and Chance, B. (1964) Oscillations of glycolytic intermediates in yeast cells, *Biochem. Biophys. Res. Comm.*, **16**, 174–181.

Giardino, I., Edelstein, D. and Brownlee, M. (1994) Nonenzymatic glycosylation in vitro and in bovine endothelial cells alters basic fibroblast growth factor activity. A model for intracellular glycosylation in diabetes, *J. Clin. Invest.*, **94**, 110–117.

Giersch, C. (1988) Control analysis of metabolic networks. 1. Homogeneous functions and the summation theorems for control coefficients, *Eur. J. Biochem.*, **174**, 509–513.

Giersch, C. (1991) Pillow Strategy, *J. Theor. Biol.*, **152**, 71.

Gitterman, M. and Weiss, G.H. (1994) Generalized theory of the kinetics of tracers in biological systems, *Bull. Math. Biol.*, **56**, 171–186.

Glansdorff, P. and Prigogine, I. (1971) *Thermodynamic Theory of Structure, Stability and Fluctuations*, Wiley-Interscience, London.

Glaser, R., Heinrich, H., Brumen, M. and Svetina, S. (1983) Ionic states and metabolism of erythrocytes, *Biomed. Biochim. Acta*, **42**, S77–S80.

Gnaiger, E. and Wyss, M. (1994) Chemical forces in the cell: calculation for the ATP system, in: Gnaiger, E., Gellerich, F.N. and Wyss, M. (eds.) *What is Controlling Life?* Innsbruck University Press, Innsbruck, pp. 207–212.

Gnaiger, E., Gellerich, F.N. and Wyss, M. (eds.) (1994) *What is Controlling Life?* Innsbruck University Press, Innsbruck.

Goldbeter, A. (1990) *Rythmes et chaos dans les systèmes biochimiques et cellulaires*, Masson, Paris.

Goldbeter, A. and Caplan, S.R. (1976) Oscillatory enzymes, *Ann. Rev. Biophys. Bioeng.*, **5**, 449–476.

Goldbeter, A. and Koshland Jr., D.E. (1984) Ultrasensitivity in biochemical systems controlled by covalent modification. Interplay between zero-order and multistep effects, *J. Biol. Chem.*, **259**, 14441–14447.

Goldbeter, A. and Lefever, R. (1972) Dissipative structures for an allosteric model. Application to glycolytic oscillations, *Biophys. J.*, **12**, 1302–1315.

Goldbeter, A. and Li, Y.-X. (1989) Frequency coding in intercellular communication, in: Goldbeter, A. (ed.) *Cell to Cell Signalling: From Experiments to Theoretical Models*, Academic Press, London, pp. 415–432.

Goldbeter, A., Dupont, G. and Berridge, M.J. (1990) Minimal model for signal-induced Ca^{2+} oscillations and for their frequency encoding through protein phosphorylation, *Proc. Nat. Acad. Sci. USA*, **87**, 1461–1465.

Goldman, D.E. (1943) Potential, impedance, and rectification in membranes, *J. Gen. Physiol.*, **27**, 37–60.

Goodwin, B.C. (1963) *Temporal Organization in Cells*, Academic Press, London.

Goodwin, B.C. (1965) Oscillatory behavior in enzymatic control processes, *Adv. Enzyme Regul.*, **3**, 425–438.

Gopalsamy, K. (1992) *Stability and Oscillations in Delay Differential Equations of Population Dynamics*, Kluwer Academic Publishers, Dordrecht.

Gottschalk, G. (1986) *Bacterial Metabolism*, Springer-Verlag, New York.

Gray, P. and Scott, S.K. (1994) *Chemical Oscillations and Instabilities. Non-Linear Chemical Kinetics*, Clarendon Press, Oxford.

Grimes, A.J. (1980) *Human Red Cell Metabolism*, Blackwell Scientific Publications, Oxford.

Groen, A.K., Wanders, R.J.A., Westerhoff, H.V., Van der Meer, R. and Tager, J.M. (1982) Quantification of the contribution of various steps of the control of mitochondrial respiration, *J. Biol. Chem.*, **257**, 2754–2757.

Groen, B.H., Berden, J.A. and Van Dam, K. (1990) Differentiation between leaks and slips in oxidative phosphorylation, *Biochim. Biophys. Acta*, **1019**, 121–127.

Groetsch, C.W. and King, J.T. (1988) *Matrix Methods and Applications*, Prentice-Hall, Englewood Cliffs, NJ.

Guckenheimer, J. and Holmes, P. (1983) *Nonlinear Oscillations, Dynamical Systems, and Bifurcations of Vector Fields*, Springer-Verlag, New York.

Guggenheim, E.A. (1967) *Thermodynamics*, North-Holland, Amsterdam.

Gutfreund, H. and Chock, P.B. (1991) Substrate channeling among glycolytic enzymes: fact or fiction, *J. Theor. Biol.*, **152**, 117–121.

Hafner, R.P., Brown, G.C. and Brand, M.D. (1990) Analysis of the control of respiration rate, phosphorylation rate, proton leak rate and protonmotive force in isolated mitochondria using the 'top-down' approach of metabolic control theory, *Eur. J. Biochem.*, **188**, 313–319.

Hahn, W. (1967) *Stability of Motion*, Springer-Verlag, Berlin.

Haldane, J.B.S. (1930) *Enzymes*, Longmans, Green and Co., London [reprinted by M.I.T. Press, Cambridge, MA, 1965].

Hänggi, P., Talkner, P. and Borkovec, M. (1990) Reaction-rate theory: fifty years after Kramers, *Rev. Mod. Phys.*, **62**, 251–342.

Hanusse, P. (1972) De l'existence d'un cycle limite dans l'évolution des systèmes chimiques ouverts, *C.R. Acad. Sc. Paris, Series C*, **274**, 1245–1247.

Hanusse, P. (1973) Étude des systèmes dissipatifs chimiques à deux et trois espèces intermédiares, *C.R. Acad. Sc. Paris, Series C*, **277**, 263–266.

Hearon, J.Z. (1953) The kinetics of linear systems with special reference to periodic reactions, *Bull. Math. Biophys.*, **15**, 121–141.

Heineken, F.G., Tsuchiya, H.M. and Aris, R. (1967) On the mathematical status of the pseudo-steady state hypothesis of biochemical kinetics, *Math. Biosci.*, **1**, 95–113.

Heinrich, R. and Hoffmann, E. (1991) Kinetic parameters of enzymatic reactions in states of maximal activity. An evolutionary approach, *J. Theor. Biol.*, **151**, 249–283.

Heinrich, R. and Holzhütter, H.-G. (1985) Efficiency and design of simple metabolic systems, *Biomed. Biochim. Acta*, **44**, 959–969.

Heinrich, R. and Rapoport, T.A. (1973) Linear theory of enzymatic chains; its application for the analysis of the crossover theorem and of the glycolysis of human erythrocytes, *Acta Biol. Med. Germ.*, **31**, 479–494.

Heinrich, R. and Rapoport, T.A. (1974a) A linear steady-state treatment of enzymatic chains. General properties, control and effector strength, *Eur. J. Biochem.*, **42**, 89–95.

Heinrich, R. and Rapoport, T.A. (1974b) A linear steady-state treatment of enzymatic chains. Critique of the crossover theorem and a general procedure to identify interaction sites with an effector, *Eur. J. Biochem.*, **42**, 97–105.

Heinrich, R. and Rapoport, T.A. (1975) Mathematical analysis of multienzyme systems. II. Steady state and transient control, *BioSystems*, **7**, 130–136.

Heinrich, R. and Rapoport, T.A. (1980) Mathematical modelling of translation of mRNA in eucaryotes. Steady states, time-dependent processes and application to reticulocytes, *J. Theor. Biol.*, **86**, 279–313.

Heinrich, R. and Reder, C. (1991) Metabolic control analysis of relaxation processes, *J. Theor. Biol.*, **151**, 343–350.

Heinrich, R. and Schuster, S. (1991) Is metabolic channelling the complicated solution to the easy problem of reducing transient times?, *J. Theor. Biol.*, **152**, 57–61.

Heinrich, R. and Sonntag, I. (1982) Dynamics of non-linear biochemical systems and the evolutionary significance of time hierarchy, *BioSystems*, **15**, 301–316.

Heinrich, R., Rapoport, S.M. and Rapoport, T.A. (1977) Metabolic regulation and mathematical models, *Progr. Biophys. Mol. Biol.*, **32**, 1–82.

Heinrich, R., Holzhütter, H.-G. and Schuster, S. (1987) A theoretical approach to the evolution and structural design of enzymatic networks; linear enzymatic chains, branched pathways and glycolysis of erythrocytes, *Bull. Math. Biol.*, **49**, 539–595.

Heinrich, R., Schuster, S. and Holzhütter, H.-G. (1991) Mathematical analysis of enzymic reaction systems using optimization principles, *Eur. J. Biochem.*, **201**, 1–21.

Heinrich, R., Montero, F., Klipp, E., Waddell, T. G., and Meléndez-Hevia, E. (1996) Theoretical approaches to the evolutionary optimization of glycolysis. Thermodynamic and kinetic constraint, *submitted.*

Henri, M.V. (1902) Théorie générale de l'action de quelques diastases, *Compt. Rend. Acad. Sci.*, **135**, 916–919.

Hess, B. and Boiteux, A. (1968) Mechanism of glycolytic oscillation in yeast. I. Aerobic and anaerobic growth conditions for obtaining glycolytic oscillation, *Hoppe-Seyler's Z. Physiol. Chem.*, **349**, 1567–1574.

Higgins, J. (1964) A chemical mechanism for oscillation of glycolytic intermediates in yeast cells, *Proc. Nat. Acad. Sci. USA*, **51**, 989–994.

Higgins, J. (1965) Dynamics and control in cellular reactions, in: Chance, B., Estabrook, R.W. and Williamson, J.R. (eds.) *Control of the Energy Metabolism*, Academic Press, New York, pp. 13–46.

Higgins, J. (1967) The theory of oscillating reactions, *Ind. Eng. Chem.*, **59**, 18–62.

Hill, A.V. (1910) The possible effects of the aggregation of the molecules of hemoglobin on its dissociation curves, *J. Physiol.*, **40**, iv–vii.

Hill, T.L. (1977) *Free Energy Transduction in Biology*, Academic Press, New York.

Hochachka, P.W. and Somero, G.N. (1973). *Strategies of Biochemical Adaptation*, W.B.Saunders Company, Philadelphia.

Höfer, T. and Heinrich, R. (1993) A second-order approach to metabolic control analysis, *J. Theor. Biol.*, **164**, 85–102.

Hofmeyr, J.-H.S. (1989) Control-pattern analysis of metabolic pathways. Flux and concentration control in linear pathways, *Eur. J. Biochem.*, **186**, 343–354.

Hofmeyr, J.-H.S. and Cornish-Bowden, A. (1991) Quantitative assessment of regulation in metabolic systems, *Eur. J. Biochem.*, **200**, 223–236.

Hofmeyr, J.-H.S. and Cornish-Bowden, A. (1993) A control analysis of metabolic regulation, in: Schuster, S., Rigoulet, M., Ouhabi, R. and Mazat, J.-P. (eds.) *Modern Trends in Biothermokinetics*, Plenum Press, New York, pp. 193–198.

Hofmeyr, J.H.S. and Van der Merwe, K.J. (1986) METAMOD: software for steady-state modelling and control analysis of metabolic pathways on the BBC microcomputer, *CABIOS*, **2**, 243–249.

Hofmeyr, J.-H.S., Kacser, H. and Van der Merwe, K.J. (1986) Metabolic control analysis of moiety-conserved cycles, *Eur. J. Biochem.*, **155**, 631–641.

Hofmeyr, J.-H.S., Cornish-Bowden, A. and Rohwer, J.M. (1993) Taking enzyme kinetics out of control; putting control into regulation, *Eur. J. Biochem.*, **212**, 833–837.

Holmes, W.F. (1959) Locating sites of interactions between external chemicals and a sequence of chemical reactions, *Trans. Faraday Soc.*, **55**, 1122–1126.

Holms, W.H. (1986) The central metabolic pathways of Escherichia coli: relationship between flux and control at a branch point, efficiency of conversion to bionmass, and excretion of acetate, *Curr. Top. Cell. Regul.*, **28**, 69–105.

Holzhütter, H.-G. (1985) Compartmental analysis: theoretical aspects and application, *Biomed. Biochim. Acta*, **44**, 863–873.

Holzhütter, H.-G., Henke, W., Dubiel, W. and Gerber, G. (1985a) A mathematical model to study short-term regulation of mitochondrial energy transduction, *Biochim. Biophys. Acta*, **810**, 252–268.

Holzhütter, H.-G., Jacobasch, G. and Bisdorff, A. (1985b) Mathematical modelling of metabolic pathways affected by an enzyme deficiency. A mathematical model of glycolysis in normal and pyruvate-kinase-deficient red blood cells, *Eur. J. Biochem.*, **149**, 101–111.

Holzhütter, H.-G., Sluse-Goffart, C.M. and Sluse, F.E. (1994) Multiphase saturation curves of the oxoglutarate carrier: a mathematical model, *Math. Comput. Modell.*, **19**, 263–272.

Hopf, E. (1942) Abzweigung einer periodischen Lösung von einer stationären Lösung eines Differentialsystems, *Ber. Math. Phys. Kl. Sächs. Akad. Wiss.*, **94**, 3–22.

Horecker, B.L., Gibbs, M., Klenow, H. and Smyrniotis, P.Z. (1954) The mechanism of pentose phosphate conversion to hexose monophosphate, *J. Biol. Chem.*, **207**, 393–403.

Horn, F. and Jackson, R. (1972) General mass action kinetics, *Arch. Rational Mech. Anal.*, **47**, 81–116.

Hunding, A. (1974) Limit-cycles in enzyme-systems with nonlinear negative feedback, *Biophys. Struct. Mech.*, **1**, 47–54.

Hurwitz, A. (1895) Ueber die Bedingungen, unter welchen eine Gleichung nur Wurzeln mit negativen reellen Theilen besitzt, *Math. Ann.*, **46**, 273–284.

Jaffe, L.F. (1991) The path of calcium in cytosolic calcium oscillations: a unifying hypothesis, *Proc. Nat. Acad. Sci. USA*, **88**, 9883–9887.

Jensen, R.A. (1976) Enzyme recruitment in evolution of new function, *Ann. Rev. Microbiol.*, **30**, 409–425.

Joshi, A. and Palsson, B.O. (1989a) Metabolic dynamics in the human red cell. I. A comprehensive kinetic model, *J. Theor. Biol.*, **141**, 515–528.

Joshi, A. and Palsson, B.O. (1989b) Metabolic dynamics in the human red cell. II. Interactions with the environment, *J. Theor. Biol.*, **141**, 529–545.

Joshi, A. and Palsson, B.O. (1990a) Metabolic dynamics in the human red cell. III. Metabolic reaction rates, *J. Theor. Biol.*, **142**, 41–68.

Joshi, A. and Palsson, B.O. (1990b) Metabolic dynamics in the human red cell. IV. Data prediction and some model computations, *J. Theor. Biol.*, **142**, 69–85.

Jou, D., Casas-Vázques, J. and Lebon, G. (1993) *Extended Irreversible Thermodynamics*, Springer-Verlag, Berlin.

Kacser, H. (1983) The control of enzyme systems in vivo: elasticity analysis of the steady state, *Biochem. Soc. Trans.*, **11**, 35–40.

Kacser, H. and Beeby, R. (1984) Evolution of catalytic proteins. On the origin of enzyme species by means of natural selection, *J. Mol. Evol.*, **20**, 38–51.

Kacser, H. and Burns, J.A. (1973) The control of flux, *Symp. Soc. Exp. Biol.*, **27**, 65–104.

Kahn, D. and Westerhoff, H.V. (1991) Control theory of regulatory cascades, *J. Theor. Biol.*, **153**, 255–285.

Kahn, D. and Westerhoff, H.V. (1993a) Regulation and homeostasis in metabolic control theory: interplay between fluctuations of variables and parameter changes, in: Schuster, S., Rigoulet, M., Ouhabi, R. and Mazat, J.-P. (eds.) *Modern Trends in Biothermokinetics*, Plenum Press, New York, pp. 199–204.

Kahn, D. and Westerhoff, H.V. (1993b) The regulatory strength: how to be precise about regulation and homeostasis, *Acta Biotheor.*, **41**, 85–96.

Katchalsky, A. and Curran, P.F. (1965) *Nonequilibrium Thermodynamics in Biophysics*, Harvard University Press, Cambridge, MA.

Kedem, O. (1972) From irreversible thermodynamics to networks thermodynamics, *J. Membr. Biol.*, **10**, 213–219.

Kedem, O. and Caplan, S.R. (1965) Degree of coupling and its relation to efficiency of energy conversion, *Trans. Faraday Soc.*, **61**, 1897–1911.

Kholodenko, B.N. (1990) Metabolic control theory. New relationships for determining control coefficients of enzymes and response coefficients of system variables, *J. Nonlin. Biol.*, **1**, 107–126.

Kholodenko, B.N. and Westerhoff, H.V. (1993) Metabolic channelling and control of the flux, *FEBS Lett.*, **320**, 71–74.

Kholodenko, B.N. and Westerhoff, H.V. (1994) Control theory of one enzyme, *Biochim. Biophys. Acta*, **1208**, 294–305.

Kholodenko, B.N., Lyubarev, A.E. and Kurganov, B.I. (1992) Control of the metabolic flux in a system with high enzyme concentrations and moiety-conserved cycles. The sum of the flux control coefficients can drop significantly below unity, *Eur. J. Biochem.*, **210**, 147–153.

Kholodenko, B.N., Demin, O.V. and Westerhoff, H.V. (1993a) 'Channelled' pathways can be more sensitive to specific regulatory signals, *FEBS Lett.*, **320**, 75–78.

Kholodenko, B.N., Cascante, M. and Westerhoff, H.V. (1993b) Dramatic changes in control properties that accompany channelling and metabolic sequestration, *FEBS Lett.*, **336**, 381–384.

Kholodenko, B.N., Westerhoff, H.V. and Brown, G.C. (1994) Rate limitation within a single enzyme is directly related to enzyme intermediate levels, *FEBS Lett.*, **349**, 131–134.

Kholodenko, B.N., Molenaar, D., Schuster, S., Heinrich, R. and Westerhoff, H.V. (1995)

Defining control coefficients in non-ideal metabolic pathways, *Biophys. Chem.*, **56**, 215–226.

King, E.L. and Altman, C. (1956) A schematic method of deriving the rate laws for enzyme-catalyzed reactions, *J. Phys. Chem.*, **60**, 1375–1378.

Klonowski, W. (1983) Simplifying principles for chemical and enzyme reaction kinetics, *Biophys. Chem.*, **18**, 73–87.

Ko, Y.F., Bentley, W.E. and Weigand, W.A. (1994) A metabolic model of cellular energetics and carbon flux during aerobic Escherichia coli fermentation, *Biotechnol. Bioeng.*, **43**, 847–855.

Kohn, M.C. and Chiang, E. (1982) Metabolic network sensitivity analysis, *J. Theor. Biol.*, **98**, 109–126.

Kohn, M.C. and Chiang, E. (1983) Sensitivity to values of the rate constants in a neurochemical metabolic model, *J. Theor. Biol.*, **100**, 551–565.

Kohn, M.C., Whitley, L.M. and Garfinkel, D. (1979) Instantaneous flux control analysis for biochemical systems, *J. Theor. Biol.*, **76**, 437–452.

Kondratiev, V.N. (1969) Chain reactions, in: Bamford, C.H. and Tipper, C.F.H. (eds.) *Comprehensive Chemical Kinetics. Vol. 2*, Elsevier, Amsterdam, pp. 81–188.

Korzeniewski, B. and Froncisz, W. (1991) An extended dynamic model of oxidative phosphorylation, *Biochim. Biophys. Acta*, **1060**, 210–223.

Koshland Jr., D.E., Némethy, G. and Filmer, D. (1966) Comparison of experimental binding data and theoretical models in proteins containing subunits, *Biochemistry*, **5**, 365–385.

Kuby, S.A. (1991) *A Study of Enzymes. I. Enzyme Catalysis, Kinetics, and Substrate Binding*, CRC Press, Boca Raton, FL.

Kuchel, P.W. (1985) Kinetic analysis of multienzyme systems in homogeneous solution, in: Welch, G.R. (ed.) *Organized Multienzyme Systems*, Academic Press, Orlando, FL, pp. 303–380.

LaBaume, L.B., Merrill, D.K., Clary, G.L. and Guynn, R.W. (1987) Effect of acute ethanol on serine biosynthesis in liver, *Arch. Biochem. Biophys.*, **256**, 569–577.

Lehninger, A.L. (1982). *Principles of Biochemistry*, Worth Publishers, New York.

Leiser, J. and Blum, J.J. (1987) On the analysis of substrate cycles in large metabolic systems, *Cell Biophys.*, **11**, 123–138.

Lengyel, S. (1989) On the relationship between thermodynamics and chemical kinetics, *Z. Phys. Chem.*, **270**, 577–589.

Letellier, T., Reder, C. and Mazat, J.-P. (1991) CONTROL: software for the analysis of the control of metabolic networks, *CABIOS*, **7**, 383–390.

Letellier, T., Malgat, M. and Mazat, J.-P. (1993) Control of oxidative phosphorylation in rat muscle mitochondria: implications for mitochondrial myopathies, *Biochim. Biophys. Acta*, **1141**, 58–64.

Letellier, T., Heinrich, R., Malgat, M. and Mazat, J.-P. (1994) The kinetic basis of threshold effects observed in mitochondrial diseases: a systemic approach, *Biochem. J.*, **302**, 171–174.

Levin, J.J. and Levinson, N. (1954) Singular perturbations of non-linear systems of differential equations and an associated boundary layer equation, *J. Rational Mech. Anal.*, **3**, 247–270.

Lewis, G.N. (1925) A new principle of equilibrium, *Proc. Nat. Acad. Sci. USA*, **11**, 179–183.

Liao, J.C. and Delgado, J. (1993) Advances in metabolic control analysis, *Biotechn. Progr.*, **9**, 221–233.

Liao, J.C. and Lightfoot Jr., E.N. (1987) Extending the quasi-steady state concept to analysis of metabolic networks, *J. Theor. Biol.*, **126**, 253–273.

Liao, J.C. and Lightfoot Jr., E.N. (1988a) Characteristic reaction paths of biochemical reaction systems with time scale separation, *Biotechnol. Bioeng.*, **31**, 847–854.

Liao, J.C. and Lightfoot, Jr., E.N. (1988b) Lumping analysis of biochemical reaction systems with time scale separation, *Biotechnol. Bioeng.*, **31**, 869–879.

Liljenström, H. and Blomberg, C. (1987) Site dependent time optimization of protein synthesis with special regard to accuracy, *J. Theor. Biol.*, **129**, 41–56.

Lotka, A. (1910) Zur Theorie der periodischen Reaktionen, *Z. Phys. Chem.*, **72**, 508–511.

Lowry, O.H. and Passonneau, J.V. (1964) The relationships between substrates and enzymes of glycolysis in brain, *J. Biol. Chem.*, **239**, 31–42.

Luvisetto, S., Pietrobon, D. and Azzone, G.F. (1987) Uncoupling of oxidative phosphorylation. 1. Protonophoric effects account only partially for uncoupling, *Biochemistry*, **26**, 7332–7338.

Majewski, R.A. and Domach, M.M. (1985) Consideration of the gain, enzymatic capacity utilization, and response time properties of metabolic networks as a function of operating point and structure, *BioSystems*, **18**, 15–22.

Majewski, R.A. and Domach, M.M. (1990a) Effect of regulatory mechanism on hyperbolic reaction network properties, *Biotechnol. Bioeng.*, **36**, 166–178.

Majewski, R.A. and Domach, M.M. (1990b) Chemostat-cultivated Escherichia coli at high dilution rate: multiple steady states and drift, *Biotechnol. Bioeng.*, **36**, 179–190.

Maretzki, D., Reimann, B., Klatt, D. and Rapoport, S. (1980) A form of (Ca^{2+} + Mg^{2+})-ATPase of human red cell membranes with low affinity for Mg-ATP: a hypothesis for its function, *FEBS Lett.*, **111**, 269–271.

Markus, M. and Hess, B. (1986) Hysteresis and crises in the dynamics of glycolysis, in: Damjanovich, S., Keleti, T. and Trón, L. (eds.) *Dynamics of Biochemical Systems*, Académiai Kiadó, Budapest, pp. 11–23.

Markus, M. and Hess, B. (1990) Control of metabolic oscillations: unpredictability, critical slowing down, optimal stability and hysteresis, in: Cornish-Bowden, A. and Cárdenas, M.L. (eds.) *Control of Metabolic Processes*, Plenum Press, New York, pp. 303–313.

Mavrovouniotis, M.L. (1992) Synthesis of reaction mechanisms consisting of reversible and irreversible steps. 2. Formalization and analysis of the synthesis algorithm, *Ind. Eng. Chem. Res.*, **31**, 1637–1653.

Mavrovouniotis, M.L., Stephanopoulos, G. and Stephanopoulos, G. (1990) Computer-aided synthesis of biochemical pathways, *Biotechnol. Bioeng.*, **36**, 1119–1132.

May, R.M. (1974) *Stability and Complexity in Model Ecosystems*, Princeton University Press, Princeton, NJ.

Mayer, R. (1845) *Die organische Bewegung in ihrem Zusammenhange mit dem Stoffwechsel*, Verlag der Drechsler'schen Buchhandlung, Heilbronn.

Mazat, J.-P., Letellier, T. and Reder, C. (1990) Metabolic control theory: the geometry of the triangle, *Biomed. Biochim. Acta*, **49**, 801–810.

Meléndez-Hevia, E. (1990) The game of the pentose phosphate cycle: a mathematical approach to study the optimization in design of metabolic pathways during evolution, *Biomed. Biochim. Acta*, **49**, 903–916.

Meléndez-Hevia, E. and Isidoro, A. (1985) The game of the pentose phosphate cycle, *J. Theor. Biol.*, **117**, 251–263.

Meléndez-Hevia, E. and Torres, N.V. (1988) Economy of design in metabolic pathways: further remarks on the game of the pentose phosphate cycle, *J. Theor. Biol.*, **132**, 97–111.

Meléndez-Hevia, E., Torres, N.V., Sicilia, J. and Kacser, H. (1990) Control analysis of transition times in metabolic systems, *Biochem. J.*, **265**, 195–202.

Meléndez-Hevia, E., Waddell, T.G. and Montero, F. (1994) Optimization of metabolism: the evolution of metabolic pathways toward simplicity through the game of the pentose phosphate cycle, *J. Theor. Biol.*, **166**, 201–219.

Mendes, P. (1993) GEPASI: a software package for modelling the dynamics, steady states and control of biochemical and other systems, *CABIOS*, **9**, 563–571.

Mendes, P., Kell, D.B. and Westerhoff, H.V. (1992) Channelling can decrease pool size, *Eur. J. Biochem.*, **204**, 257–266.

Meyer, T. (1991) Cell signalling by second messenger waves, *Cell*, **64**, 675–678.

Meyer, T. and Stryer, L. (1988) Molecular model for receptor-stimulated calcium spiking, *Proc. Nat. Acad. Sci. USA*, **85**, 5051–5055.

Meyer, T. and Stryer, L. (1991) Calcium spiking, *Ann. Rev. Biophys. Biophys. Chem.*, **20**, 153–174.

Michaelis, L. and Menten, M.L. (1913) Die Kinetik der Invertinwirkung, *Biochem. Z.*, **49**, 333–369.

Minorsky, N. (1962) *Nonlinear Oscillations*, Van Nostrand, New York.

Mitchell, P. (1961) Coupling of phosphorylation to electron and hydrogen transfer by a chemi-osmotic type of mechanism, *Nature*, **191**, 144–148.

Mitchell, P. and Moyle, J. (1965) Stoichiometry of proton translocation through the respiratory chain and adenosine triphosphatase systems of rat liver mitochondria, *Nature*, **208**, 147–151.

Moelwyn-Hughes, E.A. (1964) *Physical Chemistry*, Pergamon Press, Oxford.

Monod, J., Wyman, J. and Changeux, J.-P. (1965) On the nature of allosteric transitions: a plausible model, *J. Mol. Biol.*, **12**, 88–118.

Moore, W.J. (1972) *Physical Chemistry*, Prentice Hall, Englewood Cliffs, NJ.

Morales, M. and McKay, D. (1967) Biochemical oscillations in "controlled" systems, *Biophys. J.*, **7**, 621–625.

Morris, A.O. (1992) *Linear Algebra. An Introduction*, Chapman & Hall, London.

Nicolis, G. and Prigogine, I. (1977) *Self-Organization in Nonequilibrium Systems*, John Wiley & Sons, New York.

Novak, B. and Tyson, J.J. (1993) Modeling the cell division cycle: M-phase trigger, oscillations, and size control, *J. Theor. Biol.*, **165**, 101–134.

Nožičk, F., Guddat, J., Hollatz, H. and Bank, B. (1974) *Theorie der linearen parametrischen Optimierung*, Akademie-Verlag, Berlin.

Onsager, L. (1931) Reciprocal relations in irreversible processes. I, *Phys. Rev.*, **37**, 405–426.

Othmer, H.G. (1976) The qualitative dynamics of a class of biochemical control circuits, *J. Math. Biol.*, **3**, 53–78.

Othmer, H.G. (1981) The interaction of structure and dynamics in chemical reaction networks, in: Ebert, K.H., Deuflhard, P. and Jäger, W. (eds.) *Modelling of Chemical Reaction Systems. Proceedings of an International Workshop, Heidelberg, Sept. 1–5, 1980*, Springer-Verlag, Berlin, pp. 2–19.

Otto, M., Heinrich, R., Kühn, B. and Jacobasch, G. (1974) A mathematical model for the influence of fructose 6-phosphate, ATP, potassium, ammonium and magnesium on the phosphofructokinase from rat erythrocytes, *Eur. J. Biochem.*, **49**, 169–178.

Otto, M., Heinrich, R., Jacobasch, G. and Rapoport, S. (1977) A mathematical model for the influence of anionic effectors on the phosphofructokinase from rat erythrocytes, *Eur. J. Biochem.*, **74**, 413–420.

Ovádi, J. (1991) Physiological significance of metabolic channelling, *J. Theor. Biol.*, **152**, 1–22.

Overholser, K.A., Lomangio, N.A., Harris, T.R., Bradley, J.D. and Bosan, S. (1994) Deduction of pulmonary microvascular hematocrit from indicator dilution curves, *Bull. Math. Biol.*, **56**, 225–247.

Palsson, B.O., Jamier, R. and Lightfoot, E.N. (1984) Mathematical modelling of dynamics and control in metabolic networks. II. Simple dimeric enzymes, *J. Theor. Biol.*, **111**, 303–321.

Palsson, B.O., Palsson, H. and Lightfoot, E.N. (1985) Mathematical modelling of dynamics and control in metabolic networks. III. Linear reaction sequences, *J. Theor. Biol.*, **113**, 231–259.

Park, D.J.M. (1974) An algorithm for detecting non-steady state moieties in steady state subnetworks, *J. Theor. Biol.*, **48**, 125–131.

Park, Jr., D.J.M. (1986) The complete stoichiometer, *Comput. Meth. Prog. Biomed.*, **22**, 293–301.

Park, Jr., D.J.M. (1988) Positive compositional algorithms in chemical reaction systems, *Comput. Chem.*, **12**, 175–188.

Patel, S.S., Wong, I. and Johnson, K.A. (1991) Pre-steady-state kinetic analysis of processive DNA replication including complete characterization of an exonuclease-deficient mutant, *Biochemistry*, **30**, 511–525.

Patte, J.-C., Truffa-Bachi, P. and Cohen, G.N. (1966) The threonine-sensitive homoserine dehydrogenase and aspartokinase activities of Escherichia coli. I. Evidence that the two activities are carried by a single protein, *Biochim. Biophys. Acta*, **128**, 426–439.

Peller, L. and Alberty, R.A. (1959) Multiple intermediates in steady state enzyme kinetics. I. The mechanism involving a single substrate and product, *J. Am. Chem. Soc.*, **81**, 5907–5914.

Peschel, M. and Mende, W. (1986) *The Predator-Prey Model: Do We Live in a Volterra World?*, Akademie-Verlag, Berlin.

Pettersson, G. (1989) Effect of evolution on the kinetic properties of enzymes, *Eur. J. Biochem.*, **184**, 561–566.

Pettersson, G. (1991) Why do many Michaelian enzymes exhibit an equilibrium constant close to unity for the interconversion of enzyme-bound substrate and product?, *Eur. J. Biochem.*, **195**, 663–670.

Pettersson, G. (1992) Evolutionary optimization of the catalytic efficiency of enzymes, *Eur. J. Biochem.*, **206**, 289–295.

Petzold, L. (1983) Automatic selection of methods for solving stiff and nonstiff systems of ordinary differential equations, *SIAM J. Sci. Comput.*, **4**, 136–148.

Pietrobon, D. and Caplan, S.R. (1985) Flow-force relationships for a six-state proton pump model: intrinsic uncoupling, kinetic equivalence of input and output forces, and domain of approximate linearity, *Biochemistry*, **24**, 5764–5776.

Pietrobon, D., Zoratti, M., Azzone, G.F. and Caplan, S.R. (1986) Intrinsic uncoupling of mitochondrial proton pumps. 2. Modeling studies, *Biochemistry*, **25**, 767–775.

Poincaré, H. (1880–1890) *Mémoire sur les courbes définies par les équations différentielles I-VI*, Gauthier-Villars, Paris.

Poincaré, H. (1899) *Les méthodes nouvelles de la mécanique céleste*, Gauthier-Villars, Paris.

Possmayer, F.E. and Gräber, P. (1994) The pH_{in} and pH_{out} dependence of the rate of ATP synthesis catalyzed by the chloroplast H^+-ATPase CF_0F_1 in proteoliposomes, *J. Biol. Chem.*, **269**, 1896–1904.

Prigogine, I. and Defay, R. (1954) *Chemical Thermodynamics*, Longmans, London.

Pye, E.K. (1969) Biochemical mechanisms underlying the metabolic oscillations in yeast, *Can. J. Bot.*, **47**, 271–285.

Quant, P.A. (1993) Experimental application of top-down control analysis to metabolic systems, *Trends Biochem. Sci.*, **18**, 26–30.

Raïs, B., Heinrich, R., Malgat, M. and Mazat, J.P. (1993) Control of threonine biosynthesis in E. coli, in: Schuster, S., Rigoulet, M., Ouhabi, R. and Mazat, J.-P. (eds.) *Modern Trends in Biothermokinetics*, Plenum Press, New York, pp. 269–274.

Rapoport, I., Elsner, R., Müller, M., Dumdey, R. and Rapoport, S. (1979) NADPH-production in the oxidative pentose phosphate pathway as source of reducing equivalents in glycolysis of human red cells in vitro, *Acta Biol. Med. Germ.*, **38**, 901–908.

Rapoport, T.A. and Heinrich, R. (1975) Mathematical analysis of multienzyme systems. I. Modelling of the glycolysis of human erythrocytes, *BioSystems*, **7**, 120–129.

Rapoport, T.A., Heinrich, R., Jacobasch, G. and Rapoport, S. (1974) A linear steady-state treatment of enzymatic chains. A mathematical model of glycolysis of human erythrocytes, *Eur. J. Biochem.*, **42**, 107–120.

Rapoport, T.A., Heinrich, R. and Rapoport, S.M. (1976) The regulatory principles of glycolysis in erythrocytes in vivo and in vitro. A minimal comprehensive model describing steady states, quasi-steady states and time-dependent processes, *Biochem. J.*, **154**, 449–469.

Rapp, P.E. (1987) Why are so many biological systems periodic?, *Progr. Neurobiol.*, **29**, 261–273.

Rapp, P.E., Mees, A.I. and Sparrow, C.T. (1981) Frequency encoded biochemical regulation is more accurate than amplitude dependent control, *J. Theor. Biol.*, **90**, 531–544.

Ray, Jr., W.J. (1983) Rate-limiting step: a quantitative definition. Application to steady-state enzymic reactions, *Biochemistry*, **22**, 4625–4637.

Rédei, L. (1967) *Algebra. Vol. 1*, Pergamon Press, Oxford.

Reder, C. (1986) Mimodrame mathématique sur les systèmes biochimiques, *Université Bordeaux I, Mathématiques Appliquées*, Report No. 8608.

Reder, C. (1988) Metabolic control theory: a structural approach, *J. Theor. Biol.*, **135**, 175–201.

Reich, J.G. (1983) Zur Ökonomie im Proteinhaushalt der lebenden Zelle, *Biomed. Biochim. Acta*, **42**, 839–848.

Reich, J.G. (1985) On optimum properties in the design of metabolic and epigenetic systems, *Biomed. Biochim. Acta*, **44**, 845–852.

Reich, J.G. and Selkov, E.E. (1975) Time hierarchy, equilibrium and non-equilibrium in metabolic systems, *BioSystems*, **7**, 39–50.

Reich, J.G. and Selkov, E.E. (1981) *Energy Metabolism of the Cell*, Academic Press, London.

Reimann, B., Klatt, D., Tsamaloukas, A.G. and Maretzki, D. (1981) Membrane phosphorylation in intact human erythrocytes, *Acta Biol. Med. Germ.*, **40**, 487–493.

Ricard, J. (1978) Generalized microscopic reversibility, kinetic co-operativity of enzymes and evolution, *Biochem. J.*, **175**, 779–791.

Ricard, J. and Noat, G. (1986) Catalytic efficiency, kinetic co-operativity of oligomeric enzymes and evolution, *J. Theor. Biol.*, **123**, 431–451.

Rigoulet, M., Fraisse, L., Ouhabi, R., Guérin, B., Fontaine, E. and Leverve, X. (1990) Flux-dependent increase in the stoichiometry of charge translocation by mitochondrial ATPase/ATP synthase induced by almitrine, *Biochim. Biophys. Acta*, **1018**, 91–97.

Rockafellar, R.T. (1970) *Convex Analysis*, Princeton University Press, Princeton, NJ.

Rössler, O.E. (1979) Continuous chaos—Four prototype equations, in: Gurel, O. and Rössler, O.E. (eds.) *Bifurcation Theory and Applications in Scientific Disciplines*, The New York Academy of Sciences, New York, pp. 376–392.

Rosen, R. (1967) *Optimality Principles in Biology*, Butterworths, London.

Rosen, R. (1986) Optimality in biology and medicine, *J. Math. Anal. Appl.*, **119**, 203–222.

Rottenberg, H. (1973a) The mechanism of energy-dependent ion transport in mitochondria, *J. Membr. Biol.*, **11**, 117–137.

Rottenberg, H. (1973b) The thermodynamic description of enzyme-catalyzed reactions, *Biophys. J.*, **13**, 503–511.

Sauro, H.M. (1993) SCAMP: a general-purpose simulator and metabolic control analysis program, *CABIOS*, **9**, 441–450.

Sauro, H.M. (1994) Moiety-conserved cycles and metabolic control analysis: problems in sequestration and metabolic channelling, *BioSystems*, **33**, 55–67.

Sauro, H.M. and Fell, D.A. (1991) SCAMP: A metabolic simulator and control analysis program, *Math. Comput. Modell.*, **15**, 15–28.

Sauro, H.M. and Kacser, H. (1990) Enzyme-enzyme interactions and control analysis. 2. The case of non-independence: heterologous associations, *Eur. J. Biochem.*, **187**, 493–500.

Sauro, H.M., Small, J.R. and Fell, D.A. (1987) Metabolic control and its analysis. Extensions to the theory and matrix method, *Eur. J. Biochem.*, **165**, 215–221.

Savageau, M.A. (1969) Biochemical systems analysis. I. Some mathematical properties of the rate law for the component enzymatic reactions, *J. Theor. Biol.*, **25**, 365–369.

Savageau, M.A. (1975) Optimal design of feedback control by inhibition. Dynamic considerations, *J. Mol. Evol.*, **5**, 199–222.

Savageau, M.A. (1976) *Biochemical Systems Analysis*, Addison-Wesley Publishing Company, Reading, MA.

Savageau, M.A., Voit, E.O. and Irvine, D.H. (1987a) Biochemical systems theory and metabolic control theory. 1. Fundamental similarities and differences, *Math. Biosci.*, **86**, 127–145.

Savageau, M.A., Voit, E.O. and Irvine, D.H. (1987b) Biochemical systems theory and metabolic control theory. 2. The role of summation and connectivity relationships, *Math. Biosci.*, **86**, 147–169.

Sawaragi, Y., Nakayama, H. and Tanino, T. (1985) *Theory of Multiobjective Optimization*, Academic Press, Orlando, FL.

Schauer, M. and Heinrich, R. (1979) Analysis of the quasi-steady-state approximation for an enzymatic one-substrate reaction, *J. Theor. Biol.*, **79**, 425–442.

Schauer, M. and Heinrich, R. (1983) Quasi-steady-state approximation in the mathematical modeling of biochemical reaction networks, *Math. Biosci.*, **65**, 155–171.

Schauer, M., Heinrich, R. and Rapoport, S.M. (1981a) Mathematische Modellierung der Glykolyse und des Adeninnukleotidstoffwechsels menschlicher Erythrozyten. I. Reaktionskinetische Ansätze, Analyse des in vivo-Zustandes und Bestimmung der Anfangsbedingungen für die in vitro-Experimente, *Acta Biol. Med. Germ.*, **40**, 1659–1682.

Schauer, M., Heinrich, R. and Rapoport, S.M. (1981b) Mathematische Modellierung der Glykolyse und des Adeninnukleotidstoffwechsels menschlicher Erythrozyten. II. Simulation des Adeninnukleotidabbaus bei Glukoseverarmung, *Acta Biol. Med. Germ.*, **40**, 1683–1697.

Schellenberger, W. and Hervagault, J.-F. (1991) Irreversible transitions in the 6-phosphofructokinase/fructose 1,6-bisphosphatase cycle, *Eur. J. Biochem.*, **195**, 109–113.

Schellenberger, W., Kretschmer, M., Eschrich, K. and Hofmann, E. (1988) Dynamic structures in the fructose-6-phosphate/fructose-1,6-bisphosphatase cycle, in: *Thermodynamics and Pattern Formation in Biology*, Walter de Gruyter & Co., Berlin, pp. 205–222.

Schuster, P. (1995) How to search for RNA structures. Theoretical concepts in evolutionary biotechnology, *J. Biotechnol.*, **41**, 239–257.

Schuster, R. and Holzhütter, H.-G. (1995) Use of mathematical models for predicting the metabolic effect of large-scale enzyme activity alterations. Application to enzyme deficiencies of red blood cells, *Eur. J. Biochem.*, **229**, 403–418.

Schuster, R. and Schuster, S. (1991) Relationships between modal analysis and rapid-equilibrium approximation in the modelling of biochemical networks, *Syst. Anal. Model. Simul.*, **8**, 623–633.

Schuster, R. and Schuster, S. (1993) Refined algorithm and computer program for calculating all non-negative fluxes admissible in steady states of biochemical reaction systems with or without some flux rates fixed, *CABIOS*, **9**, 79–85.

Schuster, R., Holzhütter, H.-G. and Jacobasch, G. (1988) Interrelations between glycolysis and the hexose monophosphate shunt in erythrocytes as studied on the basis of a mathematical model, *BioSystems*, **22**, 19–36.

Schuster, R., Jacobasch, G. and Holzhütter, H.-G. (1989) Mathematical modelling of metabolic pathways affected by an enzyme deficiency. Energy and redox metabolism of glucose-6-phosphate-dehydrogenase-deficient erythrocytes, *Eur. J. Biochem.*, **182**, 605–612.

Schuster, R., Schuster, S. and Holzhütter, H.-G. (1992) Simplification of complex kinetic models used for the quantitative analysis of nuclear magnetic resonance or radioactive tracer studies, *J. Chem. Soc. Faraday Trans.*, **88**, 2837–2844.

Schuster, S. (1989) Time hierarchy in enzymatic reaction networks as derived from an extremum principle, in: Ebeling, W. and Peschel, M. (eds.) *Dynamical Networks*, Akademie-Verlag, Berlin, pp. 173–183.

Schuster, S. and Heinrich, R. (1987) Time hierarchy in enzymatic reaction chains resulting from optimality principles, *J. Theor. Biol.*, **129**, 189–209.

Schuster, S. and Heinrich, R. (1991) Minimization of intermediate concentrations as a suggested optimality principle for biochemical networks. I. Theoretical analysis, *J. Math. Biol.*, **29**, 425–442.

Schuster, S. and Heinrich, R. (1992) The definitions of metabolic control analysis revisited, *BioSystems*, **27**, 1–15.

Schuster, S. and Hilgetag, C. (1994) On elementary flux modes in biochemical reaction systems at steady state, *J. Biol. Syst.*, **2**, 165–182.

Schuster, S. and Hilgetag, C. (1995) What information about the conserved-moiety structure of chemical reaction systems can be derived from their stoichiometry?, *J. Phys. Chem.*, **99**, 8017–8023.

Schuster, S. and Höfer, T. (1991) Determining all extreme semi-positive conservation relations in chemical reaction systems: a test criterion for conservativity, *J. Chem. Soc. Faraday Trans.*, **87**, 2561–2566.

Schuster, S. and Mazat, J.-P. (1993) A model study on interrelation between the transmembrane potential and pH difference across the mitochondrial inner membrane, in: Schuster, S., Rigoulet, M., Ouhabi, R. and Mazat, J.-P. (eds.) *Modern Trends in Biothermokinetics*, Plenum Press, New York, pp. 39–44.

Schuster, S. and Schuster, R. (1989) A generalization of Wegscheider's condition. Implications for properties of steady states and for quasi-steady-state approximation, *J. Math. Chem.*, **3**, 25–42.

Schuster, S. and Schuster, R. (1991) Detecting strictly detailed balanced subnetworks in open chemical reaction networks, *J. Math. Chem.*, **6**, 17–40.

Schuster, S. and Schuster, R. (1992) Decomposition of biochemical reaction systems according to flux control insusceptibility, *J. Chim. Phys.*, **89**, 1887–1910.

Schuster, S., Schuster, R. and Heinrich, R. (1991) Minimization of intermediate concentrations as a suggested optimality principle for biochemical networks. II. Time hierarchy, enzymatic rate laws, and erythocyte metabolism, *J. Math. Biol.*, **29**, 443–455.

Schuster, S., Kahn, D. and Westerhoff, H.V. (1993a) Modular analysis of the control of complex metabolic pathways, *Biophys. Chem.*, **48**, 1–17.

Schuster, S., Rigoulet, M., Ouhabi, R. and Mazat, J.-P. (eds.) (1993b) *Modern Trends in Biothermokinetics*, Plenum Press, New York.

Segel, I.H. and Martin, R.L. (1988) The general modifier ("allosteric") unireactant enzyme mechanism: redundant conditions for reduction of the steady state velocity equation to one that is first degree in substrate and effector, *J. Theor. Biol.*, **135**, 445–453.

Segel, L.A. (1988) On the validity of the steady state assumption of enzyme kinetics, *Bull. Math. Biol.*, **50**, 579–593.

Selkov, E.E. (1968) Self-oscillations in glycolysis. 1. A simple kinetic model, *Eur. J. Biochem.*, **4**, 79–86.

Selkov, E.E. (1975a) Analysis of stoichiometric regulation of energy metabolism, in: Rapoport, S. and Jung, F. (eds.) *VII. Internationales Symposium über Struktur und Funktion der Erythrozyten*, Akademie-Verlag, Berlin, pp. 17–19.

Selkov, E.E. (1975b) Stabilization of energy charge, generation of oscillations and multiple steady states in energy metabolism as a result of purely stoichiometric regulation, *Eur. J. Biochem.*, **59**, 151–157.

Selkov, E.E. (1980) Instability and self-oscillations in the cell energy metabolism, *Ber. Bunsenges. Phys. Chem.*, **84**, 399–402.

Seressiotis, A. and Bailey, J.E. (1988) MPS: An artificially intelligent software system for the analysis and synthesis of metabolic pathways, *Biotechnol. Bioeng.*, **31**, 587–602.

Shields, R.W. and Pearson, J.B. (1976) Structural controllability of multiinput linear systems, *IEEE Trans. Autom. Contr.*, **AC-21**, 203–212.

Shiraishi, F. and Savageau, M.A. (1993) The tricarboxylic acid cycle in Dictyostelium discoideum. Systemic effects of including protein turnover in the current model, *J. Biol. Chem.*, **268**, 16917–16928.

Skulachev, V.P. (1988) *Membrane Bioenergetics*, Springer-Verlag, Berlin.

Small, J.R. and Fell, D.A. (1989) The matrix method of metabolic control analysis: its validity for complex pathway structures, *J. Theor. Biol.*, **136**, 181–197.

Small, J.R. and Fell, D.A. (1990) Metabolic control analysis. Sensitivity of control coefficients to elasticities, *Eur. J. Biochem.*, **191**, 413–420.

Small, J.R. and Kacser, H. (1993) Responses of metabolic systems to large changes in enzyme activities and effectors. 1. The linear treatment of unbranched chains, *Eur. J. Biochem.*, **213**, 613–624.

Smith, W.R. and Missen, R.W. (1991) *Chemical Reaction Equilibrium Analysis. Theory and Algorithms*, Krieger, Malabar, FL.

Somogyi, R. and Stucki, J.W. (1991) Hormone-induced calcium oscillations in liver cells can be explained by a simple one pool model, *J. Biol. Chem.*, **266**, 11068–11077.

Sorribas, A. and Savageau, M.A. (1989a) A comparison of variant theories of intact biochemical systems. I. Enzyme–enzyme interactions and biochemical systems theory, *Math. Biosci.*, **94**, 161–193.

Sorribas, A. and Savageau, M.A. (1989b) A comparison of variant theories of intact biochemical systems. II. Flux-oriented and metabolic control theories, *Math. Biosci.*, **94**, 195–238.

Srere, P.A. (1987) Complexes of sequential metabolic enzymes, *Ann. Rev. Biochem.*, **56**, 89–124.

Srivastava, D.K. and Bernhard, S.A. (1986) Enzyme–enzyme interactions and the regulation of metabolic reaction pathways, *Curr. Top. Cell. Regul.*, **28**, 1–68.

Stackhouse, J., Nambiar, K.P., Burbaum, J.J., Stauffer, D.M. and Benner, S.A. (1985) Dynamic transduction of energy and internal equilibria in enzymes: A reexamination of pyruvate kinase, *J. Am. Chem. Soc.*, **107**, 2757–2763.

Stoner, C.D. (1992) An investigation of the relationships between rate and driving force in simple uncatalysed and enzyme-catalysed reactions with applications of the findings to chemiosmotic reactions, *Biochem. J.*, **283**, 541–552.

Stucki, J.W. (1980) The optimal efficiency and the economic degrees of coupling of oxidative phosphorylation, *Eur. J. Biochem.*, **109**, 269–283.

Stucki, J.W., Compiani, M. and Caplan, S.R. (1983) Efficiency of energy conversion in model biological pumps. Optimization by linear nonequilibrium thermodynamic relations, *Biophys. Chem.*, **18**, 101–109.

Termonia, Y. and Ross, J. (1981) Oscillations and control features in glycolysis: numerical analysis of a comprehensive model, *Proc. Nat. Acad. Sci. USA*, **78**, 2952–2956.

Thèze, J., Kleidman, L. and Saint Girons, I. (1974) Homoserine kinase from Escherichia coli K-12: properties, inhibition by L-threonine, and regulation of biosynthesis, *J. Bacteriol.*, **118**, 577–581.

Thomas, A.P., Renard, D.C. and Rooney, T.A. (1991) Spatial and temporal organization of calcium signalling in hepatocytes, *Cell Calcium*, **12**, 111–126.

Thomas, S. and Fell, D.A. (1993) A computer program for the algebraic determination of control coefficients in metabolic control analysis, *Biochem. J.*, **292**, 351–360.

Thomas, S. and Fell, D.A. (1994) Metabolic control analysis: sensitivity of control coefficients to experimentally determined variables, *J. Theor. Biol.*, **167**, 175–200.

Tikhonov, A.N. (1948) O zavisimostii reshenij differencial'nykh uravnenij ot malogo parametra, *Mat. Sborn.*, **22** (**64**), No. 2, 193–204.

Topham, C.M. (1990) A generalized theoretical treatment of the kinetics of an enzyme-catalysed reaction in the presence of an unstable irreversible modifier, *J. Theor. Biol.*, **145**, 547–572.

Torres, N.V. (1994) Modeling approach to control of carbohydrate metabolism during citric acid accumulation by Aspergillus niger. 1. Model definition and stability of the steady state, *Biotechnol. Bioeng.*, **44**, 104–111.

Torres, N., Regalado, C., Sorribas, A. and Cascante, M. (1993) Quality assessment of a metabolic model and systems analysis of citric acid production by Aspergillus niger, in: Schuster, S., Rigoulet, M., Ouhabi, R. and Mazat, J.-P. (eds.) *Modern Trends in Biothermokinetics*, Plenum Press, New York, pp. 115–124.

Umbarger, H.E. (1956) Evidence for a negative-feedback mechanism in the biosynthesis of isoleucine, *Science*, **123**, 848.

Vajda, S. and Rabitz, H. (1994) Identifiability and distinguishability of general reaction systems, *J. Phys. Chem.*, **98**, 5265–5271.

Viniegra-Gonzalez, G. and Martinez, H. (1969) Stability of biochemical feedback systems, *Biophys. J.—Soc. Abstr.*, **9**, A-210.

Voet, D. and Voet, J.G. (1990) *Biochemistry*, John Wiley & Sons, New York.

Vol'pert, A.I. and Khudyaev, S.I. (1975) *Analiz v klassakh razryvnykh funkcij i uravnenija matematicheskoj fiziki*, Nauka, Moscow.

Volterra, V. (1931) *Leçons sur la théorie mathématique de la lutte pour la vie*, Gauthier-Villars, Paris.

von Bertalanffy, L. (1953) *Biophysik des Fließgleichgewichtes. Einführung in die Physik offener Systeme und ihre Anwendung in der Biologie*, Vieweg, Braunschweig.

Wakil, S.J., Stoops, J.K. and Joshi, V.C. (1983) Fatty acid synthesis and its regulation, *Ann. Rev. Biochem.*, **52**, 573–579.

Wald, G. (1964) The origins of life, *Proc. Nat. Acad. Sci. USA*, **52**, 595–611.

Waley, S.G. (1964) A note on the kinetics of multi-enzyme systems, *Biochem. J.*, **91**, 514–517.

Walz, D. and Caplan, S.R. (1988) Energy coupling and thermokinetic balancing in enzyme kinetics, *Cell Biophys.*, **12**, 13–28.

Wang, Z.-X. and Tsou, C.-L. (1987) Kinetics of substrate reaction during irreversible modification of enzyme activity for enzymes involving two substrates, *J. Theor. Biol.*, **127**, 253–270.

Wasow, W.R. (1965) *Asymptotic Expansions for Ordinary Differential Equations*, John Wiley & Sons, New York.

Wegscheider, R. (1900) Über die allgemeinste Form der Gesetze der chemischen Kinetik homogener Systeme, *Z. Phys. Chem.*, **35**, 513–587.

Wegscheider, R. (1902) Über simultane Gleichgewichte und die Beziehungen zwischen Thermodynamik und Reaktionskinetik homogener Systeme, *Z. Phys. Chem.*, **39**, 257–303.

Wei, J. (1962) Axiomatic treatment of chemical reaction systems, *J. Chem. Phys.*, **36**, 1578–1584.

Welch, G.R., Keleti, T. and Vertessy, B. (1988) The control of cell metabolism for homogeneous vs. heterogeneous enzyme systems, *J. Theor. Biol.*, **130**, 407–422.

Werner, A. and Heinrich, R. (1985) A kinetic model for the interaction of energy metabolism and osmotic states of human erythrocytes. Analysis of the stationary "in vivo" state and of time dependent variations under blood preservation conditions, *Biomed. Biochim. Acta*, **44**, 185–212.

Westerhoff, H.V. (1982) Should irreversible thermodynamics be applied to metabolic systems? Yes. Kinetics alone are impracticable, *Trends Biochem. Sci.*, **7**, 275–279.

Westerhoff, H.V. (1989) Control, regulation and thermodynamics of free-energy transduction, *Biochimie*, **71**, 877–886.

Westerhoff, H.V. and Chen, Y.-D. (1984) How do enzyme activities control metabolite concentrations? An additional theorem in the theory of metabolic control, *Eur. J. Biochem.*, **142**, 425–430.

Westerhoff, H.V. and Kell, D.B. (1987) Matrix method for determining steps most rate-limiting to metabolic fluxes in biotechnological processes, *Biotechnol. Bioeng.*, **30**, 101–107.

Westerhoff, H.V. and Van Dam, K. (1987) *Thermodynamics and Control of Biological Free-Energy Transduction*, Elsevier, Amsterdam.

Westerhoff, H.V., Groen, A.K. and Wanders, R.J.A. (1983) The thermodynamic basis for the partial control of oxidative phosphorylation by the adenine-nucleotide translocator, *Biochem. Soc. Trans.*, **11**, 90–91.

Westerhoff, H.V., Koster, J.G., Van Workum, M. and Rudd, K.E. (1990) On the control of gene expression, in: Cornish-Bowden, A. and Cárdenas, M.L. (eds.) *Control of Metabolic Processes*, Plenum Press, New York, pp. 399–412.

Westerhoff, H.V., Jensen, P.R., Kahn, D., Kholodenko, B.N. and Richard, P. (1996) Nonlinear control and self-organisation, in: Lakshmikantham, V. (ed.) *Proceedings of the 1st World Congress of Nonlinear Analysts, Tampa, Florida, USA August 19–26, 1992*, Walter de Gruyter, Berlin, pp. 3245–3253.

Wilhelm, T. and Heinrich, R. (1995) Smallest chemical reaction system with Hopf bifurcation, *J. Math. Chem.*, **17**, 1–14.

Wilhelm, T., Hoffmann-Klipp, E. and Heinrich, R. (1994) An evolutionary approach to enzyme kinetics: optimization of ordered mechanisms, *Bull. Math. Biol.*, **56**, 65–106.

Wilson, D.F. (1982) Should irreversible thermodynamics be applied to metabolic systems? Not in most cases, *Trends Biochem. Sci.*, **7**, 275–279.

Winfree, A.T. (1990) *The Geometry of Biological Time*, Springer-Verlag, Berlin.

Wonham, W.M. (1974) *Linear Multivariable Control: A Geometric Approach*, Springer-Verlag, Berlin.

Wood, H.G. and Katz, J. (1958) The distribution of C^{14} in the hexose phosphates and the effect of recycling in the pentose cycle, *J. Biol. Chem.*, **233**, 1279–1282.

Woods, N.M., Cuthbertson, K.S.R. and Cobbold, P.H. (1986) Repetitive transient rises in cytoplasmic free calcium in hormone-stimulated hepatocytes, *Nature*, **319**, 600–602.

Woods, N.M., Cuthbertson, K.S.R. and Cobbold, P.H. (1987) Agonist induced oscillations in cytoplasmic free calcium concentration in single rat hepatocytes, *Cell Calcium*, **8**, 79–100.

Wright, B.E., Butler, M.H. and Albe, K.R. (1992) Systems analysis of the tricarboxylic acid cycle in Dictyostelium discoideum. I. The basis for model construction, *J. Biol. Chem.*, **267**, 3101–3105.

Wu, X., Gutfreund, H., Lakatos, S. and Chock, P.B. (1991) Substrate channeling in glycolysis: a phantom phenomenon, *Proc. Nat. Acad. Sci. USA*, **88**, 497–501.

Yčas, M. (1974) On earlier states of the biochemical system, *J. Theor. Biol.*, **44**, 145–160.

Zeleny, M. (1974) *Linear Multiobjective Programming*, Springer-Verlag, Berlin.

Index

P

Q

R

S